Emerging Smart Agricultural Practices Using Artificial Intelligence

IEEE Press
445 Hoes Lane
Piscataway, NJ 08854

Emerging Smart Agricultural Practices Using Artificial Intelligence

Edited by

Ashish Kumar
School of Computer Science Engineering and Technology
Bennett University
Greater Noida
Uttar Pradesh, India

Jai Prakash Verma
Department of Computer Science and Engineering, Institute of Technology
Nirma University
Ahmadabad
Gujarat, India

Rachna Jain
IT Department
Bhagwan Parshuram Institute of Technology
Delhi
New Delhi, India

IEEE Press
WILEY

About the Editors

Dr. Ashish Kumar, PhD, has been working as an Associate Professor at Bennett University, Greater Noida, U.P., India, since 2022. Prior to this, he worked at Bharati Vidyapeeth's College of Engineering (affiliated with GGS Indraprastha University) from August 2009 to July 2022. He obtained his PhD in Computer Science and Engineering from Delhi Technological University (formerly DCE), New Delhi, India, in 2020. He received the Best Researcher award from the Delhi Technological University for his contribution to the computer vision domain. He received his MTech with distinction in Computer Science and Engineering from GGS Indraprastha University, New Delhi. He has published more than 45 research papers in various reputed national and international journals and conferences. He has published 15+ book chapters in various Scopus-indexed books and has authored/edited several books in the fields of AI, computer vision, and healthcare with reputed publishers. He is an active member of various international societies and clubs. He is a reviewer for many reputed journals, and he is on the technical program committee of various national/international conferences. Dr. Kumar also served as a session chair at many international and national conferences. His current research interests include object tracking, image processing, artificial intelligence, and medical imaging analysis.

Dr. Jai Prakash Verma has been working as an Associate Professor in the Computer Science and Engineering Department at Institute of Technology, Nirma University, and he has been associated with the department since July 2006. He received his BSc (PCM) and MCA degrees from the University of Rajasthan, Jaipur, and his PhD from Charusat University, Changa. His PhD subject area was Text Data Summarization and Big Data Analytics. His research interests include data mining, big data analytics, and machine learning. He has contributed to research in these domains through several international conferences and journal publications. He is a recognized PhD guide at Nirma University and currently guides four PhD scholars. He has also been serving as the IQAC Coordinator for the School of Technology at Nirma University and actively contributes to all ranking-related tasks for the institute, such as NAAC, NBA, and NIRF. He is actively involved in various training programs, such as customized training on Big Data Analytics for naval officers at INS Valsura, Indian Navy, and SAC-ISRO, Ahmedabad. He, with other faculty members of the department, coordinates activities of the Center of Excellence in Data Science, in collaboration with SUNY Binghamton, New York.

Dr. Rachna Jain has been working as an Associate Professor at Bhagwan Parshuram Institute of Technology (affiliated with GGSIPU) since August 2021. She has worked as an Assistant Professor in Computer Science Department at Bharati Vidyapeeth's College of Engineering (affiliated with GGSIPU) from August 2007 to August 2021. She completed her PhD in Computer Science from Banasthali Vidyapith in 2017. She received her ME degree in 2011 from the Delhi College of Engineering (Delhi University) with a specialization in Computer Technology and Applications. She did her BTech in Computer Science in 2006 from the N.C. College of Engineering, Kurukshetra University. Her current research interests include cloud computing, fuzzy logic, network and information security, swarm intelligence, big data and IoT, deep learning, and machine learning. She has contributed more than 10 book chapters to various books. She has also served as Session Chair in various international conferences. She was Co-PI of a DST project titled "Design an Autonomous Intelligent Drone for City Surveillance." With over 15 years of academic and research experience, Dr. Jian has authored more than 50 publications in various national and international conferences and high-repute international journals such as Scopus, ISI, and SCI.

List of Contributors

Neera Agarwal
Department of Electronics and Communication, Bharati Vidyapeeth's College of Engineering (BVCOE)
GGSIPU, Delhi
New Delhi, India

Naman Agrawal
School of Computer Science Engineering and Technology, Bennett University, Greater Noida
Uttar Pradesh, India

Yuvraj Ahuja
Department of Electronics and Communication, Bharati Vidyapeeth's College of Engineering (BVCOE)
GGSIPU, Delhi, New Delhi, India

Shalom Akhai
Professor, Department of Mechanical Engineering, Maharishi Markandeshwar (Deemed to be University) Mullana, Ambala
Haryana, India

Yogita Arora
Department of Electronics and Communication, Bharati Vidyapeeth's College of Engineering (BVCOE),
GGSIPU, Delhi, New Delhi, India

Supriya Bajpai
IIT Bombay-Monash Research Academy, IIT Bombay, Mumbai
Maharashtra, India

Himanshi Bansal
IT Department, Guru Tegh Bahadur Institute of Technology
New Delhi, India

Abhinav Bhardwaj
Department of Computer Science and Engineering, Bharati Vidyapeeth College of Engineering, Delhi
New Delhi, India

B.S Bhatia
Pro Vice Chancellor, Management Department, RIMT University
FatehgarhSahib, Punjab, India

Jitendra Bhatia
Department of Computer Science and Engineering, Institute of Technology
Nirma University, Ahmedabad,
Gujarat, India

Murat Can Cardak
School of Electronic Engineering and Computer Science, Queen Mary University of London, London, UK

Heet Dave
Department of Computer Science and Engineering, Institute of Technology, Nirma University
Ahmadabad, Gujarat, India

Vimal Gaur
CSE Department, Maharaja Surajmal Institute of Technology, Delhi
New Delhi, India

Milind Gautam
Department of Computer Science and Engineering, Bharati Vidyapeeth College of Engineering, Delhi
New Delhi, India

Sukhpal Singh Gill
School of Electronic Engineering and Computer Science
Queen Mary University of London
London, UK

Muhammed Golec
School of Electronic Engineering and Computer Science, Queen Mary University of London
London, UK

Manya Gupta
Department of Computer Science and Engineering, Bharati Vidyapeeth College of Engineering
Delhi, New Delhi, India

Neha Gupta
School of Computer Applications, Manav Rachna International Institute of Research & Studies
Faridabad, Haryana, India

Rachna Jain
IT Department, Bhagwan Parshuram Institute of Technology, Delhi
New Delhi, India

V. Kanakaris
MLV Research Group, Department of Informatics, Democritus University of Thrace, Kavala, Greece

Department of Economics, Democritus University of Thrace
Komotini, Greece

Anil Kumar
Bharati Vidyapeeth's College of Engineering, Delhi
New Delhi 110063 India

Ashish Kumar
School of Computer Science Engineering and Technology
Bennett University, Greater Noida
Uttar Pradesh, India

Manoranjan Kumar
School of Computer Science Engineering and Technology
Bennett University, Greater Noida
Uttar Pradesh, India

Malaram Kumhar
Department of Computer Science and Engineering, Institute of Technology
Nirma University, Ahmedabad
Gujarat, India

Gargi Mishra
Department of Computer Science and Engineering, Bharati Vidyapeeth College of Engineering, Delhi
New Delhi, India

G. A. Papakostas
MLV Research Group, Department of Informatics, Democritus University of Thrace, Kavala, Greece

Prasoon Kumar Pandey
School of Computer Science
Engineering and Technology
Bennett University, Greater Noida
Uttar Pradesh, India

Bhavin Patel
Research Scholar, Gujarat
Technological University
Ahmedabad, Gujarat, India

Department of Computer Engineering
Vishwakarma Government
Engineering College, Ahmedabad
Gujarat, India

Priya
School of Computer Applications,
Manav Rachna International Institute
of Research & Studies, Faridabad
Haryana, India

Shipra Raheja
IT Department, Guru Tegh Bahadur
Institute of Technology, New Delhi
India

Suraj Ranga
Department of Computer Science &
Engineering, Indira Gandhi University
Meerpur, Rewari, Haryana, India

Jaya Saini
School of Computer Science
Engineering and Technology
Bennett University
Greater Noida
Uttar Pradesh, India

Savita Kumari Sheoran
Department of Computer Science
& Engineering, Indira Gandhi
University Meerpur, Rewari
Haryana, India

Divya Singh
School of Computer Science
Engineering and Technology
Bennett University, Greater Noida
Uttar Pradesh, India

Ghanapriya Singh
Department of Electronics &
Communication Engineering
NIT, Kurukshetra
Haryana, India

Jaspreet Singh
Professor, Computer Science
Department, Christ University
Ghaziabad, Uttar Pradesh, India

Tanu Taneja
Research Scholar, Department of Civil
Engineering, RIMT University
FatehgarhSahib
Punjab, India

Shashi Tanwar
Professor, Computer Science &
Engineering, Anangpuria School
of Management & Technology
Faridabad, Haryana, India

Gautmi Tomar
Department of Electronics and
Communication, Bharati Vidyapeeth's
College of Engineering (BVCOE),
GGSIPU, Delhi, New Delhi, India

Jai Prakash Verma
Department of Computer Science and Engineering, Institute of Technology, Nirma University, Ahmadabad Gujarat, India

E. Vrochidou
MLV Research Group, Department of Informatics, Democritus University of Thrace, Kavala, Greece

Preface

In today's rapidly evolving world, agriculture stands at a crossroads, facing the dual challenge of feeding a growing global population while mitigating the environmental impact of farming. The integration of Artificial Intelligence (AI) into agriculture has emerged as a pivotal solution, promising increased efficiency, sustainability, and productivity.

This book is a comprehensive exploration of the transformative power of AI in agriculture. It delves deep into the innovative practices and technologies that are reshaping the way we cultivate crops, raise livestock, and manage agricultural resources. It provides a roadmap for harnessing the full potential of AI to optimize farming operations, minimize waste, and adapt to the dynamic demands of the modern agricultural landscape. "Emerging Smart Agricultural Practices Using Artificial Intelligence" envisions a future where AI contributes to a more sustainable, high-yield, and environmentally responsible agricultural sector. By embracing these emerging practices, one can pave the way for a smarter, more efficient, and more resilient future of farming.

The book comprises 15 chapters contributing to diverse aspects of AI for reshaping traditional farming and providing resolution to contemporary agricultural challenges. The first chapter titled "Agricultural Resilience: Water Quality and Human Well-Being" emphasizes the need of sustainable water management strategies for producing healthy crops for improved human health. The second chapter titled "Precision Farming: A Technological Revolution for Sustainable Agriculture" highlights the technological advancement for precision farming along with livestock monitoring and tracking. The third chapter titled "Precision Farming and Smart Crop Management" investigates the industrial advancements in precision farming and smart crop management techniques for sustainable practices in agriculture. The fourth chapter titled "Empowering Smart Agriculture with Artificial Intelligence" highlights the benefits of AI in agriculture for resource optimization and executing data-driven decisions. The fifth chapter titled "Foundations of Agricultural AI" discusses the fundamentals of artificial intelligence for monitoring and managing crop, data quality, and their ethical and social implications. The sixth chapter titled "AI in Agriculture: A Comprehensive Exploration of Technological Transformation" explores the role of AI in crop, disease, and weed management. The seventh chapter titled "Integrating AI and Climate-Smart Agricultural Mechanization: Strategies for Enhancing Productivity and Sustainability in a Changing Climate" discusses the government's efforts for automating farming processes to enhance productivity without contributing to the adverse effects of climate change.

The eighth chapter titled "Harvesting Tomorrow: Exploring Real-World Applications of AI in Agriculture" explores the innovations in supply chain, revolution in livestock monitoring, soil analysis, and management for improving future prospects of contemporary farming using AI. The ninth chapter titled "Smart Agriculture: Predictive Modeling of Fertilizer Requirements Using Neural Networks" analyzes the role of neural networks in identifying the requirement of fertilizer to enhance productivity. The tenth chapter titled "Reviewing Advances in Image-Based Plant Disease Detection" reviews the imaging techniques for predicting plant diseases using AI. The eleventh chapter titled "Leveraging ResNeXt50 and LSTM for Enhanced Plant Disease Detection: A Hybrid Model Proposal" examines plant diseases and automates the detection process using deep learning models to prevent severe plant damage at a later stage. The twelfth chapter titled "FarmTechAI: Artificial-Intelligence-Based Modern Farmer Management System" highlights the functional and non-functional requirements in AI-based modern farming along with legal, social, and ethical issues for global sustainability. The thirteenth chapter titled "Livestock Monitoring and Welfare" explores the innovation in livestock monitoring using AI and other recent technologies. The fourteenth chapter titled "Smart Crop Management: Harnessing Green IoT Tomorrow" discusses the green agriculture techniques using IoT for remote monitoring of crops and precision agriculture. The last chapter titled "Current Progress of Sustainable Smart Agriculture Using Internet of Things" elaborates on the progress of AI in agriculture to ensure sustainability by analyzing the current status of various sustainable development goals worldwide.

Throughout these pages, the reader discovers real-world case studies, expert insights, and practical guidance on how AI can revolutionize various facets of agriculture, from precision farming and livestock monitoring to sustainable resource management. We aim to bridge the gap between agricultural experts, data scientists, technology enthusiasts, and policymakers, ensuring that the knowledge shared here is accessible and beneficial to a diverse audience.

This book provides advanced learners with limitless possibilities to explore what AI offers to the world of agriculture. This is suitable for a farmer seeking ways to optimize crop yields, a researcher on the cutting edge of AI innovation, or a policymaker shaping the future of food production. This book aims to equip them with the knowledge and insights needed to navigate the exciting intersection of AI and agriculture.

1

Agricultural Resilience: Water Quality and Human Well-Being

Tanu Taneja[1], B.S Bhatia[2], and Shalom Akhai[3]

[1] *Research Scholar, Department of Civil Engineering, RIMT University, FatehgarhSahib, Punjab, India*
[2] *Pro Vice Chancellor, Management Department, RIMT University, FatehgarhSahib, Punjab, India*
[3] *Professor, Department of Mechanical Engineering, Maharishi Markandeshwar (Deemed to be University) Mullana, Ambala, Haryana, India*

1.1 Introduction

The water consumed by humans directly or in the form of food products harvested or prepared using water directly affects human health, depending on the quality of the water being used [1]. When the water used for agriculture is contaminated/polluted, it reduces crop yields and lowers the nutritional value due to unacceptable levels of pollutants such as heavy metals, herbicides, and pathogens [2]. This long-term contamination in soil due to harmful elements that may come into it by polluted water harms ecosystems [3]. When using sensor-based and artificial intelligence (AI)-driven water monitoring precision devices, farmers have access to real-time irrigation water quality data, which aids to address and prevent harmful contamination levels of pollutants [4]. Farmers can also save water by avoiding over-irrigation based on data analysis of the data collected using sensors, drones, and AI-based systems [5]. It is also useful for disease prevention by identification of waterborne pathogens at early stages [6]. Good water quality is required not only by humans and food crops but also by livestock for consumption [7]. Water safety in animal health is crucial for agricultural resilience because livestock health affects sustainable and profitable farming [8]. Agricultural resilience requires water management that goes beyond conservation, efficiency, and ecological harmony [9]. Rainwater harvesting, efficient irrigation, and agroecology reduce farmers' dependency on conventional water sources and optimize

Emerging Smart Agricultural Practices Using Artificial Intelligence, First Edition.
Edited by Ashish Kumar, Jai Prakash Verma, and Rachna Jain.

Published 2025 by John Wiley & Sons, Inc.

water application and resource utilization [10]. Heavy metals, pesticides, and aquatic illnesses may enter the food chain from unclean water-irrigated crops [11]. Community-protecting programs need to understand how water quality affects health [12]. Sustainable and cost-effective water filtration techniques solve agricultural water quality challenges by filtering and cleansing water naturally [13]. Farmers may improve water quality, agricultural profitability, and environmental and community well-being by using AI, precision irrigation, and sustainable water management [14].

1.2 The Nexus of Water Quality and Agriculture

Crop growth, production, and nutrition depend on water quality. Due to its complex chemical and biological makeup, its relationship to agriculture is elaborated [15]. Pollination, flower loss, and yield are affected by contaminated water [16]. Lack of moisture delays seed germination, whereas enough water stimulates flowering and fruiting for a rich harvest. Water quality is important during irrigation because pollutants may affect soil structure and plant health, reducing agricultural yield. Water quality affects nutrition because plant roots absorb nutrients and contaminants [17]. Polluted water harms soil fertility's diverse microbial communities by spreading illnesses and poisons. Diseases, economic hardship, and agricultural instability may result from contaminants [18]. Technology like AI-driven water monitoring systems, precision agriculture, and waterborne disease diagnosis in farming are improving water quality and agriculture [20]. Farmers may decrease over-irrigation, pollutant leaching, and pollution using these technologies [20]. Water quality agriculture management also involves waterborne pathogen identification, which helps farmers identify harmful bacteria [21]. Ethical, animal welfare, and agricultural profitability depend on livestock water safety [22]. Sustainable water management technologies help farmers gather and store rain, control water application, and optimize resource usage. Water is purified chemical-free by phytoremediation and wetland filtration [23]. Table 1.1 briefs the impact of water quality on agriculture.

In conclusion, water quality and soil health are interrelated, and innovative water filtration technology can help address these issues.

1.3 Impact of Contaminated Water on Crop Health

Water is crucial for crop health, transporting nutrients needed for development [31]. However, when polluted with heavy metals, pesticides, and viruses, it becomes a silent enemy, stunting growth, reducing yields, and lowering

Table 1.1 Impact of water quality on agriculture.

Aspect of agriculture	Effects of poor water quality	Effects of good water quality	Research needs
Crop development and productivity [24–26]	• Delays or prevents seed germination. Affects pollination, leading to flower loss and reduced yield • Contaminants can damage plant health, leading to lower yield	• Promotes seed germination and healthy growth • Encourages pollination and blooming, resulting in higher yields • Supports strong plant development, leading to healthy and abundant crops	• Develop cost-effective methods for on-farm water quality testing • Research on improving seed tolerance to contaminants • Develop crop varieties resilient to poor water quality
Nutrition [27, 28]	• Plants absorb pollutants alongside minerals, affecting the nutritional value of crops • Contaminated water can introduce diseases into crops, impacting food safety	• Plants absorb essential minerals, contributing to the nutritional value of crops • Promotes healthy crop growth, leading to safe and nutritious food	• Develop strategies to reduce contaminant uptake by plants • Research on biofortification techniques to enhance crop nutrient content
Soil health [3, 29, 30]	• Contaminated water damages soil structure, hindering water drainage and root growth • Disrupts the delicate balance of beneficial microbes, affecting soil fertility	• Supports healthy soil structure, allowing for proper water drainage and root growth • Creates an environment for diverse microbial populations, contributing to soil fertility	• Research on soil remediation techniques to mitigate the effects of contaminants • Develop methods to improve soil organic matter content for better water retention
Economic impacts [31, 32]	• Can lead to crop loss, requiring replanting and increasing economic burden • May contribute to the spread of diseases, impacting marketability and causing financial losses	• Supports healthy crop growth, reducing the need for replanting and associated costs • Contributes to higher yields and improved crop quality, leading to increased profitability	• Develop economic models to assess the financial impact of water quality on agriculture • Research on cost-effective water treatment solutions for farms

nutrition [33]. Heavy metals like lead, cadmium, and mercury disrupt plant physiological systems, impairing nutrition uptake and transport, leading to stunted development [34, 35]. Pesticides, used in agriculture to control pests, can also pollute water, causing stunted growth, leaf discoloration, and lower photosynthetic efficiency [36]. They can also build up in the soil, threatening agricultural land [37]. Waterborne pathogens, including bacteria, viruses, and fungus, can enter plants, producing wilting, lesions, and rotting, causing farmers significant economic losses [38]. Water contamination impacts crops via root absorption since soil pollutants circulate in their vascular systems and inhibit nutrient synthesis. This stunts growth, lowers yields, and may impact farmers, downstream consumers, and agriculture-dependent firms [39]. Heavy metals and pesticides in contaminated water limit nutrient absorption and plant synthesis, lowering agricultural nutrition and possibly affecting human health [2, 11]. Soil contaminants worsen the issue, diminishing the land's ability to support many crop cycles. Water quality's various impacts must be understood to build resilient agricultural systems [33]. Monitoring water sources and soil quality, precision irrigation utilizing real-time data, and investments in water treatment equipment, including filtration and purification systems, may prevent agricultural contamination [40, 41]. Table 1.2 summarizes contaminants and their effects.

Polluted water slows crop growth, lowers yields, and reduces crop nutrition and land sustainability. To enhance agricultural systems and feed rising populations, comprehensive agricultural water management is needed.

1.4 AI-Driven Water Monitoring Systems

AI-driven water monitoring systems are a crucial tool in agriculture, providing real-time insights, predictive capabilities, and proactive water quality solutions [52]. With modern sensors, these systems gather and analyze vast amounts of data in real time to help farmers manage water, irrigation, and soil [19, 53]. These sensors monitor water pH, nutrient content, temperature, and contaminants while powerful data analytics and machine learning algorithms detect patterns, irregularities, and contamination events. Farmers can immediately identify water quality issues and take preventative measures using real-time data [54]. These systems' sensor networks encompass the agricultural area to increase water quality information and enable personalized treatments [55]. Farmers may reduce pollution by changing irrigation, target treatments, or water sources using machine learning [56]. New data may teach AI-driven systems to change their models, making flexibility crucial in changing agricultural environments [57, 58]. In climate change, temperature, precipitation, and severe weather may alter water quality, making adaptability essential [59]. AI systems with anomaly detection

Table 1.2 Impacts of pollutants on crop health and yield.

Pollutant category	Specific pollutant examples	Uptake pathway	Effects on plants	Effects on yield andquality	Additional notes
Heavy metals [42, 43]	Lead (Pb), cadmium (Cd), and mercury (Hg)	Root uptake	• Disrupt enzyme activity and nutrient uptake • Damage cell membranes and organelles • Inhibit photosynthesis and respiration	• Stunted growth • Reduced seed production • Deformed plant organs	• Accumulate in edible parts, posing health risks to consumers
Herbicides [44, 45]	Glyphosate, atrazine, and 2,4-D	Root uptake and foliar absorption	• Disrupt plant growth hormones • Inhibit photosynthesis and protein synthesis • Cause leaf chlorosis and necrosis	• Reduced growth and development • Lower crop yield • Decreased crop quality	• Can persist in soil and water for extended periods
Waterborne pathogens [46, 47]	**Bacteria:** *Escherichiacoli* and *Pseudomonas* **Viruses:** tomato mosaic virus and cucumber mosaic virus **Fungi:** Fusarium wilt and Phytophthora blight	Stomata, wounds, and lenticels	• Cause tissue death and necrosis • Disrupt water and nutrient transport • Interfere with plant defense mechanisms	• Wilting, stunting, and death of plants • Reduced crop yield • Decreased fruit quality	• Some pathogens can be transmitted to humans through contaminated food or water
Salinity (high levels of dissolved salts) [48, 49]	Sodium ($Na+$), chloride ($Cl-$), and sulfates (SO_4^{2-})	Root uptake	• Disrupts water uptake by plants • Creates osmotic stress • Damages cell membranes and organelles	• Reduced plant growth • Leaf scorching and wilting • Decreased crop yield and quality	• Can render some land unsuitable for agriculture
Industrial pollutants [50, 51]	Polycyclic aromatic hydrocarbons (PAHs), pharmaceuticals, and personal care products (PPCPs)	Root uptake and foliar absorption	• Can disrupt various plant physiological processes • May mimic hormones and affect plant growth • Can accumulate in plant tissues	• Reduced plant growth and development • Potential negative impacts on human health through food consumption	• Emerging contaminants with limited research on long-term effects

can quickly discover water quality abnormalities, minimizing economic losses and ensuring agricultural sustainability. AI-driven water monitoring is proactive, like precision agriculture, allowing farmers to optimize irrigation operations using real-time data to ensure crops receive the proper quantity of water at the right time [60]. This saves water and prevents over-irrigation, which may cause waterlogging, nutrient leaching, and soil contamination [61]. AI-driven water monitoring systems improve agriculture's resilience and sustainability while upgrading technology. However, installation costs, training requirements, and data privacy concerns must be addressed. As technology evolves and becomes more accessible, AI-driven water monitoring becomes an essential part of contemporary agriculture [53, 62]. Table 1.3 presents AI-driven water monitoring systems that are revolutionizing water management in agriculture.

These smart systems enhance agricultural productivity while promoting efficient water usage and environmental sustainability. Thus, the extensive sensor network monitors water supplies, irrigation systems, and soil, providing real-time analytics and machine learning for predictive risk management.

1.5 Research Gaps and Research Dimensions

The study highlights the need for interdisciplinary research, advanced detection technologies, and innovative crop production methods to sustain agricultural water quality and protect water resources for future generations. Tables 1.2 and 1.3 indicate valuable research areas in agricultural water management. Research on the impact of pollutants on crop health and the benefits of AI-powered water monitoring systems in agriculture has identified gaps. These include understanding the long-term effects of contaminants, integrating emerging contaminants, and holistically assessing water quality monitoring systems. Sustainable agriculture requires studying contaminants' long-term effects on soil health, crop productivity, and food safety. Future research should focus on the detection, accumulation, and ecological effects of contaminants by data analysis as it is yet understudied. Encourage adoption of smart monitoring technologies, and high installation costs, data privacy concerns, and technological barriers must be addressed.

The management of agricultural water quality requires interdisciplinary techniques, longitudinal research, sophisticated detection technologies, policy and governance, capacity development, and environmental issues. This study explores the link between water quality, soil health, and agricultural yield through longitudinal research across multiple growing seasons. It uses sensor platforms, molecular diagnostic tools, and remote sensing to improve water quality monitoring systems. Policy and governance are crucial for effective water quality management and regulatory compliance.

Table 1.3 Benefits of AI-powered water monitoring in agriculture.

Feature	Description	Benefits
Data collection and analysis [63, 64]	• Extensive sensor network monitors water supplies, irrigation systems, and soil • Sensors track parameters such as pH, nutrient content, temperature, and pollutants • Advanced analytics and machine learning process data in real time	• Identifies trends, abnormalities, and contamination events • Allows quick recognition of deviations from ideal water quality
Predictive capabilities [65–67]	• Machine learning predicts risks based on historical data and current trends	• Enables proactive risk management by adjusting irrigation practices, targeting treatments, and shifting water sources to avoid contamination
Adaptability and flexibility [68–70]	• Systems learn and modify models with new data, adapting to changing conditions	• Crucial for responding to climate change and unforeseen circumstances
Anomaly detection [71–73]	• Quickly detects water quality abnormalities	• Minimizes economic losses and ensures agricultural sustainability
Precision irrigation [74, 75]	• Real-time data optimizes irrigation, ensuring proper water delivery	• Minimizes water loss and avoids logging off water nutrient loss and soil adulteration

1.6 Precision Irrigation Techniques

Automation in agriculture is enhancing food production and reducing environmental impact by optimizing precise irrigation for diverse crops. Using sensors, drones, and automated systems, agriculture addresses rising food demand, population growth, and water shortages by adjusting water consumption to meet crop needs at various stages, thereby conserving water and reducing over-irrigation. Precision irrigation relies on real-time monitoring and data-driven decision-making. Field sensors measure soil moisture, temperature, and crop health to help farmers save resources. Soil moisture sensors let farmers adjust irrigation schedules to match crop and soil demands in real time. Drones help farmers see water shortages, crop health concerns, and irrigation system efficiency from above. Automated irrigation systems improve water efficiency and uniformity by watering precisely and adaptably without operator intervention. Environmental advantages of precision irrigation include reduced water stress during development, plant growth, and environmental resilience. Reducing over irrigation minimizes soggy roots and nutrient loss [76–78].

1.7 Waterborne Pathogens in Farming

Waterborne pathogens pose a significant threat to agriculture, as water supplies from natural or artificial sources can harbor bacteria, viruses, and parasites that threaten crops and cattle. These infections must be detected quickly and accurately to minimize disease spread, protect agricultural systems, and ensure food supply chain safety. Traditional techniques for detecting waterborne infections are laborious and slow, delaying hazard detection. New molecular methods and diagnostic tools aid farmers in detecting hazardous bacteria, ensuring crop health, and reducing waterborne illness risks [78, 79]. Molecular methods like Polymerase Chain Reaction (PCR) and Next-Generation Sequencing (NGS) have revolutionized pathogen identification in water sources, allowing for high sensitivity and specificity in identifying harmful bacteria.[80]. Rapid diagnostic technologies speed up identification and reveal water source microbial composition in real-time, allowing for real-time monitoring to prevent disease spread and safeguard crops [81]. Waterborne diseases affect cattle as well as crops, and farmers can avoid cattle infections by quickly identifying pathogens in water sources. The importance of aquatic pathogen detection extends beyond agricultural production and animal health, as it also impacts human health [82]. A comprehensive water monitoring and pathogen detection system requires strategic sampling and continuous water source surveillance, with automated sensors and remote monitoring systems improving this procedure [83]. Farmers must be aware of waterborne pathogen detection and best practices shared by agricultural extension services, research institutions, and government agencies to integrate advanced detection technologies into agricultural practices.

1.8 Livestock Health and Water Safety

Water quality is crucial for livestock health, causing infections, reproductive problems, and economic losses. Farmers must implement comprehensive management techniques and sustainable water management and educate farmers about water safety to maintain agricultural systems' resilience and prevent waterborne illnesses [83–85]. The following key points emphasize the importance of livestock health, water safety, and water quality in agricultural settings:

- Water quality is crucial for cattle health and well-being in agricultural settings.
- Contaminated water can lead to infections, impaired development, reproductive issues, and producer losses.
- Waterborne diseases can harm cattle's development, weight gain, and reproduction.
- Farmer management should include periodic testing for water safety.
- Livestock farms can reduce dependence on polluting external water sources and maintain water supply systems.
- Regular cleaning and maintenance of water storage facilities protect water quality and livestock health.
- Farmers should be educated on livestock water safety and equipped with tools to prevent animal health concerns.

1.9 Sustainable Water Management Strategies

Water management in agricultural systems must be sustainable to maintain water supplies and promote environmental sustainability [86]. Rainwater collection, irrigation, soil moisture management, and agroecology are essential [87]. Modern drip irrigation and sprinkler systems decrease runoff and evaporation, while rainwater collecting lowers water dependence [88]. Real-time soil moisture management ensures crops receive the right amount of water at the right time [89]. Environmental balance and natural resource conservation support sustainable water management in agroecological systems. Cover cropping, agroforestry, and rotational grazing improve soil structure, erosion, and water retention [90]. Riparian buffer zones and wetlands in agricultural landscapes remove sediments and contaminants before reaching water sources [91]. Crop rotation and diversity optimize water use and limit resource depletion risk. Deep root systems in certain crops improve soil structure and water retention [92]. Water-efficient crop selection, such as drought-tolerant or water-efficient cultivars, connects agricultural operations with water availability, boosting sustainability [93]. Education and outreach are essential for implementing sustainable water management strategies. Table 1.4 presents a comprehensive list of agricultural practices that focus on water management and soil health, each offering unique benefits for sustainable agriculture.

Table 1.4 Sustainable water conservation practices in agriculture.

Practice	Description	Benefits
Rainwater harvesting [94, 95]	Capturing and storing rainwater for later use in irrigation	Reduces reliance on conventional water sources.
Modern irrigation systems (drip, sprinkler) [95–97]	Targeted and efficient application of water directly to crops.	Reduces water runoff and evaporation
Real-time soil moisture management [98, 99]	Utilizing sensors and data to monitor soil moisture and adjust irrigation accordingly	Ensures crops receive the right amount of water at the right time
Agroecology: cover cropping [100, 101]	Planting additional crops between main crops to improve soil health and water retention	Promotes environmental balance and natural resource conservation
Agroforestry: integrating trees and shrubs within agricultural landscapes [102, 103]	Improving soil structure and water infiltration	Enhances soil quality and water storage capacity
Rotational grazing [104, 105]	Rotating livestock grazing patterns to prevent overgrazing and improve soil health	Promotes healthy soil ecosystems
Riparian buffer zones and wetlands [106, 107]	Creating buffer areas between agricultural land and water bodies to filter sediments and pollutants	Protects water quality and aquatic ecosystems
Crop rotation and diversity [108, 109]	Planting different crops in sequence or together to improve soil health, nutrient availability, and water use efficiency	Reduces the risk of resource depletion and promotes healthy soil ecosystems
Deep-rooted crops [110, 111]	Selecting crops with deep root systems to improve soil structure and water retention	Adapts agricultural practices to local water conditions and promotes sustainable use
Water-efficient crop varieties [101, 112]	Selecting drought-tolerant or water-efficient cultivars based on local water availability	Adapts agricultural practices to local water conditions and promotes sustainable use

1.10 Human Health Implications

Clean air, water, and nutritious food are essential requirements for all living beings, including humans. Both indoor and outdoor air quality significantly impact human health, similar to the importance of clean water [113–118]. Water quality and agriculture are interconnected, affecting human health in various ways [119, 120]. Contaminated crops and livestock products can introduce harmful pollutants into the food chain, leading to infections, heavy metals, pesticides, and other contaminants [121]. To assess food safety and reduce health hazards, it is crucial to understand water pollutants and their effects on crops. Livestock may also absorb toxins from polluted water, increasing health hazards in meat, milk, and eggs [122, 123]. Infected water sources can transmit illnesses via agricultural goods, with pathogenic bacteria, viruses, and parasites infecting crops through irrigation or contact with polluted surfaces [124]. Consuming contaminated agricultural goods can cause diarrhea, gastrointestinal infections, and other waterborne disorders [125]. Bioaccumulation/toxin exposure is essential for understanding the health effects of agricultural water contamination, as waterborne heavy metals, herbicides, and other pollutants accumulate in crops and animals, potentially causing long-term health problems due to toxin buildup [126]. Figure 1.1 illustrates the significant effects of water pollution on both human health and the environment.

Community health and clean water are crucial for agriculture-dependent populations, as contaminated water sources increase waterborne illness and toxin exposure [127]. Safe and clean water for agriculture and homes is essential for community well-being [128]. Pregnant women and small children are especially vulnerable to agricultural water contamination, as toxin exposure may cause birth abnormalities or developmental difficulties. Protecting mother and child health requires specific measures to protect agricultural water quality [129]. E-waste is a major environmental hazard, including its contribution to water pollution [130]. Agriculture pollution concerns include cancer, neurological issues, and endocrine disruption. Monitoring and control are crucial to prevent chronic health impacts from agricultural pesticides while addressing human health consequences of water quality in agriculture is essential for a resilient and sustainable system [131].

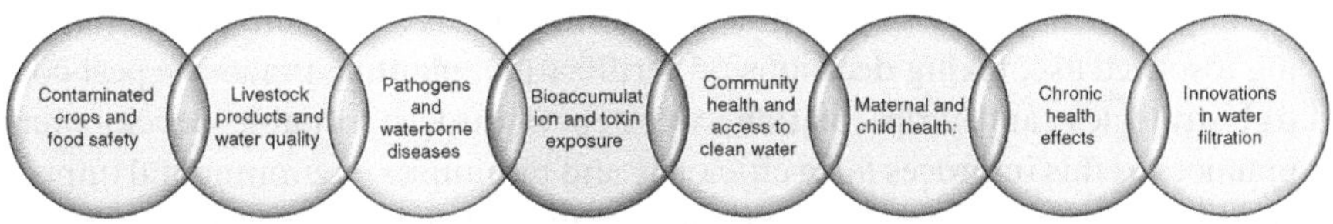

Figure 1.1 Impacts of water pollution on human health and environment.

1.11 Significance of Research in Agricultural Water Quality

AI-powered water monitoring systems are a boon for agricultural communities, as they monitor the water quality via sensors and may even actuate alarming signals based on the data collected and thus, they provide quality water and food security, help in the direction toward better human health, environmental sustainability, agricultural resilience practices, and economic prosperity [132]. AI provides reliable, real-time information [133, 134]. This information can be further studied to predict trends, allowing legislators, water resource managers, and farmers to make educated decisions. AI systems thus act as early warning systems for contamination levels and pollution, accordingly, directing data-driven decision-making and supporting environmentally conscious efforts [135, 136]. Modern agriculture is therefore undergoing a transition shift by employing smart sensors, IoT devices, and machine learning algorithms to improve water efficiency, minimize waste, and raise crop yields [137–140]. So, understanding the link between water pollution and crop vitality is crucial as it allows these AI systems to avoid hazards, assuring a prosperous future for agricultural enterprises.

1.12 Conclusion

- Data-centric techniques in modern agriculture have many advantages/benefits as well as challenges to overcome. The practice in use includes data monitoring and decision-making for precise water quality control and management, proactive crop health monitoring, data-driven sustainability, reduced environmental footprint building agricultural resilience for sustainable future.
- Modern water management uses AI-powered sensors to monitor water quality in real time, identifying salinity, pH, and nutrient levels. This aids to protect food and agriculture crops from deterioration/damage due to the water pollution levels. Practical crop health monitoring by using precise smart sensors and data analysis using machine learning algorithms aid to predict nutrient deficiencies, detect disease by enabling early detection, and thereby provide instant treatment to agriculture crops by early detecting issues.
- By making decisions using practical data-driven information, i.e. data analysis for implementing data-driven practices from various sources thereby optimizing resource use, taking decisions on fertilization, planting practices, pest control strategies, and crop rotation, we step toward sustainable development practices as this improves farm efficiency and minimizes environmental impact.
- Data analysis-based knowledge of water quality related to agriculture facilitates the information sharing of data among farmers, researchers, and agribusinesses

promoting cooperation among them for accelerating agriculture and better practices. Secure open-source data sharing drives improvement in agriculture technology practices and promotes cooperation among farmers, researchers, and agribusinesses by promoting digital literacy.

- Agricultural firms can greatly benefit from data-driven insights, which are essential for building long-term agricultural resilience. These insights enable informed planning and risk management, helping firms to mitigate the impacts of shifting weather patterns, water and air pollution, and other environmental challenges. The collection of data and its advanced data analytics for predictive modeling must be prioritized. In the long run, focusing on data collected from sensor networks can help food production systems become more resilient to environmental change and resource scarcity.

References

1 Thoradeniya, B., Pinto, U., and Maheshwari, B. (2019). Perspectives on impacts of water quality on agriculture and community well-being – a key informant study from Sri Lanka. *Environmental Science and Pollution Research* 26 (3): 2047–2061.

2 Alengebawy, A., Abdelkhalek, S.T., Qureshi, S.R., and Wang, M.Q. (2021). Heavy metals and pesticide toxicity in agricultural soil and plants: Ecological risks and human health implications. *Toxics* 9 (3): 42.

3 Morgado, R.G., Loureiro, S., and González-Alcaraz, M.N. (2018). Changes in soil ecosystem structure and functions due to soil contamination. In: *Soil Pollution* (ed. A.C. Duarte, A. Cachada, and T. Rocha-Santos), 59–87. Academic Press.

4 Mana, A.A., Allouhi, A., Hamrani, A. et al. (2024). Sustainable AI-based production agriculture: exploring AI applications and implications in agricultural practices. *Smart Agricultural Technology* 7: 100416.

5 Anjum, M.N., Cheema, M.J.M., Hussain, F., and Wu, R.S. (2023). Precision irrigation: challenges and opportunities. *Precision Agriculture* 1: 85–101.

6 Dong, Y., Jiang, Z., Hu, Y. et al. (2024). Pathogen contamination of groundwater systems and health risks. *Critical Reviews in Environmental Science and Technology* 54 (4): 267–289.

7 Miller, J.D., Workman, C.L., Panchang, S.V. et al. (2021). Water security and nutrition: current knowledge and research opportunities. *Advances in Nutrition* 12 (6): 2525–2539.

8 Sekaran, U., Lai, L., Ussiri, D.A. et al. (2021). Role of integrated crop-livestock systems in improving agriculture production and addressing food security–a review. *Journal of Agriculture and Food Research* 5: 100190.

9 Srivastav, A.L., Dhyani, R., Ranjan, M. et al. (2021). Climate-resilient strategies for sustainable management of water resources and agriculture. *Environmental Science and Pollution Research* 28 (31): 41576–41595.

10 Patle, G.T., Kumar, M., and Khanna, M. (2020). Climate-smart water technologies for sustainable agriculture: a review. *Journal of Water and Climate Change* 11 (4): 1455–1466.

11 Sharma, A., Grewal, A.S., Sharma, D., and Srivastav, A.L. (2023). Heavy metal contamination in water: consequences on human health and environment. In: *Metals in Water: Global Sources, Significance, and Treatment* (ed. S.K. Shukla, S. Kumar, et al.), 39–52. Elsevier.

12 Babuji, P., Thirumalaisamy, S., Duraisamy, K., and Periyasamy, G. (2023). Human health risks due to exposure to water pollution: a review. *Water* 15 (14): 2532.

13 Razali, M.C., Wahab, N.A., Sunar, N., and Shamsudin, N.H. (2023). Existing filtration treatment on drinking water process and concerns issues. *Membranes* 13 (3): 285.

14 Srivastava, A., Jain, S., Maity, R., and Desai, V.R. (2022). Demystifying artificial intelligence amidst sustainable agricultural water management. *Current Directions in Water Scarcity Research* 7: 17–35.

15 Akhtar, N., SyakirIshak, M.I., Bhawani, S.A., and Umar, K. (2021). Various natural and anthropogenic factors responsible for water quality degradation: a review. *Water* 13 (19): 2660.

16 Vohra, R. and Kumar, A. (Eds.). (2024). *Advanced Geospatial Practices in Natural Environment Resource Management*. IGI Global.

17 Bhattacharya, A. (2021). Effect of soil water deficit on growth and development of plants: a review. In: *Soil Water Deficit and Physiological Issues in Plants*. Singapore: Springer https://doi.org/10.1007/978-981-33-6276-5_5.

18 Singh, N.S., Sharma, R., Parween, T., and Patanjali, P.K. (2018). Pesticide contamination and human health risk factor. In: *Modern Age Environmental Problems and their Remediation* (ed. M. Oves, M. Zain Khan, M.I. Ismail, and I.). Cham: Springer https://doi.org/10.1007/978-3-319-64501-8_3.

19 Tsolakis, N., Aivazidou, E., and Srai, J.S. (2019). Sensor applications in agrifood systems: current trends and opportunities for water stewardship. *Climate* 7 (3): 44.

20 Pérez-Lucas, G., Vela, N., El Aatik, A., and Navarro, S. (2019). Environmental risk of groundwater pollution by pesticide leaching through the soil profile. *Pesticides-use and Misuse and their Impact in the Environment* 17: 1–28.

21 Pal, M., Ayele, Y., Hadush, M. et al. (2018). Public health hazards due to unsafe drinking water. *Air Water Borne Dis 7* (1000138): 2.

22 Scanes, C.G. (2018). Impact of agricultural animals on the environment. In: *Animals and Human Society* (ed. C.G. Scanes and S.R. Toukhsati), 427–449. Academic Press.

23 Lata, S. (2021). Sustainable and eco-friendly approach for controlling industrial wastewater quality imparting succour in water-energy nexus system. *Energy Nexus* 3: 100020.

24 Shivanna, K.R., Tandon, R., and Koul, M. (2020). 'Global pollinator crisis' and its impact on crop productivity and sustenance of plant diversity. In: *Reproductive Ecology of Flowering Plants: Patterns and Processes* (ed. R. Tandon, K. Shivanna, and M. Koul). Singapore: Springer https://doi.org/10.1007/978-981-15-4210-7_16.

25 Yanagi, M. (2024). Climate change impacts on wheat production: reviewing challenges and adaptation strategies. *Advances in Resources Research* 4 (1): 89–107.

26 Ahmed, N., Zhang, B., Chachar, Z. et al. (2024). Micronutrients and their effects on horticultural crop quality, productivity and sustainability. *Scientia Horticulturae* 323: 112512.

27 Bisht, N. and Chauhan, P.S. (2020). Excessive and disproportionate use of chemicals cause soil contamination and nutritional stress. In: *Soil Contamination-threats and Sustainable Solutions*, 1–10.

28 Zhao, F.J., Tang, Z., Song, J.J. et al. (2022). Toxic metals and metalloids: uptake, transport, detoxification, phytoremediation, and crop improvement for safer food. *Molecular Plant* 15 (1): 27–44.

29 Usharani, K.V., Roopashree, K.M., and Naik, D. (2019). Role of soil physical, chemical and biological properties for soil health improvement and sustainable agriculture. *Journal of Pharmacognosy and Phytochemistry* 8 (5): 1256–1267.

30 Gavrilescu, M. (2021). Water, soil, and plants interactions in a threatened environment. *Water* 13 (19): 2746.

31 Savary, S. and Willocquet, L. (2020). Modeling the impact of crop diseases on global food security. *Annual Review of Phytopathology* 58: 313–341.

32 Rosa, L., Chiarelli, D.D., Rulli, M.C. et al. (2020). Global agricultural economic water scarcity. *ScienceAdvances* 6 (18): eaaz6031.

33 Alengebawy, A., Abdelkhalek, S.T., Qureshi, S.R., and Wang, M.Q. (2021). Heavy metals and pesticides toxicity in agricultural soil and plants: ecological risks and human health implications. *Toxics* 9 (3): 42.

34 Lawal, K.K., Ekeleme, I.K., Onuigbo, C.M. et al. (2021). A review on the public health implications of heavy metals. *World Journal of Advanced Research and Reviews* 10 (3): 255–265.

35 Nuray, D.A.Ğ. and ARICI, Ö. K. (2021). Heavy metals in soils Pb (Lead), Hg (Mercury), Cd (Cadmium), As (Arsenic) effects on human health. *International Journal of Environmental Trends (IJENT)* 5 (2): 48–59.

36 Singh, D. and Kumar, A. (2024). A deep recurrent neural network for plant disease classification. *SN Computer Science* 5 (8): 1053.

37 Muñoz, M., Behnke, L.E., Bridges, W.C. et al. (2025). Postharvest calcium chloride dips. An effective strategy to reduce Botrytis blight severity and increase petal strength in cut roses. *Postharvest Biology and Technology 219*: 113292.

38 Allaq, A.A.A., Mahid, H.H., Yahya, E.B. et al. (2023). Emerging drinking water borne diseases: a review on types, sources and health precaution. *Journal of Pharmaceutical Research International* 35 (31): 1–17.

39 Dotaniya, M.L., Dotaniya, C.K., Solanki, P. et al. (2020). Lead contamination and its dynamics in soil–plant system. In: *Lead in Plants and the Environment*, 83–98.

40 Majsztrik, J.C., Fernandez, R.T., Fisher, P.R. et al. (2017). Water use and treatment in container-grown specialty crop production: a review. *Water, Air, & Soil Pollution* 228: 1–27.

41 Wato, T., Amare, M., Bonga, E. et al. (2020). The agricultural water pollution and its minimization strategies–a review. *Journal of Resources Development and Management* 64: 10–22.

42 Ishtiyaq, S., Kumar, H., Varun, M. et al. (2018). Heavy metal toxicity and antioxidative response in plants: an overview. In: *Plants under Metal and Metalloid Stress: Responses, Tolerance and Remediation*, 77–106.

43 Noor, I., Sohail, H., Sun, J. et al. (2022). Heavy metal and metalloid toxicity in horticultural plants: tolerance mechanism and remediation strategies. *Chemosphere* 303: 135196.

44 Verdeguer, M., Sánchez-Moreiras, A.M., and Araniti, F. (2020). Phytotoxic effects and mechanism of action of essential oils and terpenoids. *Plants* 9 (11): 1571.

45 Varanasi, A., Prasad, P.V., and Jugulam, M. (2016). Impact of climate change factors on weeds and herbicide efficacy. *Advances in Agronomy* 135: 107–146.

46 Patel, R., Mitra, B., Vinchurkar, M. et al. (2023). Plant pathogenicity and associated/related detection systems. A review. *Talanta* 251: 123808.

47 Yashwant, Y.S., Sharma, V., Barupal, T., and Banu, J. (2023). Efficacy of nanoparticles in plant disease control and their phytotoxicity: nanoparticles in phytopathology. *Biomaterials Journal* 2 (2): 1–15.

48 Baranova, E.N. and Gulevich, A.A. (2021). Asymmetry of plant cell divisions under salt stress. *Symmetry* 13 (10): 1811.

49 Arif, Y., Singh, P., Siddiqui, H. et al. (2020). Salinity induced physiological and biochemical changes in plants: an omic approach towards salt stress tolerance. *Plant Physiology and Biochemistry* 156: 64–77.

50 Wei, H., Tang, M., and Xu, X. (2023). Mechanism and influence factors of plant uptake, accumulation, transport, metabolism pathways of pharmaceuticals and personal care products and their phytotoxicity: a review. *Science of the Total Environment* 892: 164413.

51 Olowoyo, J.O. and Mugivhisa, L.L. (2019). Evidence of uptake of different pollutants in plants harvested from soil treated and fertilized with organic materials as source of soil nutrients from developing countries. *Chemical and Biological Technologies in Agriculture* 6: 1–11.

52 Pérez-Beltrán, C.H., Robles, A.D., Rodriguez, N.A. et al. (2024). Artificial intelligence and water quality: from drinking water to wastewater. *TrAC Trends in Analytical Chemistry* 172: 117597.

53 Bhangar, N.A. and Shahriyar, A.K. (2023). IoT and AI for next-generation farming: opportunities, challenges, and outlook. *International Journal of Sustainable Infrastructure for Cities and Societies* 8 (2): 14–26.

54 Vohra, R., Kumar, A., and Rongali, G. (2024). Multi-scale extraction and spatial analysis of growth pattern changes in urban water bodies using sentinel-2 MSI imagery: a study in the central part of India. *Stochastic Environmental Research and Risk Assessment* 38 (6): 2397–2423.

55 Manjakkal, L., Mitra, S., Petillot, Y.R. et al. (2021). Connected sensors, innovative sensor deployment, and intelligent data analysis for online water quality monitoring. *IEEE Internet of Things Journal* 8 (18): 13805–13824.

56 Krishnan, S.R., Nallakaruppan, M.K., Chengoden, R. et al. (2022). Smart water resource management using Artificial Intelligence – a review. *Sustainability* 14 (20): 13384.

57 González-Rodríguez, V.E., Izquierdo-Bueno, I., Cantoral, J.M. et al. (2024). Artificial intelligence: a promising tool for application in phytopathology. *Horticulturae* 10 (3): 197.

58 Babakhouya, A., Naji, A., Daaif, A., and Hnini, A. (2023). Leveraging artificial intelligence in the agri-food industry: a comprehensive review. In *E3S Web of Conferences* (Vol. 469, p. 00079). EDP Sciences, 23–24 November 2023, Morocco.

59 Vohra, R., Kumar, A., Jain, R., and Hemanth, D.J. (2024). Analysis and prediction of land surface temperature with increasing urbanisation using satellite imagery. Heliyon, 10 (22).

60 Abioye, E.A., Hensel, O., Esau, T.J. et al. (2022). Precision irrigation management using machine learning and digital farming solutions. *AgriEngineering* 4 (1): 70–103.

61 Sultan, M. and Ahmad, F. (Eds.). (2023). Irrigation and drainage: recent advances.

62 Araújo, S.O., Peres, R.S., Barata, J. et al. (2021). Characterising the agriculture 4.0 landscape – emerging trends, challenges and opportunities. *Agronomy* 11 (4): 667.

63 Paepae, T., Bokoro, P.N., and Kyamakya, K. (2021). From fully physical to virtual sensing for water quality assessment: a comprehensive review of the relevant state-of-the-art. *Sensors* 21 (21): 6971.

64 Afridi, W.A., Akhter, F., Vitoria, I., and Mukhopadhyay, S.C. (2022). A technology review and field testing of a soil water quality monitoring system. In Nagender Kumar Suryadevara, Boby George, Krishanthi P. Jayasundera, Subhas Chandra Mukhopadhyay *International Conference on Sensing Technology* (pp. 460–475). Cham: Springer Nature Switzerland.

65 Drogkoula, M., Kokkinos, K., and Samaras, N. (2023). A comprehensive survey of machine learning methodologies with emphasis in water resources management. *Applied Sciences* 13 (22): 12147.

66 Li, L., Rong, S., Wang, R., and Yu, S. (2021). Recent advances in artificial intelligence and machine learning for nonlinear relationship analysis and process control in drinking water treatment: a review. *Chemical Engineering Journal* 405: 126673.

67 Shaikh, T.A., Rasool, T., and Lone, F.R. (2022). Towards leveraging the role of machine learning and artificial intelligence in precision agriculture and smart farming. *Computers and Electronics in Agriculture* 198: 107119.

68 Chen, L., Chen, Z., Zhang, Y. et al. (2023). Artificial intelligence-based solutions for climate change: a review. In: *Environmental Chemistry Letters*, 1–33.

69 Camarena, S. (2020). Artificial intelligence in the design of the transitions to sustainable food systems. *Journal of Cleaner Production* 271: 122574.

70 Shaikh, T.A., Mir, W.A., Rasool, T., and Sofi, S. (2022). Machine learning for smart agriculture and precision farming: towards making the fields talk. *Archives of Computational Methods in Engineering* 29 (7): 4557–4597.

71 Bhattacharjee, P., Moy Ghosh, A., and Indu, P. (2022). A study on the social and economic impact of artificial intelligence-based environmental forecasts. In: *Environmental Informatics: Challenges and Solutions* (ed. P.K. Paul, A. Choudhury, A. Biswas, and B.K. Singh), 67–95. Springer Nature Singapore: Singapore.

72 Nova, K. (2023). AI-enabled water management systems: an analysis of system components and interdependencies for water conservation. *Eigenpub Review of Science and Technology* 7 (1): 105–124.

73 Kose, U., Prasath, V.S., Mondal, M.R.H. et al. (2022). *Artificial Intelligence and Smart Agriculture Technology*. CRC Press.

74 Ahmad, L. and Nabi, F. (2021). *Agriculture 5.0: Artificial Intelligence, IoT and Machine Learning*. CRC Press.

75 Raj, E.F.I., Appadurai, M., and Athiappan, K. (2022). Precision farming in modern agriculture. In: *Smart Agriculture Automation Using Advanced Technologies: Data Analytics and Machine Learning, Cloud Architecture, Automation and IoT* (ed. A. Choudhury, A. Biswas, T.P. Singh, and S.K. Ghosh), 61–87. Springer Singapore: Singapore.

76 Bwambale, E., Abagale, F.K., and Anornu, G.K. (2022). Smart irrigation monitoring and control strategies for improving water use efficiency in precision agriculture: a review. *Agricultural Water Management* 260: 107324.

77 Nhamo, L., Magidi, J., Nyamugama, A. et al. (2020). Prospects of improving agricultural and water productivity through unmanned aerial vehicles. *Agriculture* 10 (7): 256.

78 Bridle, H. (ed.) (2020). *Waterborne Pathogens: Detection Methods and Applications*. Academic Press.

79 Oon, Y.L., Oon, Y.S., Ayaz, M. et al. (2023). Waterborne pathogens detection technologies: advances, challenges, and future perspectives. *Frontiers in Microbiology* 14: 1286923.

80 Zhang, S., Li, X., Wu, J. et al. (2021). Molecular methods for pathogenic bacteria detection and recent advances in wastewater analysis. *Water* 13 (24): 3551.

81 Walper, S.A., LasarteAragonés, G., Sapsford, K.E. et al. (2018). Detecting biothreat agents: from current diagnostics to developing sensor technologies. *ACS Sensors* 3 (10): 1894–2024.

82 da Silva, A.D., Paschoalino, W.J., Neto, R.C., and Kubota, L.T. (2022). Electrochemical point-of-care devices for monitoring waterborne pathogens: protozoa, bacteria, and viruses–An overview. *Case Studies in Chemical and Environmental Engineering* 5: 100182.

83 Katko, T. and Højris, B. (ed.) (2019). *Microbiological Sensors for the Drinking Water Industry*. IWA Publishing.

84 Elahi, E., Abid, M., Zhang, L., and Alugongo, G.M. (2017). The use of wastewater in livestock production and its socioeconomic and welfare implications. *Environmental Science and Pollution Research* 24: 17255–17266.

85 Walper, S.A., LasarteAragonés, G., Sapsford, K.E. et al. (2018). Detecting biothreat agents: from current diagnostics to developing sensor technologies. *ACS Sensors* 3 (10): 1894–2024.

86 Yadav, V., Singh, S., Garg, M.P., and Akhai, S. (2024). Effect of natural Sapindusmukorossi treatment process on bio-waste bagasse fibers for biocomposite fabrication and application purposes. *Industrial Crops and Products 219*: 119066.

87 Anantha, K.H., Garg, K.K., Moses, D.S. et al. (2021). Impact of natural resource management interventions on water resources and environmental services in different agroecological regions of India. *Groundwater for Sustainable Development* 13: 100574.

88 Kremsa, V.Š. (2021). Sustainable management of agricultural resources (agricultural crops and animals). In: *Sustainable Resource Management* (ed. C.M. Hussain and J.F. Velasco-Muñoz), 99–145. Elsevier.

89 Liao, R., Zhang, S., Zhang, X. et al. (2021). Development of smart irrigation systems based on real-time soil moisture data in a greenhouse: proof of concept. *Agricultural Water Management* 245: 106632.

90 Sarvade, S., Upadhyay, V.B., Kumar, M., and Imran Khan, M. (2019). Soil and water conservation techniques for sustainable agriculture. In: *Sustainable Agriculture, Forest and Environmental Management*, 133–188.

91 Cole, L.J., Stockan, J., and Helliwell, R. (2020). Managing riparian buffer strips to optimise ecosystem services: a review. *Agriculture, Ecosystems & Environment* 296: 106891.

92 Peng, H., Xiong, J., Zhang, J. et al. (2023). Water requirements and comprehensive benefit evaluation of diversified crop rotations in the Huang-Huai plain. *Sustainability* 15 (13): 10229.

93 Chen, Y., Zhang, J.H., Chen, M.X. et al. (2023). Optimizing water conservation and utilization with a regulated deficit irrigation strategy in woody crops: a review. *Agricultural Water Management* 289: 108523.

94 Richards, S., Rao, L., Connelly, S. et al. (2021). Sustainable water resources through harvesting rainwater and the effectiveness of a low-cost water treatment. *Journal of Environmental Management* 286: 112223.

95 Velasco-Muñoz, J.F., Aznar-Sánchez, J.A., Batlles-delaFuente, A., and Fidelibus, M.D. (2019). Rainwater harvesting for agricultural irrigation: an analysis of global research. *Water* 11 (7): 1320.

96 Eisenhauer, D.E., Martin, D.L., Heeren, D.M., and Hoffman, G.J. (2021). *Irrigation Systems Management*. American Society of Agricultural and Biological Engineers (ASABE).

97 Bhavsar, D., Limbasia, B., Mori, Y. et al. (2023). A comprehensive and systematic study in smart drip and sprinkler irrigation systems. *Smart Agricultural Technology* 5: 100303.

98 Zheng, L. and Deng, Y. (2024). Advancing rainwater treatment technologies for irrigation of urban agriculture: a pathway toward innovation. *Science of the Total Environment* 916: 170087.

99 Wanyama, J., Bwambale, E., Kiraga, S. et al. (2024). A systematic review of fourth industrial revolution technologies in smart irrigation: constraints, opportunities, and future prospects for Sub-Saharan Africa. *Smart Agricultural Technology* 7: 100412.

100 Sharma, P., Singh, A., Kahlon, C.S. et al. (2018). The role of cover crops towards sustainable soil health and agriculture – a review paper. *American Journal of Plant Sciences 9* (9): 1935–1951.

101 Islam, R. and Sherman, B. (ed.) (2021). *Cover Crops and Sustainable Agriculture*. CRC Press.

102 Fahad, S., Chavan, S.B., Chichaghare, A.R. et al. (2022). Agroforestry systems for soil health improvement and maintenance. *Sustainability 14* (22): 14877.

103 Zhu, X., Liu, W., Chen, J. et al. (2020). Reductions in water, soil and nutrient losses and pesticide pollution in agroforestry practices: a review of evidence and processes. *Plant and Soil 453*: 45–86.

104 Xu, S., Jagadamma, S., and Rowntree, J. (2018). Response of grazing land soil health to management strategies: a summary review. *Sustainability 10* (12): 4769.

105 Teague, R. and Kreuter, U. (2020). Managing grazing to restore soil health, ecosystem function, and ecosystem services. *Frontiers in Sustainable Food Systems* 157: 534187.

106 Boyd, C.E. and Boyd, C.E. (2020). Water quality protection. In: *Water Quality: An Introduction*, 379–409.

107 Liu, C., Qu, L., Clausen, J. et al. (2023). Impact of riparian buffer zone design on surface water quality at the watershed scale, a case study in the jinghe watershed in China. *Water* 15 (15): 2696.

108 Renwick, L.L., Bowles, T.M., Deen, W., and Gaudin, A.C. (2019). Potential of increased temporal crop diversity to improve resource use efficiencies: exploiting water and nitrogen linkages. In: *Agroecosystem Diversity*, 55–73. Academic Press.

109 Bodner, G., Mentler, A., and Keiblinger, K. (2021). Plant roots for sustainable soil structure management in cropping systems. In: *The Root Systems in Sustainable Agricultural Intensification*, 45–90.

110 Pierret, A., Maeght, J.L., Clément, C. et al. (2016). Understanding deep roots and their functions in ecosystems: an advocacy for more unconventional research. *Annals of Botany 118* (4): 621–635.

111 Maurya, S.K., Kalhapure, A., Verma, V.K. et al. (2024). Irrigation scheduling and cultivar management for increasing water productivity under dryland condition: a review. *International Journal of Environment and Climate Change* 14 (1): 461–470.

112 Ullah, H., Santiago-Arenas, R., Ferdous, Z. et al. (2019). Improving water use efficiency, nitrogen use efficiency, and radiation use efficiency in field crops under drought stress: a review. *Advances in Agronomy 156*: 109–157.

113 Akhai, S., Mala, S., and Jerin, A.A. (2021). Understanding whether air filtration from air conditioners reduces the probability of virus transmission in the environment. *Journal of Advanced Research in Medical Science & Technology* 8 (1): 36–41.

114 Akhai, S., Mala, S., and Jerin, A.A. (2020). Apprehending air conditioning systems in context to COVID-19 and human health: a brief communication. *International Journal of Healthcare Education &Medical Informatics* 7 (1 and 2): 28–30.

115 Akhai, S., Singh, V.P., and John, S. (2016). Human performance in industrial design centers with small unit air conditioning systems. *Journal of Advanced Research in Production Industrial Engineering* 3 (2): 5–11.

116 Akhai, S., Singh, V.P., and John, S. (2016). Investigating indoor air quality for the split-type air conditioners in an office environment and its effect on human performance. *Journal of Mechanical Civil Engineering* 13 (6): 113–118.

117 Tanwar, N. and Akhai, S. (2017). Survey analysis for quality control comfort management in air conditioned classroom. *Journal of Advanced Research in Civil and Environmental Engineering* 4 (1 and 2): 20–23.

118 Akhai, S., Thareja, P., and Singh, V.P. (2017). Assessment of indoor environment health sustenance in air conditioned class rooms. *Advanced Research in Civil and Environmental Engineering* 4 (1 and 2): 1–9.

119 Von Schneidemesser, E., Driscoll, C., Rieder, H.E., and Schiferl, L.D. (2020). How will air quality effects on human health, crops and ecosystems change in the future? *Philosophical Transactions of the Royal Society A* 378 (2183): 20190330.

120 Kumar, A. and Jha, R. (2022). IoT in smart farming: a boon to farmers. In: *Internet of Everything: Smart Sensing Technologies*. Nova Publisher.

121 Alengebawy, A., Abdelkhalek, S.T., Qureshi, S.R., and Wang, M.Q. (2021). Heavy metals and pesticides toxicity in agricultural soil and plants: Ecological risks and human health implications. *Toxics* 9 (3): 42.

122 Scutarașu, E.C. and Trincă, L.C. (2023). Heavy metals in foods and beverages: global situation, health risks and reduction methods. *Foods* 12 (18): 3340.

123 Ullah, A.A., Afrin, S., Hosen, M.M. et al. (2021). Concentration, source identification, and potential human health risk assessment of heavy metals in chicken meat and egg in Bangladesh. In: *Environmental Science and Pollution Research*, 1–12.

124 Adegoke, A.A., Amoah, I.D., Stenström, T.A. et al. (2018). Epidemiological evidence and health risks associated with agricultural reuse of partially treated and untreated wastewater: a review. *Frontiers in Public Health* 6: 337.

125 Fazal-ur-Rehman, M. (2019). Polluted water borne diseases: symptoms, causes, treatment and prevention. *Journal of Medicinal and Chemical Sciences* 2 (1): 21–26.

126 Rani, R., Sharma, P., Kumar, R., and Hajam, Y.A. (2022). Effects of heavy metals and pesticides on fish. In: *Bacterial Fish Diseases*, 59–86. Academic Press.

127 Das, A., Joardar, M., De, A. et al. (2021). Pollution index and health risk assessment of arsenic through different groundwater sources and its load on soil-paddy-rice system in a part of Murshidabad district of West Bengal, India. *Groundwater for Sustainable Development* 15: 100652.

128 Young, S.L., Frongillo, E.A., Jamaluddine, Z. et al. (2021). Perspective: the importance of water security for ensuring food security, good nutrition, and well-being. *Advances in Nutrition* 12 (4): 1058–1073.

129 Dar, B.A., Yaqoob, A., and Mir, G.J. (2023). Effect of environmental contaminants on female reproductive health: a narrative review: environmental contaminants & female reproductive health. *BME Horizon 1* (3): https://doi.org/10.37155/2972-449X-vol1(3)-86.

130 Aggarwal, P., Rana, M., and Akhai, S. (2022). Briefings on e-waste hazard until COVID era in India. *Materials Today Proceedings* 71: 389–393.

131 Sabarwal, A., Kumar, K., and Singh, R.P. (2018). Hazardous effects of chemical pesticides on human health–cancer and other associated disorders. *Environmental Toxicology and Pharmacology* 63: 103–114.

132 Thangamani, R., Sathya, D., Kamalam, G.K., and Lyer, G.N. (2024). AI green revolution: reshaping agriculture's future. In: *Intelligent Robots and Drones for Precision Agriculture*, 421–461. Springer Nature Switzerland: Cham.

133 Akhai, S. (2024). Towards trustworthy and reliable AI the next frontier. *Explainable Artificial Intelligence (XAI) in Healthcare 1*: 119–129.

134 Akhai, S. and Kumar, V. (2024). Quantum resilience and distributed trust: the promise of blockchain and quantum computing in defense. In: *Sustainable Security Practices Using Blockchain, Quantum and Post-Quantum Technologies for Real Time Applications*, 125–153.

135 Rajitha, A., Aravinda, K., Nagpal, A.et al. (2024). Machine learning and AI-driven water quality monitoring and treatment. *E3S Web of Conferences*, Vol. 505, p. 03012. EDP Sciences.

136 Akhai, S. (2024). Trends and environmental impact of paper consumption: a prognostic scenario for the Indian market by 2030-a case study. *International Conference on Interdisciplinary Approaches in Civil Engineering for Sustainable Development*, Vol. 464, pp. 11–18. Springer Nature Singapore.

137 Ahmad, L. and Nabi, F. (2021). *Agriculture 5.0: Artificial intelligence, IoT and machine learning*. CRC Press.

138 Kumar, A.S., Sharma, P., Kaur, S. et al. (2023). AI-equipped IoT applications in high-tech agriculture using machine learning. In: *Handbook of Research on AI-Equipped IoT Applications in High-Tech Agriculture*, 38–64. IGI Global.

139 Reddy, K.S., Ahmad, S.S., and Tyagi, A.K. (2024). Artificial intelligence and the internet of things-enabled smart agriculture for the modern era. In: *AI Applications for Business, Medical, and Agricultural Sustainability*, 68–99. IGI Global.

140 Karn, S., Kotecha, R., and Pandey, R.K. (2024). Towards sustainable farming: leveraging AIoT for precision water management and crop yield optimization. *Procedia Computer Science* 233: 772–781.

2

Precision Farming: A Technological Revolution for Sustainable Agriculture

Ashish Kumar, Prasoon Kumar Pandey, and Divya Singh

School of Computer Science Engineering and Technology, Bennett University, Greater Noida, Uttar Pradesh, India

2.1 Introduction

Artificial intelligence (AI) has a wide range of applications in various fields that includes tracking [1, 2], education [3, 4], tourism [5], and disease prediction [6, 7]. In agriculture, AI with its machine learning (ML)- and deep learning (DL)-based algorithms have been explored to improve productivity. With the advent of technology, conventional farming techniques are improved, such as precision farming. Precision farming is a local management system that adopts, analyzes, and handles the variability in fields using information technology and the Internet of Things (IoT) [8]. The key focus of precision farming is to increase profitability, by increasing productivity, preventing land resources and preserving sustainable resources [9]. Precision farming improves various aspects of crop production and helps farmers in taking intelligent decisions for better crop yield. This method considers the natural fluctuation in various farming fields and tries to handle it efficiently to obtain maximum efficiency. The idea is not to get the same yields all over the field but to adopt a way to manage and distribute inputs on the basis of a site-specific manner to achieve long-term cost-effectiveness.

Conventional farming techniques utilize the same methods and procedures uniformly throughout the field, which is an ineffective method of production. On the other hand, precision farming is considered by farmers all over the world to boost the output of crops in their fields. It is typically associated with technologies such as variable-rate farming or GPS (global positioning system) agriculture [10].

Emerging Smart Agricultural Practices Using Artificial Intelligence, First Edition.
Edited by Ashish Kumar, Jai Prakash Verma, and Rachna Jain.

Published 2025 by John Wiley & Sons, Inc.

These tools can increase the productivity by helping farmers in applying the received information effectively. Farmers who apply information successfully will, in most cases, be able to get more returns than those who don't.

Precision farming emphasizes dividing the whole field into small areas and treats them as individual units for their proper management [11]. This method will increase the level of management, which clearly shows how important it is to have good agronomic practices and is proof of the necessity of these practices. In order to effectively implement precision agriculture, farmers must first have a strong farm management system in place. Precision agriculture is a farming method that is based on systems and offers benefits to the environment, economy, and practical aspects of field-level management and technology [12]. The adoption of the right agricultural techniques, the use of various crop varieties, strategic crop rotation, precise application of chemicals and fertilizers, as well as the careful monitoring of crops, help the farmers to achieve high yields, reduce inputs, and maximize profits.

Gaining a better understanding of the evolution of precision agriculture throughout the years, it is necessary to first of all, mention some basic aspects of this concept. In the bygone days, when family-owned small-scale farms were the leading type of organization, farmers were able to see the spatial variability of soil and its effect on crop production. Hence, here crop yields are according to the changes in the weather [13]. The mechanization of agriculture was later used to make the agriculture economical in the large fields with uniform input usage. A farmer who currently bears the responsibility for the management of larger areas has less agronomic information than the ten farmers who previously used to work in the same area. GPS enabling the turning of this process around. The GPS technology that is used along with the specialized equipment that is capable of measuring variability and applying inputs such as fertilizers and herbicides is mandatory for the progress of precision agriculture.

Precision farming is a new trend in the agricultural world that is created by the progress of technology. The main goal of this study is to analyze the widely used concepts of this kind of farming and to determine the technical and economic effectiveness of various technologies through a systematic literature review mentioned in this research paper. The key contributions of this chapter are as follows:

- The comprehensive description about the principles of precision farming is highlighted to under the components, advantages, and challenges.
- The key technologies and benefits of robots in precision farming are explored to understand the requirements.
- The limitations and benefits are also discussed to explore the future of precision farming in agricultural practices.

The rest of the chapter is organized as follows. Section 2.2 discusses the principles and procedures of precision farming. The key technologies in precision farming are elaborated in Section 2.3. The role and benefits of robots in precision farming are reviewed in Section 2.4. The limitations and advantages of precision farming in revolutionizing conventional farming are highlighted in Section 2.5. Finally, the conclusion and future directions are sketched in Section 2.6.

2.2 Principles of Precision Farming

The basic principle of precision farming is to initially capture the data using various sensors such as soil moisture sensors, field temperature, and humidity measurements. The next step is to analyze this sensor data for better analysis [14]. Figure 2.1 represents the basic principles of precision farming.

The basic concepts of precision farming are as follows:

1) **Data Collection**: Precision farming is all about the collection of different kinds of data, including soil type, weather patterns, crop growth stages, and yield potential [11]. The gathered data are obtained by a variety of methods, such as sensors, satellites, drones, and other monitoring devices.
2) **Data Analysis**: After the data has been collected, it is combed through to find the patterns, trends, and correlations [12]. The new high-tech analytics, including ML algorithms, are being employed to get unprecedented benefits. IoT devices are also used for analysis of the data collected from the sensors.
3) **Agricultural Database and Real-Time Monitoring**: The collected data from the different sensors acts as an input to various IoT devices [8]. These devices are responsible for continuous and real-time monitoring of various information such as soil moisture, water temperature, fertilizer quality, soil

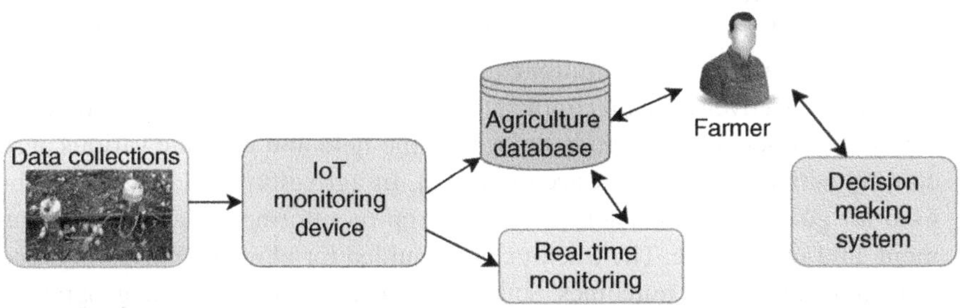

Figure 2.1 Basic procedure steps for data analysis and monitoring in precision farming.

nutrient level, and many more. The data getting added to the agricultural database so that accurate and effective decisions can be provided to farmers to increase crop yield.

4) **Decision-Making System**: The so-called knowledge gathered from data analysis is used to make decisions about the management of crops, irrigation, fertilization, pest control, and harvesting. The decision support systems help the farmers to keep the inputs optimum, the yields to be maximum, and the costs and environmental impact to be as low as possible.

These steps are crucial in precision farming to improve the overall productivity, and hence, to increase the farmer's income. The data collected from sensors is integrated with the data recorded in the agricultural database to identify the individual needs of the farmer. This strategy is helpful in providing personalized decisions on the choice of pesticides, fertilizers, and manure in the field. It will also prevent water wastage and excessive use of pesticides and weedicides to prevent the growth of germs and weeds in the field. This step not only improves the crop yield but also prevents the environment from the negative impact of harmful chemicals. However, this sort of application is limited to large farming fields where the financial capital is quite high. The benefits of precision farming for small farms and farming cooperatives are yet to be explored for better outcomes.

2.3 Technologies in Precision Farming

Precision agriculture is an amalgamation of several technologies that are employed to acquire data, monitor crops, and carry out farming activities [15]. Key technologies include:

Global Positioning System (GPS): GPS utilizes a constellation of satellites through which the farmers can be accurately positioned and navigate. They can map fields with precision, monitor equipment movements as well as apply variable rates effectively. This technology records the geographical coordinates in terms of latitude and longitude to localize the field and navigate various agricultural activities with great accuracy [15]. In agriculture fields, GPS can be used for guiding tractors, soil sampling, crop monitoring, irrigation management, and many more. The tractor can be guided for plowing, harvesting, and pest control for better outcomes. Soil samples can be collected, and geographic coordinates are mapped using GPS and the mapping software. These samples can be used for soil treatment and providing region-related information. This will save money by providing variable rate applications depending on the requirements of certain regions. Also, GPS can monitor and assess the crop by dividing the farms into small zones or regions. The information can be used for understanding the soil requirements, irrigation management, and planting samplings in the near future.

Remote Sensing: The usage of technologies, such as satellites, drones, and aerial imagery, is very helpful in providing different kinds of information about crop health, soil moisture levels, pest problems, and yield fluctuations [16]. These technologies utilize visible light, infrared, and near-infrared light to collect agricultural data without making physical connections with any object. Satellites can provide real-time data for monitoring the field and can guide tractors for better management of various agricultural activities. Drones are helpful in monitoring the large fields. With the help of drones' pest control and chemical spraying in large fields can be done in a faster and economical way. There are variety of drones available fulfilling the farmers need. Aerial imagery is helpful in the assessment of soil quality and moisture. Aerial imagery can also be helpful in examining crop quality, soil humidity, and air pressure.

Remote Sensors: These sensors collect useful information suitable for determining the soil quality, moisture, humidity, and nutrients. Depending on the utilized technologies, sensors in precision farming are categorized as ground sensors, aerial sensors, and satellite sensors. Ground sensors collect data by establishing a physical connection with objects. These sensors record and monitor the soil quality, moisture, and humidity [17]. These sensors can monitor the information from a small geographical area. For large fields, aerial and satellite sensors are preferred. These sensors collect data with the help of remote sensing techniques for better management of the agricultural fields.

IoT: IoT connects many physical devices with the help of the Internet to collect and share data globally. IoT in the agricultural field not only improves the crop yield but also makes it profitable for the farmers [18]. IoT has various applications in agriculture, such as automated irrigation systems, weather stations, and soil moisture sensors, which provide real-time agricultural operation, monitoring, and control [19]. These techniques modernize the conventional agricultural practices by preserving the precious natural resources. Automated and intelligent irrigation systems check the station keep a check of on the dynamic weather conditions and inform farmers beforehand so that timely decisions can be taken to prevent the crops from the calamity.

Autonomous Machinery: Autonomous tractors, harvesters, and drones powered by AI and computer vision technologies can carry out tasks such as planting, spraying, and monitoring with the least human involvement. This machinery can cover a large area at a faster speed. Also, the transporting of crops from the field to the storage warehouses for supplying to local markets is crucial and can be done in an efficient way with the help of tractors and other machineries [20]. Technology is essential to prevent wastage during supply and delivery by maintaining the correct temperature and moisture. Predictive analysis must be performed to ensure high profitability and long-term viability.

2.3.1 Disease Prediction

Identification of plant disease at an early stage is very important for taking appropriate and timely measures to minimize the loss. With the advancement in technology, AI-based disease prediction algorithms are very popular. These algorithms utilized either ML [21] or DL [22–25] techniques in their predictive models. Figure 2.2 depicts the necessary steps in classifying the leaves either as healthy or infected. Initially, the data can be captured either with the help of a mobile device or the sensors mounted in the field. This data act as an input to the ML or DL algorithms to predict the disease. After this, the decision-making system will classify the input as healthy or infected leaves. Based on the information, the infected leaves can be provided with the disease type and the corresponding cure and treatment.

Table 2.1 tabulates the salient features of the techniques utilized for disease prediction. The plants are suffering from various infections such as fungal, bacterial, viruses, and other diseases [32]. Conventional techniques require high expertise knowledge for examining the plant infection and is time-consuming too. On the other hand, AI-based algorithms are superior in predicting plant diseases at an early stage. These algorithms can diagnose diseases in grains plant such as wheat, maize, and rice, in vegetable plants such as tomato, peanuts, and in spices plant such as pepper and also in kiwi and orchids plant too. The early

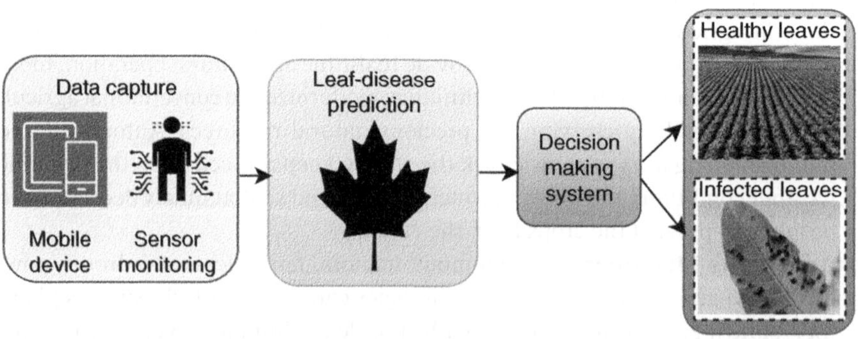

Figure 2.2 Technology-based disease prediction and monitoring in precision farming. *Source:* Kaggle, Inc./https://www.kaggle.com/datasets/emmarex/plantdisease/suggestions?status=pending&yourSuggestions=true// last accessed on January 10, 2024.

Table 2.1 Salient features of recent techniques for disease prediction models in precision farming.

SN	Citation	Technique	Methodology	Disease prediction	Summary
1.	Joseph et al. [26]	DL	Eight CNN models	Fungal and bacterial disease in rice, maize, and wheat plant	• Obtained ACC upto 98% among various models • Suitable to predict disease in three food grains efficiently
2.	Kini et al. [27]	DL and TL	Inception V3, GoogleNet, SqueezeNet, and Resnet18	Prominent diseases in black pepper leaves	• Obtained ACC up to 99.7% using various DL models • Utilized trainable network to predict newly developed diseases in pepper leaves
3.	Maurya et al. [23]	DL and IoT	MLP and LSTM	Cotton, maize, and derived plant village	• Computationally efficient with a smaller number of trainable parameters • Achieved ACC of upto 98% in most of DB for disease prediction
4.	Khalid and Karan [22]	DL and XAI	CNN, MobileNet, and GradCAM	Diseased and healthy leaves from Kaggle DB	• Automated and precise detection of diseased leaves for food security • Achieved 89% ACC for diseased leaves classification
5.	Pereira et al. [28]	Hyperspectral sensing	Automated spectral band analysis tool	Bacterial disease in tomato and kiwi plants	• Obtained overall ACC up to 75% bacterial disease prediction • Feature reduction for better computational performance
6.	Pahlawanto et al. [24]	DL	VGG16	Rice plant disease	• Up to 93% ACC in test DB • Utilized data augmentation technique to increase DB size

(Continued)

Table 2.1 (Continued)

SN	Citation	Technique	Methodology	Disease prediction	Summary
7.	Kothari et al. [29]	DL and GPT-3.5 Turbo	ResNet50	Plant disease	• AI-driven chatbot for detection of plant disease • Provided plant disease prediction, cure, and preventive measures
8.	Kotwal et al. [30]	DL	UNet	Plant disease	• Extracted robust features using GLCM, SIFT, and Gabor filter • Segmented image using UNet to extract the infected region
9.	Tsai and Hsu [25]	DL	DL, edge computing, and IoT	Orchid plant disease	• Extracted low-level features using edge computing • IoT for obtaining high-level features ignoring environmental changes
10.	Routis et al. [21]	ML	ASIC	Plant disease	• IoT-based application to classify leaf image either as healthy or diseased • ML-based algorithms for better efficiency and ACC
11.	Guo et al. [31]	DL	CNN and LSTM	Peanut leaf spot prediction	• Captured spatial temporal features for capturing disease dynamics • SE network for key feature extraction and improved performance

ACC, Accuracy; AI, Artificial Intelligence; ASIC, Application specific integrated circuits; CNN, Convolutional neural network; DB, Database; DL, Deep learning; GLCM, Gray level co-occurrence matrix; IoT, Internet of Things; LSTM, Long-short term memory; ML, Machine learning; SE, Squeeze and excitation network; SIFT, Scale invariant feature transform; TL, Transfer learning identification of infection can not only prevent the damage to the farming but also enhances productivity for improved profitability. In this direction, authors exploited CNN (convolutional neural network) for predicting the infection in rice plants to improve the grain quality and quantity [24]. The algorithm was able to diagnose bacterial infections in rice plants, such as leaf blight, brown spot, and smut. It analyzed the shape, size, and color of the leaf to predict the correct disease and avoid yield loss. On the other hand, Maurya et al. [23] have examined a lightweight meta-ensemble approach consisting of MLP-mixer (multilayer perceptron) and LSTM (long-short term memory) for predicting plant diseases in cotton, maize, and plant village dataset (DB). Initially, the model was trained, and the classifier was used for predicting the plant diseases. The model was suitable for predicting bacterial and virus infection in cotton, tomato, potato, pepper, and maize leaves. In [27], authors presented a transfer learning (TL) along with a DL model for predicting fungal and bacterial infections in pepper plant leaves. The methodology can diagnose the infection at an early stage and in real time.

To increase the yield, it is essential to prevent the plant from infections and insects. For this, authors utilized CNN and other DL models to predict plant diseases with high accuracy [22]. Explainable AI (XAI) provided the model interpretation using GradCAM for understanding the model decision-making. However, the authors examined the hyperspectral data for diagnosing bacterial infection in the tomato and kiwi plants [28]. The automated spectral band analysis tool discriminated the tomato and kiwi leaves into healthy categories and two types of bacterial suspensions. This tool utilized parametric spectral vegetation indices and Gaussian process classification for analyzing the leaf diseases for discriminating the nonsymptomatic tissues from the symptomatic one. The authors designed an AI-based chatbot application named as CropGuard for recognizing various plant diseases [29]. Initially, DB was preprocessed to modify the images using scaling, normalization, and standard deviation steps. DB was augmented to increase the training size by subjecting images to rotation, shearing, flipping, and zooming. These steps enhanced the classification accuracy of the model. However, authors constantly monitored plant health by using ML-based algorithms in IoT environments [21]. To make the model computationally efficient, single board computers were used for disease identification. The correlation between the input and output was generated to improve the learning process accuracy.

Time series data was considered suitable for accurate and timely management of critical diseases in plants. In this direction, authors validated the peanut leaf spot disease in peanut plants by investigating the time series data using DL models [31]. Spatial-temporal patterns in leaves were captured and analyzed using CNN and LSTM models. Also, the squeeze and excitation network was exploited to analyze the dynamic features for enhancing the model's stability and accuracy. However, Tsai et al. [25] exploited edge computing in IoT technology for identifying various fungal, bacterial, and virus infections in orchid plants. Global and local features such as primitive, logical inference, and semantic content were extracted from orchid images to analyze the dynamic features for accurate classification of disease. The features were processed using a DL model to monitor the health of the orchid plant. The authors proposed a hybrid and optimized strategy for predicting plant diseases automatically [30]. The dynamic features such as gray level co-occurrence matrix, scale-invariant feature transform, and Gabor filter were utilized for robust feature extraction to enhance classification accuracy in the EfficientNet DL model. Hunterprey optimization was utilized during image segmentation to improve the computational performance of the model.

To summarize, automatic plant disease prediction and detection at an early stage is crucial for improving agricultural yield. Agricultural plants must be monitored continuously for accurate diagnosis of the disease. Plant diseases are a major threat not only to productivity but also to the quality of the crop and the economy of the country. Hence, to extend suitable preventive measures and cure the disease in the plants effective methods are required for its identification to minimize the loss.

2.3.2 Livestock Monitoring and Tracking

Livestock monitoring and cattle tracking are paramount in sustainable agriculture [33]. The animals are mounted with GPS devices, and their location can be tracked in the fields. AI and sensors are helpful in the social welfare of animals by considering the ethical limitations of managing product quality, disease management, and environmental parameters. These techniques have revolutionized precision farming by providing compelling solutions to real-time livestock and environment monitoring.

GPS-based trackers are helpful for real-time tracking and monitoring of livestock. GPS-based geotrackers were discussed in agricultural research as a benefit to precision farming for industrial growth [34]. There are different types of GPS devices, such as FindMy, Digitanimal, and FarmRanger, available for monitoring livestock. These devices are country specific and require mobile phones with satellite connections for their functioning.

For enhanced monitoring and surveillance of the livestock, sensors were integrated with IoT for efficient resource management in decision-making [35]. Initially, the challenges in current production system were analyzed and the segments were identified to integrate livestock management for betterment. Concerns in livestock were detected, and variables were selected as indicators. Sensors were selected for appropriate data transmission with selected software and data analysis. Teams were trained, and applications were managed by fine-tuning the applications for monitoring the application for cost benefits.

Precision farming monitors, manages and controls the real-time assessment of livestock farming with the help of sensors and advanced algorithms [36]. The technologies include RFID (radio frequency identification) for noncontact and remote monitoring of cattle. The RFID tags were tied with the animals and information such as breed, age, type, and health status was stored in it. These RFID tags ensured the continuous monitoring of the animals and were helpful in growth and medical treatment when required. GPS systems were also embedded in the cattle fields. These systems are helpful for monitoring the grazing activities of the animals. There exist other sensors such as thermal and weight, which are helpful in continuous monitoring and assessment of livestock for their betterment and welfare.

2.4 Role of Drones in Precision Farming

Drones are currently being used in agriculture to monitor every inch of farmland using special cameras. Drones have changed the face of modern farming, and they are saving costs and reducing the need for human intervention at the same time [37]. In modern farming, drones have a wide range of applications that include weed removal, smart irrigation, fertilization, and pest control. Drones can automate several farming processes and can reduce the farmer's workload. In addition, drones can acquire and monitor real-time data so that timely actions can be taken to prevent farming loss due to various environmental impacts. Figure 2.3 illustrates the key benefits of robots in precision farming.

Precision Agriculture: Unmanned vehicles, which are fitted with different kinds of sensors, like multispectral or hyperspectral cameras, can gather data in high resolution about crop health, moisture levels, and soil conditions. Therefore, farmers can take precise actions regarding irrigation, fertilization, and pest control that will result in the best yields and optimization of the resources [38].

Cost Efficiency: Drone-based data collection is cheaper than the traditional ways of data collection like manned aircraft and ground surveys. They can cover big areas in no time and need less manpower and fuel, thus, it brings the farmers saving [39].

Crop Monitoring: Drones provide a bird's-eye view of the whole fields; thus, farmers are able to observe the crop growth and discover the problems in disease outbreaks, nutrient deficiencies, or weed infestations early on [40]. The

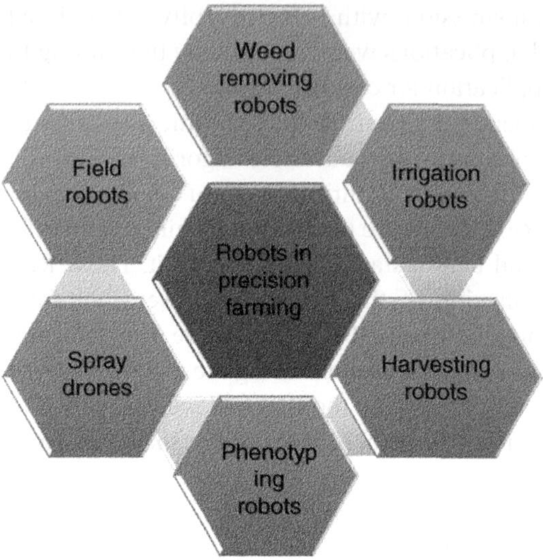

Figure 2.3 Benefits of robots in precision farming.

instant information obtained on the fields through this technology will allow the intervention to be done at the right time which will in turn prevent crop losses and will also maximize the productivity.

Spraying and Pest Control: The drones that are fitted with the spraying system can exactly spray the pesticides, herbicides, or fertilizers to specific areas, which helps in the reduction of chemical usage and minimizes the environmental impact [41]. They are able to go through the difficult roads more comfortably than the conventional machines, to the remote or unreachable places they can reach more easily.

Time Savings: Through drones, farmers can do activities like field scouting, mapping, and spraying in a fraction of the time that it would take them to work the same tasks with the traditional methods. This way, the most precious time is freed up and can be assigned to other farming activities, thus improving the overall efficiency of the farm management [42].

Safety: Through the decrease of the necessity of manual work in hazardous areas, like highlands or steep lands, drones make the safety of the farm people better. Apart from that, they also decrease exposure to the chemicals that are harmful by applying them only on the places where they are needed, hence, they reduce the health risks [43].

These advantages altogether show that drones are the key factor in the modern farming, and this is what is making the farmers to make the decisions that are based on the data, increase the productivity, and to adopt the sustainable agriculture practices.

2.5 Benefits of Precision Farming

Precision farming collects the data with the help of sensors and analyzes it to take intelligent and accurate decisions for better farming results. Precision farming offers several advantages and benefits that include enhanced productivity, economic efficiency, irrigation management, sustainable farming, and prevention of excessive use of fertilizers and chemicals. These steps contribute to enhanced productivity and efficiency of the farming results. The key benefits of precision farming for farmers, consumers, and the environment are as follows.

Enhanced Productivity: Precision agriculture makes use of technology to optimize the use of pesticides, fertilizers, and water, resulting in higher yields and higher-quality crops. Recent technologies such as ML and DL have designed algorithms that are helpful in determining the quality and quantity of pesticides and chemicals that are crucial for high productivity [15].

Economic Efficiency: Precision agriculture is a method that helps farmers to efficiently use the resources, which consequently reduces the input costs and increases the profits of the farmers. Real-time monitoring of soil and crops by the sensors enables IoT devices to take intelligent decisions and prevent the excessive usage of potential resources [32]. These innovative technologies not only determine the suitability of the resources but also helpful in conservating the precious environmental resources.

Sustainable Agriculture: By cutting the chemical intake, soil erosion, and water conservation, precision agriculture is the key to environmental good and sustainability. To increase productivity, harmful chemicals, fertilizers, and pest control mechanisms are used in conventional farming [35]. Nitrogen fertilizers, phosphate, and urea are used by farmers to increase agricultural production. However, these chemicals are not only affecting the fertility of the soil but also lead to soil and water pollution. Precision farming effectively utilizes these chemicals and by monitoring the quality and nutritional value of the soil and helpful in substantiable development.

Data-Driven Decision-Making: Using real-time data to inform decisions, precision agriculture enables farmers to manage their crops more effectively while lowering risks. This helps farmers to take better judgments. The real-time information should be shared with the policy related to yield management, disease prevention, and storage management [36]. This information is helpful in planning marketing strategies for efficient distribution of the crops. Many recent technologies are utilized in precision farming that are not only helpful in yield prediction but also determine the diseased plant. These techniques enable farmers to take timely decisions to prevent the farming loss at later stages.

2.5.1 Advantages and Limitations of Precision Farming

Precision farming is not only advantageous to farmers but also to consumers, policymakers, and the environment. [15]. The advantages and limitations of precision farming are detailed as follows:

Enhanced Productivity: Precision farming makes the optimum use of resources like water, fertilizer, and pesticides to maximize crop quality and yield the highest yield.

Cost-Effectiveness: Through the proper use of resources, precision farming, which is the farming of crops and other agricultural products on a small scale, decreases the input costs and, hence, the profitability of farmers.

Sustainable Practices: Precision farming is the way of farming that maximizes both productivity and sustainability by using less chemicals, decreasing soil erosion, and saving water.

Informed Decision-Making: Through real-time data, farmers can make wise decisions, thus, precision farming improves the management of crops and reduces the risks.

2.5.2 Possibilities and Opportunities of Precision Farming

The future of precision farming is very bright, and it will be great news for further advancement and innovation, which will be driven by the ongoing progress in sensors, robotics, AI, and data analytics [38]. Hence, the unceasing advancements in technology will enable precision farming to be more affordable and more efficient.

Through the combination of precision farming with sustainable agricultural practices such as organic farming, agroforestry, and conservation agriculture, we can strengthen environmental protection and resilience [9]. Thus, this integration shall be the main factor that will lead to the sustainability of the whole agricultural area.

The rise of digital agriculture platforms and marketplaces will be the key factor in data sharing, collaboration, and value creation throughout the entire agricultural value chain. These platforms will provide a way for the farmers and stakeholders to access the data that will be the key to good decisions and better productivity [19].

As more and more people are becoming aware of the advantages of precision farming, we can foresee a considerable rise in its worldwide use and commonness. The developing countries, to be specific, will be the ones that will use these new technologies to deal with the food security problems they have and to improve their agricultural ways. Numerous definitions and research studies have demonstrated that improved decision-making in agriculture may lead to a variety of benefits [17]. Precision farming has a number of benefits, including higher agricultural yields, more efficient use of inputs, and higher-quality labor and management.

There are many advantages of precision farming that includes higher crop yield, better process control, and less environmental exposure to agrochemicals. This ensures better input management to enhance the gross margins for the betterment of farmers. The adoption of modern technology in conventional fields increases the farmer's profits [20]. The yield monitoring in precision farming is done to maintain the style, size, and impact of crop type on financial outcomes.

The experiments, site-specific fertilizer response functions, and expert farmer assessments serve as the foundation for economic research on precision fertilizer management [32]. Soil sampling is an innovative way to determine differential lime treatment to reduce the production cost, and spatial variability of crop yield which are significant part of the precision farming. In addition, sensor-based fertilizer distribution in the field ensured the variable treatment based on the high and low requirements of the zones to prevent excessive usage of fertilizers [12]. Nitrogen fertilization and precision irrigation are widely used because they can increase yields while conserving water and nutrients. Various studies are conducted to analyze the spatial patterns in water bodies to monitor the variations to monitor the climate changes [44, 45].

2.5.3 Exploration of Precision Farming at Farm Level

Most of the steps and procedures highlighted in precision farming are either explored at large farms or cooperative farms. In order to promote precision farming at the small farm level, the following strategies should be considered [46]. The key highlights are as follows:

- Widespread use of precision farming technology should be promoted among farmers who are creative, innovative, and have a high-risk tolerance; this technology often requires a lot of money.
- Determine the areas that will be suitable for planting organic crops that are specific to the area.
- Adoption of the water accounting protocols on farms should be emphasized for better outcomes.
- Microlevel irrigation systems and water-saving techniques should be adopted to save environmental and improve productivity.
- The research should be carried out to study the spatial and temporal variability of the input parameters at the field level, using field-level primary data.
- The policy should be designed that is helpful in smooth and effective transition of technology to farmers.
- Technical support for the development of pilot projects or models that can be scaled up should be extended to farmers. It is the key to implement the innovative idea of the food production.

- Procurement pricing and the formation of cooperative or self-help groups should be created to help small-scale farmers utilize the benefits of technology efficiently.
- Export promotion zones have to be set up on the basis of precision agriculture with precision.
- The creation of export promotion zones with the requisite infrastructure, such as the cold storage, processing, and grading facilities, is the process to designate the export promotion zones.

2.6 Conclusion

The precision agriculture makes the farmers to use the crop inputs like fertilizers, pesticides, tillage, and irrigation water in the most efficient way to optimize the farming process. The input maximization will enable the farmers to get the higher crop yield and quality as well as the reduction of environmental pollution. Nevertheless, the cost benefits of precision agriculture management were examined, and it was found to be a tough task. A lot of the technologies that are used now are still in their early stages, thus, the true price for the equipment and services cannot be determined. Thus, the economic evaluations of technologies will be deemed obsolete in the future.

Precision agriculture is considered a possible solution to the world's economic and environmental problems in modern agricultural production. Although the cost-effectivity and the best way of using the existing technological tools are still unknown, the principal idea of "doing the right thing in the right place at the right time" is still attractive. The achievement of precision agriculture is largely dependent on the speedy, knowledge-based acquisition and application of which the new technologies are guided.

References

1 Kumar, A. (2023). *Visual Object Tracking Using Deep Learning*. CRC Press.

2 Kumar, A., Jain, R., Vairamani, A.D., and Nayyar, A. (2023). *Object Tracking Technology: Trends, Challenges and Applications*. Springer.

3 Kumar, A., Nayyar, A., Sachan, R.K., and Jain, R. (2023). *AI-assisted Special Education for Students with Exceptional Needs*. IGI Global.

4 Kumar, A., Singh, D., and Vohra, R. (2023). Improving learning abilities using AI-based education systems. In: *AI-Assisted Special Education for Students With Exceptional Needs* (ed. A. Kumar, A. Nayyar, R. Sachan, and R. Jain), 137–155. IGI Global.

5 Kumar, A. and Vohra, R. (2023). Impact of deep learning models for technology sustainability in tourism using big data analytics. In: *Deep Learning Technologies for the Sustainable Development Goals: Issues and Solutions in the Post-COVID Era* (ed. V. Kadyan, T.P. Singh, and C. Ugwu), 83–96. Springer.

6 Kumar, A., Aelgani, V., Vohra, R. et al. (2024). Artificial intelligence bias in medical system designs: a systematic review. *Multimedia Tools and Applications* 83 (6): 18005–18057.

7 Kumar, A., Jain, R., Gupta, M., and Islam, S.M. (2023). *6G-enabled IoT and AI for Smart Healthcare: Challenges, Impact, and Analysis*. CRC Press.

8 Ruan, J., Jiang, H., Zhu, C. et al. (2019). Agriculture IoT: emerging trends, cooperation networks, and outlook. *IEEE Wireless Communications* 26 (6): 56–63.

9 Finger, R., Swinton, S.M., El Benni, N., and Walter, A. (2019). Precision farming at the nexus of agricultural production and the environment. *Annual Review of Resource Economics* 11: 313–335.

10 Yadav, N. and Sidana, N. (2023). Precision agriculture technologies: analysing the use of advanced technologies, such as drones, sensors, and GPS, in precision agriculture for optimizing resource management, crop monitoring, and yield prediction. *Journal of Advanced Zoology* 44.

11 Paulin, M.S., Pant, R., Naveena, S.S. et al. (2024). Designing a precision seed sowing machine for enhanced crop productivity. *E3S Web of Conferences*. EDP Sciences, 11th an 12th January, 2024, AAA college of Engineering and Technology, Affiliated to Anna University, Chennai.

12 Kushwaha, M., Singh, S., Singh, V., and Dwivedi, S. (2024). Precision farming: a review of methods, technologies, and future prospects. *International Journal of Environment, Agriculture and Biotechnology* 9 (2): 242–253.

13 Pande, C.B. and Moharir, K.N. (2023). Application of hyperspectral remote sensing role in precision farming and sustainable agriculture under climate change: a review. In: *Climate Change Impacts on Natural Resources, Ecosystems and Agricultural Systems* (ed. C.B. Pande, K.N. Moharir, S.K. Singh, et al.), 503–520. Springer.

14 Mohyuddin, G., Khan, M.A., Haseeb, A. et al. (2024). *Evaluation of Machine Learning approaches for precision Farming in Smart Agriculture System-A Comprehensive Review*. IEEE Access.

15 Bajaj, S., Mishra, S., Chouhan, D. et al. (2023). A comprehensive review of application of RS, GIS and GPS in agriculture India. *International Journal of Environment and Climate Change* 13 (11): 2597–2608.

16 Prajapati, J.B., Barad, R., Patel, M.B. et al. (2023). Smart farming ingredients: IoT sensors, software, connectivity, data analytics, robots, drones, GIS-GPS. In: *Applying Drone Technologies and Robotics for Agricultural Sustainability* (ed. P. Raj, K. Saini, and V. Pacheco), 31–49. IGI Global.

17 Alahmad, T., Neményi, M., and Nyéki, A. (2023). Applying IoT sensors and big data to improve precision crop production: a review. *Agronomy* 13 (10): 2603.

18 Kumar, A. and Jha, R. (2022). IoT in smart farming: a boon to farmers. In: *Internet of Everything: Smart Sensing Technologies* (ed. T. Kavitha, V.A. Devi, S.N. Pari, and S. Ramanathan), 185–206. Nova Publisher.

19 Alazzai, W.K., Abood, B.S.Z., Al-Jawahry, H.M., and Obaid, M.K. (2024). Precision farming: the power of AI and IoT technologies. *E3S Web of Conferences*. EDP Sciences, January 11–12, 2024, Tamil Nadu, India.

20 D'Antonio, P., Mehmeti, A., Toscano, F., and Fiorentino, C. (2023). Operating performance of manual, semi-automatic, and automatic tractor guidance systems for precision farming. *Research in Agricultural Engineering* 69 (4): 179.

21 Routis, G., Michailidis, M., and Roussaki, I. (2024). Plant disease identification using machine learning algorithms on single-board computers in IoT environments. *Electronics* 13 (6): 1010.

22 Khalid, M.M. and Karan, O. (2024). Deep learning for plant disease detection. *International Journal of Mathematics, Statistics, and Computer Science* 2: 75–84.

23 Maurya, R., Mahapatra, S., and Rajput, L. (2024). A lightweight meta-ensemble approach for plant disease detection suitable for IoT-based environments. *IEEE Access* 12: 28096–28108.

24 Pahlawanto, R.D.S., Salsabila, H., and Pratiwi, K.R. (2024). Detection and prediction of rice plant diseases using convolutional neural network (CNN) method. *Journal of Student Research Exploration* 2 (1): 22–33.

25 Tsai, Y.-H. and Hsu, T.-C. (2024). An effective deep neural network in edge computing enabled internet of things for plant diseases monitoring. *Proceedings of the IEEE/CVF Winter Conference on Applications of Computer Vision*, 11th an 12th January, 2024, AAA college of Engineering and Technology, Affiliated to Anna University, Chennai.

26 Joseph, D.S., Pawar, P.M., and Chakradeo, K. (2024). *Real-time Plant Disease Dataset Development and Detection of Plant Disease Using Deep Learning.* IEEE Access.

27 Kini, A.S., Prema, K., and Pai, S.N. (2024). Early stage black pepper leaf disease prediction based on transfer learning using ConvNets. *Scientific Reports* 14 (1): 1404.

28 Reis Pereira, M., Verrelst, J., Tosin, R. et al. (2024). Plant disease diagnosis based on hyperspectral sensing: comparative analysis of parametric spectral vegetation indices and nonparametric gaussian process classification approaches. *Agronomy* 14 (3): 493.

29 Kothari, S., Bagane, P., Mishra, M. et al. (2024). CropGuard: empowering agriculture with AI driven plant disease detection chatbot. *International Journal of Intelligent Systems and Applications in Engineering* 12 (12s): 530–537.

30 Kotwal, J.G., Kashyap, R., and Shafi, P.M. (2024). Artificial driving based efficientnet for automatic plant leaf disease classification. *Multimedia Tools and Applications* 83 (13): 38209–38240.

31 Guo, Z., Chen, X., Li, M. et al. (2024). Construction and validation of peanut leaf spot disease prediction model based on long time series data and deep learning. *Agronomy* 14 (2): 294.

32 Jafar, A., Bibi, N., Naqvi, R.A. et al. (2024). Revolutionizing agriculture with artificial intelligence: plant disease detection methods, applications, and their limitations. *Frontiers in Plant Science* 15: 1356260.

33 Neethirajan, S. (2024). Artificial intelligence and sensor innovations: enhancing livestock welfare with a human-centric approach. *Human-Centric Intelligent Systems* 4 (1): 77–92.

34 Meisner, J., Bodha, B., Stothard, J.R. et al. (2024). Current approaches in livestock geotagging: assessing available technologies and applications to public health research. *medRxiv* 2024. 02. 06.24302394.

35 Nyamuryekung'e, S. (2024). Transforming ranching: precision livestock management in the internet of things era. *Rangelands* 46 (1): 13–22.

36 Tzanidakis, C., Tzamaloukas, O., Simitzis, P., and Panagakis, P. (2023). Precision livestock farming applications (PLF) for grazing animals. *Agriculture* 13 (2): 288.

37 Milella, A., Rilling, S., Rana, A. et al. (2024). *Robot-as-a-Service as a New Paradigm in Precision Farming*. IEEE Access.

38 Holtorf, L., Titov, I., Daschner, F., and Gerken, M. (2023). UAV-based Wireless data collection from underground sensor nodes for precision agriculture. *AgriEngineering* 5 (1): 338–354.

39 Zhichkin, K., Nosov, V., Zhichkina, L. et al. (2023). Efficiency of using drones in agricultural production. *E3S Web of Conferences*. EDP Sciences, 11th an 12th January, 2024, AAA college of Engineering and Technology, Affiliated to Anna University, Chennai.

40 Chin, R., Catal, C., and Kassahun, A. (2023). Plant disease detection using drones in precision agriculture. *Precision Agriculture* 24 (5): 1663–1682.

41 Kumar, S.P., Subeesh, A., Jyoti, B., and Mehta, C. (2023). Applications of drones in smart agriculture. In: *Smart Agriculture for Developing Nations: Status, Perspectives and Challenges* (ed. K. Pakeerathan), 33–48. Springer.

42 Tahir, M.N., Lan, Y., Zhang, Y. et al. (2023). Application of unmanned aerial vehicles in precision agriculture. In: *Precision Agriculture* (ed. Q. Zaman), 55–70. Elsevier.

43 Ganeshkumar, C., David, A., Sankar, J.G., and Saginala, M. (2023). Application of drone technology in agriculture: a predictive forecasting of Pest and disease incidence. In: *Applying Drone Technologies and Robotics for Agricultural Sustainability*, 50–81. IGI Global.

44 Vohra, R., Kumar, A., and Rongali, G. (2024). Multi-scale extraction and spatial analysis of growth pattern changes in urban water bodies using sentinel-2 MSI imagery: a study in the central part of India. *Stochastic Environmental Research and Risk Assessment* 38 (6): 2397–2423.

45 Vohra, R. and Kumar, A. (ed.) (2024). *Advanced Geospatial Practices in Natural Environment Resource Management*. IGI Global.

46 Mizik, T. (2023). How can precision farming work on a small scale? A systematic literature review. *Precision Agriculture* 24 (1): 384–406.

3

Precision Farming and Smart Crop Management

Bhavin Patel[1,2], Jitendra Bhatia[3], and Malaram Kumhar[3]

[1] *Research Scholar, Gujarat Technological University, Ahmedabad, Gujarat, India*
[2] *Department of Computer Engineering, Vishwakarma Government Engineering College, Ahmedabad, Gujarat, India*
[3] *Department of Computer Science and Engineering, Institute of Technology, Nirma University, Ahmedabad, Gujarat, India*

3.1 Introduction

Farming is essential for producing food and raw materials, sustaining human life, and driving the economy by contributing significantly to GDP and employing a huge population [1]. In the latter part of the 20th century, precision farming emerged as a solution to optimize farming practices using state-of-the-art technology. Initially, the Green Revolution focused on raising crop yields through the use of chemical fertilizers, high-yield variety seeds, and improved irrigation solutions [2]. Precision farming leverages contemporary technologies to optimize every aspect of crop management, surpassing the Green Revolution of the mid-20th century. The precise application of water, nutrients, and pesticides is made possible by the continuous data flow using various smart devices, which minimizes waste and lowers the impact on the environment. Moreover, intelligent crop management can reduce risks and guarantee more precise harvests by assisting farmers in anticipating and responding to insect infestations and climate effects. Thus, decision-making is further enhanced by applying AI and ML to analyze massive, complicated data sets and generate insights tailored to the requirements of certain crops. In the initial phase, it mostly relied on manual data collecting and simple analysis tools that allowed farmers to apply inputs like fertilizers and pesticides more precisely within different zones of a field as it centered on site-specific management techniques. With the usage of GPS, GIS, and variable rate technology (VRT) technologies, in this field, it brings second

Emerging Smart Agricultural Practices Using Artificial Intelligence, First Edition.
Edited by Ashish Kumar, Jai Prakash Verma, and Rachna Jain.

Published 2025 by John Wiley & Sons, Inc.

revaluation that allows for more accurate mapping and field monitoring [3]. Later on, the third stage of precision farming emerged with the integration of remote sensing (RS) technologies, such as satellite imagery and drones, which provided zone-wise real-time data on soil conditions, crop health, and pest infestations. With the emergence of big data analytics, artificial intelligence (AI)/machine learning (ML), and the Internet of Things (IoT), agriculture activities have now entered a new phase. Thus, the development of precision farming has moved from simple site-specific management to complex, data-driven farming techniques around the globe. However, adoption is hindered by high initial expenses, insufficient technological know-how, and issues with data administration, particularly in developing nations [4]. Still, there are lots of opportunities for improvement, including the creation of more accessible technologies and improved data integration frameworks. It has enormous potential to revolutionize agriculture, particularly in developing nations, as these technologies advance and become more widely available. For example, drones equipped with multispectral/hyperspectral cameras can monitor crop conditions and scout issues such as crop water stress [5].

The key contributions of this chapter include:

- An analysis of the agricultural scenarios by focusing on how precision farming and smart crop management can meet future global demands.
- A comprehensive review of existing literature to identify persisting challenges and solutions in modern agricultural practices.
- An analysis of several technologies utilized, investigation of the challenges exists in precision farming and smart crop management, along with potential site-specific applications to address these issues.

The rest of the chapter is organized as follows: Section (3.2) provides a literature review; various precision farming technologies and smart crop management techniques are discussed in Sections (3.3) and (3.4), respectively; Section (3.5) presents site-specific applications. Section (3.6) highlights the challenges and limitations, and Section (3.7) presents the conclusions and future directions.

3.2 Related Work

Precision farming and smart crop management are so important to the modernization of agriculture that a lot of research has been conducted on them in recent years. The foundation of precision farming systems is the integration of many technologies to maximize crop farming at the field level. Here, the contributions of various researchers are presented like. Kamilaris and Prenafeta-Boldu [6] have experimented on large volumes of sensors and satellites data using AI-driven models for crop management, optimizing irrigation systems, and predicting crop yields that provide farmers with useful information to increase output. Sankaran et al. [7] have explored how UAVs equipped with multispectral and hyperspectral cameras can assess crop health, detect diseases early, and monitor soil conditions to mitigate crop yield reduction issues in advance. A complex system with soil sensors that provides real-time data on soil pH, moisture content, and nutrient content was created by Adamchuk et al. [8]. These sensors' precise mapping of the soil facilitates the application of fertilizer and water at different rates, two things that are necessary for sustainable farming practices [9]. Additionally, Qazi et al. [10] looked into IoT and AI-enabled systems that use CNN-based deep learning (DL) frameworks to identify diseases utilizing intelligent farming equipment and drones. Bouali et al. [11] suggested a cloud-based IoT solution for smart irrigation in order to increase accuracy using ML algorithms and highlight the necessity for adoption across different crops. Fuzzy logic and an Arduino Nano equipped with Zigbee were employed by the system to gather data. Raj et al.'s study [12] focused on the difficulties associated with integrating ML into Agriculture 4.0's integration of IoT, drones, and AI for farm operations. Nagasubramanian et al. [13] focused on crop disease detection using IoT and ensemble classification, suggesting enhancements through larger datasets and DL methods. Liu et al. [14] presented Industry 4.0's impact on agriculture, emphasizing IoT, robotics, AI, and big data analytics integration challenges. Kour et al. [15] analyzed recent IoT developments in agriculture, proposing architectural improvements for resource management and disease detection. These studies collectively highlight advancements and challenges in integrating cutting-edge technologies to enhance agricultural practices. Here, Table 3.1 summarizes the various technologies covered in various research/review papers and their proposed solution utilized in the agriculture domain by different authors. Also, Table 3.2 presented analysis of work carried out by various researchers in this domain.

3.3 Technologies in Precision Farming

Numerous advanced technologies for agriculture are available to enhance crop productivity while maintaining efficient resource utilization, less human intervention, and reducing wastage. The main objectives are to enhance agriculture productivity in resource-constrained environments by adopting precision farming solutions such as agriculture drones, autonomous robots, smart tractors, web applications, and farm management solutions. The agricultural solutions for future farming are presented in the following section. Figure 3.1 represents the various possible technologies and use cases involved in precision farming.

Table 3.1 Summary of parameters and recent technologies covered in the literature.

Author	Year	IoT sensors	Communication technology	Drone technology	ML/DL used	Edge/fog computing	Cloud computing	SDN	Blockchain	Big data	Security	Policy and regulation	Agriculture parameter
Qazi et al. [10]	2022	Yes	Yes	Yes	Yes	Yes	Yes	No	Yes	Yes	No	No	Smart irrigation, weed detection
Bouali et al. [11]	2021	Yes	Yes	No	Yes	No	Yes	No	No	Yes	No	No	Smart irrigation
Raj et al. [12]	2021	Yes	Yes	Yes	Yes	No	Yes	No	No	Yes	No	No	Corp yielding, weed detection
Nagasubramanian et al. [13]	2021	Yes	Yes	No	Yes	No	No	No	No	No	No	No	Crop disease monitoring
Liu et al. [14]	2021	Yes	Yes	Yes	Yes	Yes	Yes	No	Yes	Yes	Yes	No	Production patterns, food supply
Kour et al. [15]	2020	Yes	Yes	Yes	Yes	Yes	Yes	No	No	Yes	Yes	No	Precision farming

Table 3.2 Analysis of various papers in the domain of precision farming based on the present study.

Authors	Key aspects	Contribution	Methodology	Scope for improvement
Qazi et al. [10]	Smart agriculture using IoT- and AI-enabled solutions	Reviewed on smart agriculture system, IoT, AI/DL technology-related use case Highlighted impact of technologies like drones and other intelligent farming equipment	CNN-based deep learning framework used to detect and classify plant diseases using annotated training images of known diseases	DL algorithm needs to be included based on agriculture application scenario
Bouali et al. [11]	Smart agriculture deployment using cloud and IoT using a renewable energy source	Integral solution for cost-effectiveness smart irrigation via real-time data collection and monitoring using a Cloud-based IoT Fuzzy logic in smart drip irrigation is being tested	System manages drip irrigation actuators as per water-table needs of plant Arduino Nano board used with Zigbee communication technology for data acquisition	Proposed smart irrigation solution not deployed for any crop to determine water schedule Scope of machine learning algorithm to improve accuracy
Raj et al. [12]	The adoption and promotion of Agriculture 4.0 and the IoT	Utilization of IoT, drone technology, data analytics, and AI-based ML/DL approach for farm-related operations Various applications and services of Agriculture 4.0 are listed	The state of Agriculture 4.0 and the future prospects in regard to technology companies Discussed ML approaches, a flowchart of big data-based processing, and agricultural datasets	ML/DL integration-related issues of IoT agriculture still persist Other use cases of IoT agriculture will be explored

(Continued)

Table 3.2 (Continued)

Authors	Key aspects	Contribution	Methodology	Scope for improvement
Nagasubramanian et al. [13]	Crop disease identification using IoT-based pattern recognition and ensemble classification	Technique uses ensemble nonlinear SVM (ENSVM) to detect infections in leaves and crops Compares different ML methods, including SVM, naïve Bayes, K-nearest neighbors, and CNN	The HySpexHD camera's high-spectral image transformed into a grayscale image to train the IoT infrastructures Use of hyperspectral images to identify the infection areas using ENSVM and CNN in crop	Scope to evaluate a large dataset of multiple crops to enhance efficiency using the DL approaches
Liu et al. [14]	State-of-the-art about industry 4.0 to agriculture 4.0	Review on industrial agriculture-based production patterns, processes, and agri-food supply chain Talk about the SAGUIN (space-air-ground-under surface integrated network) for widespread agriculture sensing and networking	Discussion of significant uses and problems for agriculture using the IoT, robotics, blockchain, A), and big data analytics	Persisting challenges and demands for valid agriculture's adoption of cutting-edge Industry 4.0 technology
Kour et al. [15]	Analysis of recent IoT developments in agriculture	Examination of IoT models based on conceptual, architectural, and implementation prospects for precision agriculture Various IoT hardware, platforms, and IoT OS discussed	Proposed architecture to observe and monitor, improve the use of resources, early disease detection, and plant growth identification, among other things	Only abstract explanation given on proposed architecture

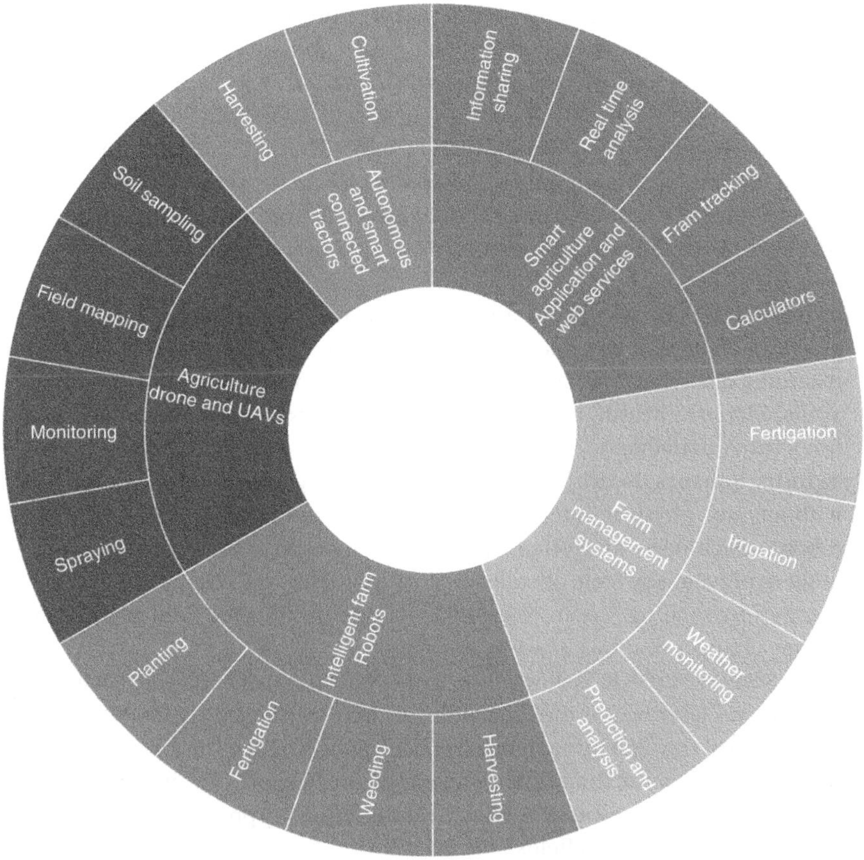

Figure 3.1 Possible technologies and its use cases for precision farming.

3.3.1 Sensors and IoT Applications

In IoT-based agriculture, various biggest technology industries such as Google, Amazon, IBM, CISCO, Intel, Oracle, Bosch, and Microsoft are now involved to provide customized solutions in different layers of IoT agriculture. IoT applications integrate data from these sensors to create comprehensive farm management systems. These systems utilize cloud computing and data analytics to process sensor data, providing actionable insights [16]. Smartphone compatibility has been built into a variety of hardware and software solutions for the agricultural sector. Here is the list of various sensors that are used in IoT applications to provide real-time data.

Soil Moisture and Temperature Sensors: Soil moisture sensors are an essential component of smart irrigation systems that automatically modify watering schedules based on real-time soil conditions. These sensors, which are available in resistive, tensiometer, and capacitive varieties, offer vital information for irrigation technique optimization. It is used to measure the volumetric water content of the soil. These sensors usually detect volumetric water content between the range of 0% and 100% [17]. By monitoring both soil and ambient temperatures, these sensors support crop health and growth. The soil and air temperatures are measured in real time using sensors such as thermocouples or thermistors, which operate within a range of −40 to 125 °C [18].

Dendrometer Sensors: These are used for continuous measurement of plant growth, which basically changes in plant diameter and stem size. It enables the recording of environmental information in a temporal manner, including water availability, tree health, and other plant parameters like minute variations in a tree's trunk diameter.

Proximity Sensors: Nearby objects and obstacle detection are mainly carried out without any physical intervention while performing operations by Drone, robotic machinery, and other autonomous tools in the agriculture field. In order to prevent any form of collision during agricultural operations, the sensor operates by sending out an electromagnetic beam into the field and scanning for any changes that take place in the signal.

LiDAR Sensor: LiDAR is known as light detection and ranging technology. LiDAR uses laser beams directed at the ground's surface to estimate distances. The distance traveled by each pulse is determined by the sensor by measuring the time it takes for the pulse to return to the sensor. This information was made into a helpful resource for applications requiring precise geospatial measurements in agricultural fields [14].

Multispectral and Hyperspectral Sensors: Different wavelengths can be used to assess crop conditions, soil properties, and plant health, providing a thorough understanding through the use of these spectral sensor types. Farmers can improve crop management procedures and maximize inputs by making informed judgments based on the analysis of particular light reflectance patterns. Depending on the application, these sensors can operate at spectral resolutions between 10 and 50 nm and spatial resolutions between 400 and 2500 nm.

Nutrient Sensors: These instruments measure essential elements in the soil, like potassium, phosphate, and nitrogen, to provide information about nutritional deficits and soil fertility. Certain nutrient concentrations, such as 0–100 ppm for nitrate, can be measured using these sensors [19]. Utilizing technology such as optical sensors and ion-selective electrodes, these instruments provide vital information for personalized fertilization procedures.

3.3.2 Farm Management System

A farm management system, which integrates WSN, GSM, and microcontroller data collection methods, enables farmers to monitor and control their whole land. For the objective of smart farming, it is a crucial component for planning, processing, and decision-making [20]. All of the farm's sensors and machines provide real-time statistics on fertilization, weather information, buffer zone width monitoring, and automatically generated records based on daily operations. For subsequent processing, the entire collection of data is saved in a computer in a common format and is available online or via a mobile device. To optimize the use of both water and chemical resources in the field, it can automate irrigation and fertigation modules [21]. As well as IoT sensors and cameras create the perfect habitat for plant monitoring and produce useful maps that identify changes if any problems are apparent [22]. In order to maximize productivity, yield, and revenue, data analytics also enables different farms to be connected and managed via a single platform. Information on scientific advancements, production, marketing, farm management, suggestions, and other relevant issues are provided from this platform [23].

3.3.3 Unmanned Aerial Vehicles (UAV)

Drone devices, or unmanned aerial vehicles, fall into a number of categories. They are equipped with modern sensors, including thermal, hyperspectral, multispectral, and LiDAR that provide extensive imagery and real-time data collection than traditional approaches of image data collection. The agility and flexibility of UAVs overcome the logistical challenges associated with employing more traditional ground-based tactics, allowing large agricultural regions to be monitored regularly and economically. Decision support and data interpretation are improved by combining UAV data with geographic information systems (GISs) and processing it with ML techniques [24]. So, farmers may apply focused treatments like precise nutrient fulfillment and irrigation control by capturing spectral bands that reflect crop vigor, nutrient deficits, and stress levels.

3.3.4 Remote Sensing

Remote sensing technologies are revolutionizing precision farming by providing comprehensive and accurate data on crop health, soil conditions, and environmental factors, which use satellite imagery, drones, and a variety of sensor technologies to provide real-time data that supports agricultural decision-making [25]. By analyzing the reflectance patterns of crops, multispectral sensors can gather data across various wavelengths of light, which helps to discover issues related to plant health, such as water stress, chlorophyll content, and crop growth stages. Farmers can minimize water stress, insect infestations, and disease outbreaks by using thermal sensors to detect variations in canopy temperature and make informed decisions regarding irrigation scheduling and pest control [26]. Some well-known satellites, such as Sentinel-2 and Landsat 8, provide high-resolution multispectral photography that enables continuous monitoring of crop conditions across time [27]. To create thorough field maps, predictive models, and decision support systems, these data can be connected with GIS and IoT platforms.

3.3.5 GIS and GNSS

GISs and global navigation satellite systems (GNSSs) basically provide accurate location-based information and advanced spatial data analysis to optimize agricultural operations. Using data from multiple sources, such as soil samples, yield maps, and remote sensing imagery, GIS creates complex computerized maps of agricultural regions. With the use of these maps, farmers can apply site-specific management techniques, such as irrigation, fertilizer, and pesticide treatment, at varying rates based on the particular requirements of each field [28]. For real-time mapping and navigation inside fields, accurate location data can be made possible by GNSS, which includes the GPS (global location system) [29]. Application of site-specific crop management that maximizes input efficiency and minimizes environmental effect is made possible by the integration of GNSS and GIS technologies.

3.3.6 Robotics and Automation

Robotics and automation technologies are revolutionizing precision farming by automating tasks and enhancing operational efficiency across agricultural operations. The design of smart and intelligent equipment is a current focus for numerous traditional farming equipment manufacturers. It links to IoT devices, smart linked tractors, and autonomous pumps that can sense their surroundings and act immediately in the event of an abnormality [30]. Harvesting the crop in a timely manner is essential, as being either early or late will significantly reduce the yield. So, for automating and designing a more specialized harvesting mechanism, the use of robots like fruit harvesting arms has increased. Even if they are partially or entirely hidden by leaves or mixed up with other fruits, the goal is to recognize the different fruits in their native habitat [13]. Robots that are used for picking and harvesting identify and locate ripe fruits and vegetables in the field while gathering them in boxes. It makes tomato, citrus, apple, and strawberry harvesting more productive for farmers [31]. These technologies encompass autonomous vehicles, robotic arms, and AI-powered systems that perform various tasks such as seeding, spraying, harvesting, and monitoring with precision and consistency. Deepfield Robotics has developed and tested an autonomous vehicle that should take weeds into account when cultivating sugar beets. This robotic weed controller utilized cameras to discriminate between crops and weeds and a GPS antenna to self-navigate throughout the field [32]. The UGV is the robotic solution for precision agriculture outfitted with a variety of sensors, including GPS, RTK–GPS, RGB, and hyperspectral cameras, LiDAR technology, wheel odometers, and others to carry out automatic weeding in the field with the integration of drone data. John Deere has created a weed-control device that can discern between crops and weeds using cameras, computers, and AI. It then employs a chemical strategy to eradicate weeds by spraying herbicides just where they are present [33].

3.3.7 Artificial Intelligence, Machine Learning, Data Analytics, Data Fusion, and Management

AI and ML technologies are widely explored in many areas [34–38]. In precision farming, AI and ML are at the forefront, driving significant advancements through sophisticated data analytics, data fusion, and management technologies. Here, we have discussed Cloud computing, fog-edge computing, ML/DL modeling, big data analytics, software-defined networking (SDN), Blockchain technology, security mechanisms technology, and its role in precision farming. AI and ML algorithms analyze vast amounts of data collected from various sources such as sensors, drones, and satellites, providing actionable insights into crop health, soil conditions, weather patterns, and pest infestations. Multilayer classifiers are used in [13], which work on SVM with ensemble nonlinear classification to project more effective classified findings on disease classification under uncertain leaf pattern recognition. Similarly to this, CNN is used to identify leaf diseases. A CNN is a kind of multilayer perceptron model that splits image patterns by performing a convolution at the beginning of the filter section. Finally, it performs deconvolution to get the integrated result. With the integration of DL networks and the IoT, Khalil et al. [39] study incorporates convolutional neural networks (CNNs) such as AlexNet, ResNet, GoogleNet, and VGG16. Agriculture has made substantial use of DA-based image processing for a variety of objectives, including the identification of infection in leaves, stems, and fruits, fruit quality and weed identification, and water management [40]. For example, ADSS mobile agricultural expert systems utilize predictive analytics to provide farmers with intelligent and precise agricultural recommendations based on big data [9]. These systems also conduct thorough risk assessments, which can help farmers manage risks associated with production, market, institutional, personal, and financial factors [41]. Advanced data management platforms facilitate real-time monitoring and historical data analysis, supporting precision agriculture by enabling detailed field mapping, variable rate applications, and efficient farm management practices [9]. Here, Table 3.3 provides a summary of the IoT integration elements of enabling technology.

Table 3.3 Summary of key enabling technologies integrated in precision farming solutions.

Technology	Description	Features	Examples of use	Specifications	Advantages	Challenges
Cloud computing [42]	Provides storage, processing, and analysis of data from IoT devices	Large network access, remote computational, on-demand service, resource pooling, centralized	Storage of sensor data and analysis results in the cloud	Reliable, cost-effective, security	Enables access to vast computing resources for data analysis	Potential for data breaches and hacking of cloud storage
Fog-Edge computing [43]	Enables data processing to occur closer to the source of data, reducing latency and network bandwidth usage	Decentralization, proximity to users, real-time interaction	Processing sensor data on the edge device to reduce latency	Low latency, bandwidth, mobility	Reduces network traffic and latency, enabling faster data processing and decision-making	Limited computing resources and processing power on edge devices
ML-DL modelling [13]	Enables devices to learn and adapt, improving their performance and automating decision-making	Feature selection, diverse algorithms	Machine learning models for plant disease detection and prediction	No human intervention, multidimensional data	Enables automated decision-making and real-time response to changes in environmental conditions	Complexity of training and deploying machine learning models
Big data analytics [44]	Analyzes large amounts of data generated by IoT devices to derive insights and make decisions	Volume, veracity, variety, velocity, knowledge discovery	Analysis of crop yield data to optimize crop growth	Cost-efficient data analysis	Enables analysis of large amounts of data for pattern detection and predictive analytics	Complexity of analyzing and interpreting large amounts of data

Software-defined networking [13]	Manage network services through the abstraction of lower-level functionality	Central control, programmability, separation of control logic from switch	Easier to provide high performance, highly available services to end-users.	Central management, dynamic configuration	Network traffic can be prioritized, directed to specific devices, and managed in a more intelligent way.	Controller compromisation, latency
Blockchain technology [45]	Provides a decentralized, tamper-proof ledger to record and verify transactions and data generated by IoT devices	Decentralized, immutable	Supply chain tracking and tracing of food products	Transparent, highly secure	Provides a secure and transparent way to track the origin and movement of food products	Limited scalability and interoperability of blockchain networks
Security mechanisms [46]	Ensures the confidentiality, integrity, and availability of data and devices, preventing unauthorized access and cyberattacks	Confidentiality, integrity, nonrepudiation, authentication, access control	Encryption of data transmitted over wireless networks	Attacks detection and mitigation, content protection	Ensures the security and privacy of data and devices	Vulnerability to cyberattacks and malicious threats

3.4 Smart Crop Management Techniques

3.4.1 Variable Rate Technology

This method adjusts the amount of input applied based on the specific needs of different field zones by using sensors, GPS, and mapping data. It enables the management of agricultural inputs, such as seeds, fertilizer, and herbicides, in accordance with the particular requirements of every location. For instance, by applying fertilizer at optimal rates in nutrient-rich and nutrient-deficient zones, VRT can increase agricultural productivity while minimizing environmental impact [47]. Moreover, VRT encourages effective water management by altering irrigation schedules in reaction to information about soil moisture [48]. Likewise, real-time data and recommendations can be obtained during crop growth stages by connecting VRT to IoT devices and DSS [49].

3.4.2 Decision Support System

This system integrates real-time data from various sources, including IoT sensors, satellite imagery, and weather forecasts, to provide targeted recommendations for actions such as irrigation, fertilization, and pest management [50]. DSS employs advanced algorithms and ML techniques to analyze this data, offering predictions and advisories on irrigation schedules, fertilization needs, pest and disease management, and harvest timings. For instance, ML algorithms are used by DSS systems such as AgroInfo DSF to analyze soil health, predict crop yields, and recommend the best times for crop planting [51]. Farmers can quickly access and comprehend difficult information thanks to the user-friendly interfaces and real-time data processing capabilities of modern DSS solutions. These tools help farmers manage resources more effectively and reduce waste. By using ML models and advanced analytics, they can identify patterns and forecast results, mitigating various challenges [52]. These systems are compatible with VRT, which maximizes efficiency and reduces environmental effects by adjusting the application rates of seeds, fertilizers, and pesticides in accordance with site-specific requirements [53].

3.4.3 Sustainable Practices and Precision Agriculture

Sustainable practices are transforming current farming by combining state-of-the-art technology with intelligent crop management techniques to improve output while minimizing environmental effects and maximizing resource use [54]. Crop rotation, cover crops, and conservation tillage are other sustainable precision agriculture techniques that enhance soil health, lower erosion, and increase biodiversity. For instance, by controlling soil moisture sensors and meteorological data, precision irrigation systems maximize water consumption, minimize water waste, and support sustainable water management [55]. Therefore, by reducing greenhouse gas emissions, minimizing chemical runoff, and maintaining soil health, these developments enhance agricultural output and promote sustainable farming practices.

3.5 Mapping to Site-Specific Applications

The use of precision farming has great potential to advance the agriculture sector by enabling real-time monitoring and detecting operations in the field. IoT devices and sensors can collect valuable information on various factors such as humidity, temperature, weather, and moisture levels. Various researchers have performed a study in the agriculture field about IoT and smart technologies deployment. The main objective is to enhance agricultural productivity in the resource-constrained environment of agriculture. Figure 3.2 gives a pictorial representation of smart agriculture with a various application perspective.

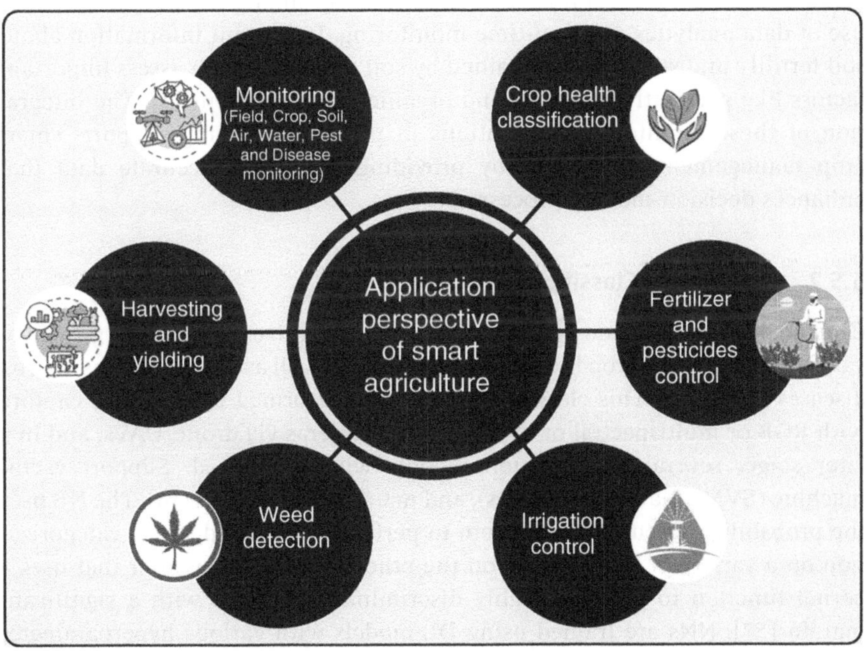

Figure 3.2 Application perspective of smart agriculture.

3.5.1 Monitoring

Precision farming integrates advanced monitoring applications for field, crop, soil, air, water, pest, and disease management to optimize agricultural practices and enhance crop productivity. Field monitoring employs GPS and GIS technologies to create detailed maps of field variability, guiding the precise application of inputs [24]. Crop monitoring utilizes remote sensing technologies, including drones and satellite imagery, to assess plant health, growth stages, and yield potential by analyzing spectral data that indicate chlorophyll content and biomass. Soil monitoring involves sensors that measure soil moisture, temperature, and nutrient levels, enabling site-specific irrigation and fertilization strategies that improve soil health and crop yields. Air monitoring, through weather stations and atmospheric sensors, provides critical data on temperature, humidity, and wind patterns, informing decisions on planting, spraying, and harvesting. Water management applications use IoT-enabled sensors and automated irrigation systems to monitor and control water usage, ensuring efficient water use and preventing over-irrigation. Pest and disease monitoring combines ML algorithms with sensor data to detect early signs of infestations and infections, enabling timely and targeted interventions that reduce crop loss and pesticide use [56]. In order to sustainably manage water resources and preserve soil health, soil and water management systems make use of data analytics and real-time monitoring. Important information about soil fertility and structure is obtained by soil sensors, which assess important factors like pH, nutrient levels, and organic matter content [3]. The integration of these monitoring applications in precision farming supports smart crop management techniques by providing real time, accurate data that enhances decision-making processes.

3.5.2 Crop Health Classification

Crop health classification systems are used to recognize healthy crop growth at various stages of the crop life cycle in the field, as well as the early signs of crop diseases and pests. This classification can be performed using image capture with RGB or multispectral or hyperspectral cameras via drone/UAVs, and in a later stage, several classification models will be applied. Support vector machine (SVM), Naïve Bayes (NBs), and neural network (NN) [6]. The NB uses the probability-based Bayes theorem to perform supervised object categorization on a variety of items. SVM, on the other hand, is a classifier that uses a kernel function to produce highly discriminating classes with a significant margin [57]. NNs are trained using DL models with various hyperparameter settings, including hidden layer selections, hidden node configurations, activation functions, and loss functions [58].

3.5.3 Yielding

Crop yield forecast is primarily influenced by a number of variables, including farming management techniques, soil properties, and environmental conditions. Information retrieval is therefore essential for these elements. Data collections obtain information via remote sensing, drones, farm vehicle-mounted devices, ground-based sensors, and satellite images. Farmers may identify issues limiting crop productivity, anticipate yields with accuracy, and gain insights by combining this data with predictive modeling and ML tools [59]. Monitoring key performance indicators such as crop health and developmental stages is facilitated by feeding the source data into software programs.

3.5.4 Fertigation and Pesticides

The use of real-time monitoring platforms and automatic fertigation systems helps optimize the delivery of fertilizers and irrigation. These platforms obtain field-level information on plant health and nutrient levels provided by remote sensing data and soil sensors without any human intervention. To guarantee that crops receive the proper amount of nutrients at the correct time, IoT-enabled control systems dynamically modify the timing and concentration of fertigation treatments. By combining fertigation, integrated pest management (IPM) systems, and VRT, the use of pesticides is optimized by applying treatments only when and where needed, minimizing chemical consumption and environmental impact [60, 61]. This approach can improve crop health and productivity by ensuring crops receive the appropriate amount of nutrients at the right time. These systems can be easily accessed remotely via mobile applications.

3.5.5 Weeding

The explosion of weeds in farmland significantly impacts crop growth and yield. To address this, various chemical and mechanical solutions are available. These days, farmers' needs are satisfied by AI-driven robotic weed cutters and chemical spraying robots [62]. These technologies use ML algorithms and computer vision to distinguish between weeds and crops, allowing for precise, targeted weed management with minimal herbicide use. By delivering accurate spectral signatures of various plant species, advanced imaging technologies such as multispectral and hyperspectral cameras improve the detection capabilities of autoloom weeding devices.

3.6 Challenges and Limitations

This section highlights the challenges and issues that remain unresolved and may be contributing to the slow adoption of IoT in the agriculture industry. Figure 3.3 reveals the research difficulties and problems related to the adoption and use of precision farming and smart crop management.

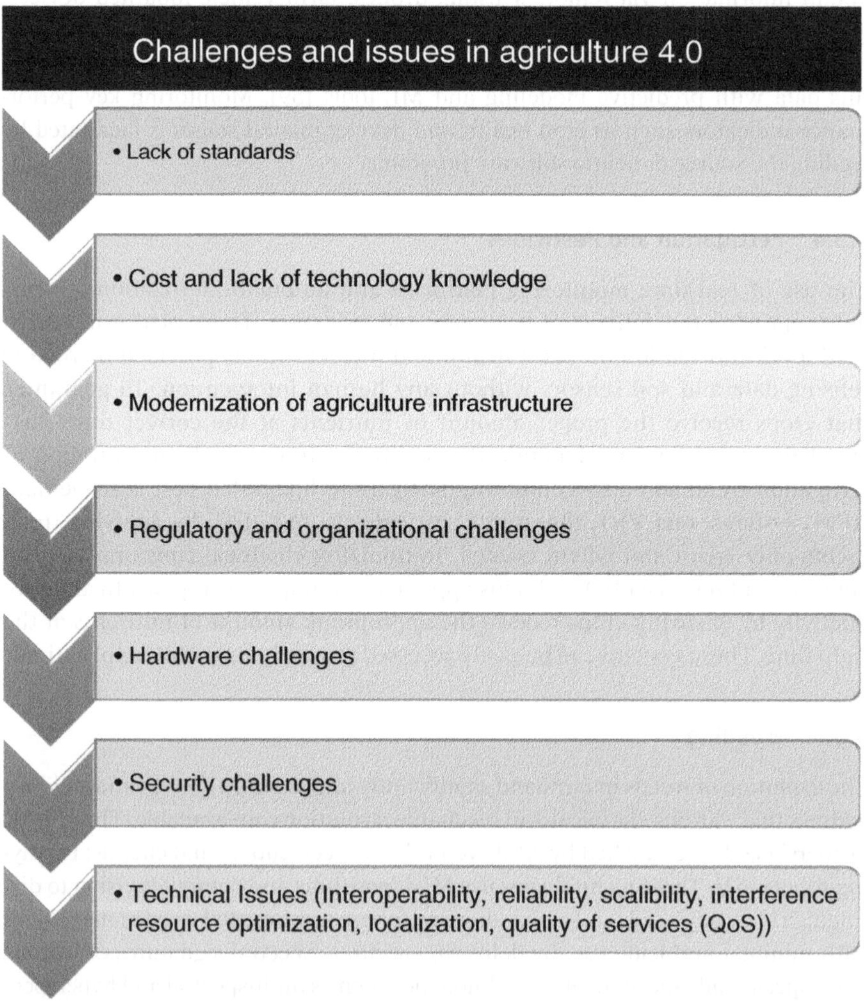

Figure 3.3 Challenges and limitations with precision agriculture and smart crop management.

3.6.1 Cost and Lack of Technology Knowledge

One of the significant challenges is the financial constraints and limited access to credit, which hinder their ability to modernize their production practices. Furthermore, there are significant variations between farmer's ability and desire to invest in new technology, which runs the risk of exploiting people and leading to significant productivity gaps between regions. When installing an IoT system, hardware implementation costs such as IoT devices, sensors, gateways, communication devices, and base station infrastructure are relatively high. The ongoing cost of using centralized services or IoT platforms that offer services like data gathering, device management, and information exchange is known as the operating cost. Costs related to data transfer between IoT devices, gateway and cloud servers, energy, and maintenance are additional operational and scheduled expenses [23]. Additionally, the main barrier to IoT in rural regions is essentially a lack of technical understanding of IoT and its uses. Therefore, it is most important to overcome these end-user-related issues.

3.6.2 Lack of Standards

Due to the lack of standardized protocols and formats for data collection, integration, and analysis. This absence complicates the seamless exchange of information between different agricultural technologies and platforms. Without universally accepted standards, compatibility issues arise, hindering the efficiency and reliability of data-driven farming practices and limiting the scalability of precision farming solutions, impacting their widespread adoption and effectiveness [2].

3.6.3 Modernization of Agriculture Infrastructure

Agriculture infrastructure must be able to ensure network Connectivity issues, energy management issues, and a centralized abstracted platform for proper applicability in rural areas. In comparison to wired networks, wireless networks are more scalable, offer flexible networking, and require less setup budget. However, it still has a lot of problems with communication breakdowns caused by field changes like plant growth, background noise production, and power source and battery drain problems in network equipment. Long-range communication technologies like low power wide area network (LPWAN) can help with some issues, but the expense to install LPWAN base stations is still expensive [63]. The capacity to exchange and analyze data on a common platform without any kind of geographic restrictions is essential for precision farming to succeed. So, such a system that can be easily adapted to support a variety of agricultural applications, from management to monitoring, needs to be developed.

3.6.4 Regulatory Hurdles and Policy Implications

Regulatory frameworks and policies often lag behind the rapid advancements in precision farming technologies, creating uncertainties and barriers to adoption. Regulatory frameworks must address issues related to data ownership, liability, and the ethical use of AI and automation in farming practices. Policymakers play a crucial role in fostering an enabling environment that encourages innovation while safeguarding consumer interests and environmental sustainability [14].

3.6.5 Regulatory and Organizational Challenges

Smart agriculture is a modern approach to agricultural practices with a global perspective that requires the implementation of standard regulations and legal frameworks for controlling and owning farm data between farmers and service providers. However, regulations and legal procedures may differ from country to country, which can lead to challenges such as resource allocation, infrastructure setup, data ownership, technical challenges, and data privacy and security. It may create a bottleneck for the application of IoT use cases such as Agri-food supply in the agriculture domain [23]. Cloud computing provides high-quality services, such as efficient storage, powerful computation, and tools for developing hardware diagnostics, at the edge of the network for data storage and processing. However, transferring data to the cloud may still incur significant latency and cost [42].

3.6.6 Data Security and Privacy Concerns

The integration of IoT devices, sensors, and data analytics in agriculture raises critical concerns about data security and privacy. These technologies generate vast amounts of sensitive data, including crop yields, soil conditions, and farm operations, which are vulnerable to cyber threats and unauthorized access. Farmers and technology providers must implement robust encryption methods, access controls, and anonymization techniques to protect confidential information from unauthorized access and cyber threats [46].

3.6.7 Technical Complexities and Skill Gap

The complexity of integrating and managing diverse technologies in precision farming underscores the need for skilled agricultural professionals and technicians. However, there exists a significant skill gap in understanding data analytics, ML, and IoT applications in agriculture. The importance of educational programs and training initiatives to empower farmers with the necessary technical competencies and enhance their capacity to adopt and implement smart crop management techniques effectively [6].

3.6.8 Reliability

In the agriculture sector, IoT devices are used in an open environment, where they may encounter problems with environmental effects, device operation-related issues, communication barriers, deterioration of deployed sensors, etc. Therefore, these expensive gadgets must be kept secure from unauthorized users, rough weather conditions, or theft. So it can work properly, and availability remains constant with different application scenarios.

3.6.9 Quality of Service

The QoS can be defined as a metric that measures the performance, quality, and efficiency of IoT devices and architectures [46]. For IoT applications, reliability, security, availability, cost, power consumption, and service time are very important and essential QoS metrics. At the IoT network layers, it still demands of some mechanism that guarantees QoS throughout which will be capable to transmit sensitive information without any barrier in precision farming.

3.6.10 Interoperability and Standardization

Achieving interoperability among diverse agricultural technologies remains a pressing challenge. Different hardware and software solutions often operate on proprietary systems, making seamless data exchange and integration difficult. Standardization efforts are crucial to establish common communication protocols and data formats that enable interoperability across various precision farming tools and platforms [51].

3.6.11 Energy Efficiency and Power Management

The deployment of precision farming technologies requires substantial energy consumption, primarily for running sensors, drones, and automated machinery. High energy consumption not only increases operational costs but also contributes to environmental impacts. Energy efficiency measures, such as the use of low-power sensors, renewable energy sources, and optimized power management systems, are essential to mitigate the environmental impact and operational costs associated with intensive energy use in agriculture [64].

3.6.12 Scalability Issues in Large-Scale Farming

Scaling precision farming practices from small pilot projects to large agricultural operations poses significant challenges. Factors such as high initial investment costs, infrastructure requirements, and the need for specialized expertise limit the widespread adoption of precision agriculture techniques. Addressing scalability issues involves developing cost-effective solutions and supportive infrastructures tailored to the unique needs of large-scale farming operations [48].

3.7 Conclusion

The development of precision farming has been presented in this study, which shows traditional methods have been replaced by modern approaches that include data analytics, remote sensing, IoT, GIS, etc. It has been covered that implementing critical technologies like sensors, drones, automated irrigation, and GPS-guided machinery together with revolutionary smart crop management strategies like real-time monitoring and predictive analytics greatly increases crop yields while maximizing resource efficiency and reducing environmental impact. In line with the ongoing recent trends, the combination of ML and AI has the potential to significantly improve smart crop management and precision farming. Future research should focus on overcoming the existing barriers to adoption, such as improving system integration, enhancing data security, and developing robust solutions for farmers.

References

1 Basso, B. and Antle, J. (2020). Digital agriculture to design sustainable agricultural systems. *Nature Sustainability* 3 (4): 254–256.

2 Gebbers, R. and Adamchuk, V.I. (2010). Precision agriculture and food security. *Science* 327 (5967): 828–831.

3 Mulla, D.J. (2013). Twenty five years of remote sensing in precision agriculture: key advances and remaining knowledge gaps. *Biosystems Engineering* 114 (4): 358–371.

4 Lowenberg-DeBoer, J. and Erickson, B. (2019). Setting the record straight on precision agriculture adoption. *Agronomy Journal* 111 (4): 1552–1569.

5 Schimmelpfennig, D. Farm Profits and Adoption of Precision Agriculture, ERR-217, U.S. Department of Agriculture, Economic Research Service, October 2016.

6 Kamilaris, A. and Prenafeta-Boldú, F.X. (2018). Deep learning in agriculture: a survey. *Computers and Electronics in Agriculture* 147: 70–90.

7 Sankaran, S., Khot, L.R., Espinoza, C.Z. et al. (2015). Low-altitude, high-resolution aerial imaging systems for row and field crop phenotyping: a review. *European Journal of Agronomy* 1 (70): 112–123.

8 Adamchuk, V.I., Hummel, J.W., Morgan, M.T., and Upadhyaya, S.K. (2004). On-the-go soil sensors for precision agriculture. *Computers and Electronics in Agriculture* 44 (1): 71–91.

9 Wolfert, S., Ge, L., Verdouw, C., and Bogaardt, M.J. (2017). Big data in smart farming – a review. *Agricultural Systems* 1 (153): 69–80.

10 Qazi, S., Khawaja, B.A., and Farooq, Q.U. (2022). *IoT-equipped and AI-enabled Next Generation Smart Agriculture: A Critical Review, Current Challenges and Future Trends*. IEEE Access.

11 Bouali, E.-T., Abid, M.R., Boufounas, E.-M. et al. (2021). Renewable energy integration into cloud & IoT-based smart agriculture. *IEEE Access* 10: 1175–1191.

12 Raj, M., Gupta, S., Chamola, V. et al. (2021). A survey on the role of internet of things for adopting and promoting agriculture 4.0. *Journal of Network and Computer Applications* 187: 103107.

13 Nagasubramanian, G., Sakthivel, R.K., Patan, R. et al. (2021). Ensemble classification and IoT-based pattern recognition for crop disease monitoring system. *IEEE Internet of Things Journal* 8 (16): 12847–12854.

14 Liu, Y., Ma, X., Shu, L. et al. (2020). From industry 4.0 to agriculture 4.0: current status, enabling technologies, and research challenges. *IEEE Transactions on Industrial Informatics* 17 (6): 4322–4334.

15 Kour, V.P. and Arora, S. (2020). Recent developments of the internet of things in agriculture: a survey. *IEEE Access* 8: 129924–129957.

16 Karunathilake, E.M., Le, A.T., Heo, S. et al. (2023). The path to smart farming: innovations and opportunities in precision agriculture. *Agriculture* 13 (8): 1593.

17 Akhter, R. and Sofi, S.A. (2022). Precision agriculture using IoT data analytics and machine learning. *Journal of King Saud University, Computer and Information Sciences* 34 (8): 5602–5618.

18 Senapaty, M.K., Ray, A., and Padhy, N. (2023). IoT-enabled soil nutrient analysis and crop recommendation model for precision agriculture. *Computers* 12 (3): 61.

19 Arrubla-Hoyos, W., Ojeda-Beltrán, A., Solano-Barliza, A. et al. (2022). Precision agriculture and sensor systems applications in Colombia through 5G networks. *Sensors* 22 (19): 7295.

20 Gardašević, G., Veletić, M., Maletić, N. et al. (2017). The IoT architectural framework, design issues and application domains. *Wireless Personal Communications* 92: 127–148.

21 Katyara, S., Shah, M.A., Zardari, S. et al. (2017). WSN based smart control and remote field monitoring of Pakistan's irrigation system using scada applications. *Wireless Personal Communications* 95 (2): 491–504.

22 Farooq, M.S., Riaz, S., Abid, A. et al. (2019). A survey on the role of IoT in agriculture for the implementation of smart farming. *IEEE Access* 7: 156237–156271.

23 Elijah, O., Rahman, T.A., Orikumhi, I. et al. (2018). An overview of internet of things (IoT) and data analytics in agriculture: benefits and challenges. *IEEE Internet of Things Journal* 5 (5): 3758–3773.

24 Zhang, C. and Kovacs, J.M. (2012). The application of small unmanned aerial systems for precision agriculture: a review. *Precision Agriculture* 13: 693–712.

25 Omia, E., Bae, H., Park, E. et al. (2023). Remote sensing in field crop monitoring: a comprehensive review of sensor systems, data analyses and recent advances. *Remote Sensing* 15 (2): 354.

26 Radočaj, D., Jurišić, M., and Gašparović, M. (2022). The role of remote sensing data and methods in a modern approach to fertilization in precision agriculture. *Remote Sensing* 14 (3): 778.

27 Singh, P., Pandey, P.C., Petropoulos, G.P. et al. (2020). Hyperspectral remote sensing in precision agriculture: Present status, challenges, and future trends. In: *Hyperspectral Remote Sensing* (ed. P.C. Pandey, P.K. Srivastava, H. Balzter, et al.), 121–146. Elsevier.

28 Sishodia, R.P., Ray, R.L., and Singh, S.K. (2020). Applications of remote sensing in precision agriculture: a review. *Remote Sensing* 12 (19): 3136.

29 Perez-Ruiz, M., Martínez-Guanter, J., and Upadhyaya, S.K. (2021). High-precision GNSS for agricultural operations. In: *GPS and GNSS Technology in Geosciences* (ed. G.P. Petropoulos and P.K. Srivastava), 299–335. Elsevier.

30 Misra, N.N., Dixit, Y., Al-Mallahi, A. et al. (2020). IoT, big data and artificial intelligence in agriculture and food industry. *IEEE Internet of Things Journal* 9: 6305.

31 Zhang, L., Jia, J., Gui, G. et al. (2018). Deep learning based improved classification system for designing tomato harvesting robot. *IEEE Access* 6: 67940–67950.

32 Evan Ackerman (2015) Bosch, deepfield robotics: we develop autonomous machines to improve conventional and organic weed management. http://www.deepfield-robotics.com/ (accessed 25 December 2021).

33 Evan Ackerman (2015) B. River, see and spray agricultural machines. http://www.bluerivertechnology.com/ (accessed 8 May 2022).

34 Kumar, A., Vohra, R., Jain, R. et al. (2024). Correlation filter based single object tracking: a review. *Information Fusion* 112: 102562.

35 Kumar, A. (2023). *Visual Object Tracking Using Deep Learning*. CRC Press.

36 Kumar, A., Walia, G.S., and Sharma, K. (2020). Recent trends in multicue based visual tracking: a review. *Expert Systems with Applications* 162: 113711.

37 Kumar, A. and Bhalla, S. (2025). Exploring brain tumor detection through artificial intelligence. In: *Revolutionizing Medical Systems using Artificial Intelligence* (ed. Kumar, A., Singh, D.), 93–120. Academic Press.

38 Kumar, A., Singh, D., and Vohra, R. (2023). Improving learning abilities using AI-based education systems. In: *AI-Assisted Special Education for Students with Exceptional Needs* (ed. A. Kumar and A. Nayyar), 137–155. IGI Global.

39 Khalil, R.A., Saeed, N., Masood, M. et al. (2021). Deep learning in the industrial internet of things: potentials, challenges, and emerging applications. *IEEE Internet of Things Journal* 8 (14): 11016–11040.

40 Cambra, C., Sendra, S., Lloret, J., and Garcia, L. (2017). An IoT service-oriented system for agriculture monitoring. *2017 IEEE International Conference on Communications (ICC)*, pp. 1–6. IEEE, 21–25 May 2017, Paris, France.

41 Komarek, A.M., De Pinto, A., and Smith, V.H. (2020). A review of types of risks in agriculture: what we know and what we need to know. *Agricultural Systems* 178: 102738.

42 Botta, A., De Donato, W., Persico, V., and Pescape, A. (2016). Integration of cloud computing and internet of things: a survey. *Future Generation Computer Systems* 56: 684–700.

43 ONS Cisco (2012). 15454 40 gbps cp-dqpsk full c-band tuneable transponder card.

44 Marjani, M., Nasaruddin, F., Gani, A. et al. (2017). Big IoT data analytics: architecture, opportunities, and open research challenges. *IEEE Access* 5: 5247–5261.

45 Bodkhe, U., Tanwar, S., Parekh, K. et al. (2020). Blockchain for industry 4.0: a comprehensive review. *IEEE Access* 8: 79764–79800.

46 Chen, L., Thombre, S., Järvinen, K. et al. (2017). Robustness, security and privacy in location-based services for future IoT: a survey. *IEEE Access* 5: 8956–8977.

47 Sharma, S. (2023). Precision agriculture: reviewing the advancements technologies and applications in precision agriculture for improved crop productivity and resource management. *Reviews in Food and Agriculture* 4 (2): 45–49.

48 Kumar, M., Sinha, A., and Kumar, N. (2023). Precision agriculture – a vital approach towards modernizing the smart farming in India. *International Journal for Research in Applied Science and Engineering Technology* 11: 848–853.

49 Smith, R.J., Raine, S.R., McCarthy, A.C., and Hancock, N.H. (2009). Managing spatial and temporal variability in irrigated agriculture through adaptive control. *Australian Journal of Multi-Disciplinary Engineering* 7 (1): 79–90.

50 Balafoutis, A.T., Evert, F.K., and Fountas, S. (2020). Smart farming technology trends: economic and environmental effects, labor impact, and adoption readiness. *Agronomy* 10 (5): 743.

51 Javed, Z., Mumtaz, I., Zia, M.A., Nawaz, Q. (July 14 2022). AgroInfo DSF: a smart decision support framework for precision agriculture and farming. *International Conference on Management Science and Engineering Management*, pp. 87–96. Cham: Springer International Publishing, 3–6 August, 2022.

52 Ryo, M. (2022). Explainable artificial intelligence and interpretable machine learning for agricultural data analysis. *Artificial Intelligence in Agriculture* 1 (6): 257–265.

53 Lindblom, J., Lundström, C., Ljung, M., and Jonsson, A. (2017). Promoting sustainable intensification in precision agriculture: review of decision support systems development and strategies. *Precision Agriculture* 18: 309–331.

54 Shafi, U., Mumtaz, R., García-Nieto, J. et al. (2019). Precision agriculture techniques and practices: from considerations to applications. *Sensors* 19 (17): 3796.

55 Adewusi, A.O., Asuzu, O.F., Olorunsogo, T. et al. (2024). AI in precision agriculture: a review of technologies for sustainable farming practices. *World Journal of Advanced Research and Reviews* 21 (1): 2276–2285.

56 Singh, D. and Kumar, A. (2024). A deep recurrent neural network for plant disease classification. *SN Computer Science* 5 (8): 1053.

57 Ma, Y. and Guo, G. (2014). *Support Vector Machines Applications*, vol. 649. Springer.

58 Shafi, U., Mumtaz, R., Iqbal, N. et al. (2020). A multi-modal approach for crop health mapping using low altitude remote sensing, internet of things (IoT) and machine learning. *IEEE Access* 8: 112708–112724.

59 Liakos, K.G., Busato, P., Moshou, D. et al. (2018). Machine learning in agriculture: a review. *Sensors* 18 (8): 2674.

60 Xiao, D., Pan, Y., Feng, J. et al. (2022). Remote sensing detection algorithm for apple fire blight based on UAV multispectral image. *Computers and Electronics in Agriculture* 1 (199): 107137.

61 Zhou, Z., Chen, X., Li, E. et al. (2019). Edge intelligence: paving the last mile of artificial intelligence with edge computing. *Proceedings of the IEEE* 107 (8): 1738–1762.

62 Wang, B., Yan, Y., Lan, Y. et al. (2023). Accurate detection and precision spraying of corn and weeds using the improved YOLOv5 model. *IEEE Access* 17 (11): 29868–29882.

63 Friha, O., Ferrag, M.A., Shu, L. et al. (2021). Internet of things for the future of smart agriculture: a comprehensive survey of emerging technologies. *IEEE/CAA Journal of Automatica Sinica* 8 (4): 718–752.

64 Singh, A.K. (2022). Precision agriculture in India–opportunities and challenges. *Indian Journal of Fertilisers* 18 (4): 308–331.

4

Empowering Smart Agriculture with Artificial Intelligence

Shipra Raheja and Himanshi Bansal

IT Department, Guru Tegh Bahadur Institute of Technology, Delhi, New Delhi, India

4.1 Introduction

The epidemic expansion of the international population, flung to reach 10 billion by 2050, places immense strain on the agrarian industry to augment vegetable generation and optimize outcomes. In response to impending food deficits, two fundamental approaches can be followed: the expansion of land usage coupled with intensive farming or the implementation of innovative methods and making use of technology developments to raise output on current acreage (driven by numerous obstacles such as dwindling soil fertility, scarce staffing, changing climates, restricted land possessions, and issues with the environment the contemporary agricultural landscape is undergoing a profound transformation, diversifying into various innovative directions) [1]. Farming has undeniably progressed significantly from the era of hand usage of machinery. Every while, it presents modern advances outlined to increment effectiveness and maximize abdicate. Be that as it may, agriculturists and businesses around the world regularly neglect the benefits of counterfeit insights (AI) for their farming [2]. By joining information into cultivating, ranchers can open more openings to alter the way they work. From prescient analytics to edit administration and cultivating procedures to optimize asset assignment, AI can increment productivity and decrease chance some time recently permaculture usage. Embracing counterfeit insights in agribusiness is an imperative step in assembly the needs of a developing populace and parallelly guaranteeing the long-term maintainability of our nourishment [3].

Emerging Smart Agricultural Practices Using Artificial Intelligence, First Edition.
Edited by Ashish Kumar, Jai Prakash Verma, and Rachna Jain.

Published 2025 by John Wiley & Sons, Inc.

As horticulture competes with issues emerging from populace development and natural impacts, the integration of information is advantageous to make the farm's farming effective, productive, and solid. Supporting AI-driven development is crucial to meet the needs of today's agribusiness and guarantee nourishment security for future generations.

4.2 Benefits of AI in Agriculture

Agriculture and farming have been transformed by AI. These advancements are essential to address issues like climate change and population growth. Environmental risk reduction, crop yield prediction, and resource optimization are among the ways that AI empowers frame AI solutions to make agriculture more productive, sustainable, and efficient, from drone monitoring to automated irrigation [4, 5]. With its capability to advance food production and innovation, this technology serves as a barrier.

4.2.1 Data-Driven Decisions

The agricultural sector uses data to gain insights ranging from individual fields to supply chains. Thus, big data analytics of which the efficiency reaches the level of AI are now available to farmers. Market trends will be provided by farmers when they incorporate the use of artificial intelligence (AI) in the forecasts of prices and the adjustment of the planting or harvesting seasons. The other productivity and sustainability enhancements include basic soil analysis with the help of AI, weather predicting and basic suggestions about fertilizers or pesticides. Through the implementation of AI in farm management software, the use of data to make decisions throughout the crops' growth period becomes possible, which in turn results in higher yields and low effects on the surrounding environment [6, 7].

4.2.2 Optimizing Resources, Boosting Profits

Farmers are always after higher yields of their crops. AI in precision farming solves this by ensuring efficient use of the resources by balancing between the smart soil and the variable rate technology and data. Improved ability to get real-time crop health information by use of AI empowers specific actions like irrigation, fertilization, and pesticide application. In this way, this active stance provides the most favorable condition for crops to grow and save losses of resources and harm to the earth, as well. Additionally, new solutions such as vertical farming present food production as a chance to be developed with little funds, with minimized herbicide applications, higher crop yield, and higher income [8]. Through the union of agriculture with AI, the farming process is being revolutionized. There are new levels of productivity only exhausting the resources, as well as profitability, attainable with the help of AI. It can then be seen that with the help of AI solutions, farmers can enhance land productivity and thereby contribute toward better and future-proof food security for the world [9].

4.2.3 Automation: A Helping Hand

Traditionally, farming has been observed to involve much effort, and at the moment, there are negative supply shocks affecting the labor force. However, it is replaced with automation and does not entail the need for any more workers. Cultivation and other farming activities were earlier hampered by manual work and the use of animals as a means of mechanization. A relatively newer form of automation that is gradually changing the face of this industry is now in vogue. These are self-driven tractors, smart irrigation systems, agriculture IoT, and AI-enabled drones, and so on, as well as AI-based robots for crop picking. AI's tools are exceptionally accurate and efficient when compared to human effort. Such technologies assist in mitigating the problems of labor shortage needed in agriculture as well as enhancing the sector's efficiency and sustainability. Automated tractors enhance efficiency during field operations and at the same time reduce on mishaps. Sensitization is done through the use of smart technology in irrigation and fertilizer methods to reduce wastage [10]. They use drones with better sensors monitoring crops and are in a position to identify certain issues that may occur. Vertical farming software also helps in the utilization of available space and other resources for year-round production. AI-powered robots excel at tedious tasks like harvesting, saving labor, increasing consistency, and minimizing crop damage. These technologies also give farmers the ability to gather and evaluate enormous volumes of data, enabling data-driven decision-making for enhanced crop care as well as utilization of resources see (Fig. 4.1).

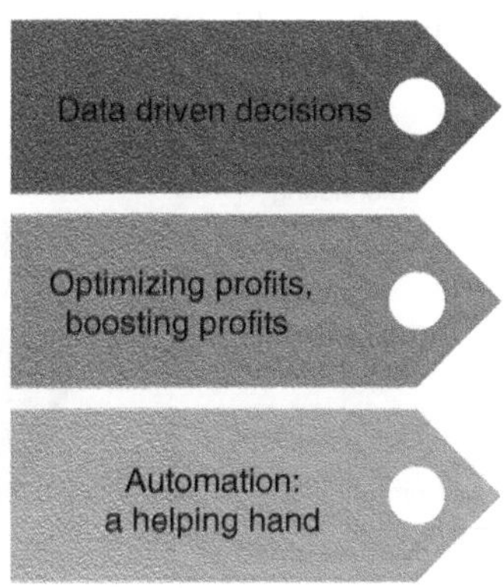

Figure 4.1 Benefits of artificial intelligence in agriculture.

In general, the utilization of automation in agriculture promotes efficiency, sustainability, and innovation throughout the industry while also addressing labor issues. Farmers can accommodate the growing demand for sustenance in a world that is changing quickly by optimizing output, cutting costs, and utilizing AI-driven tools and digital technology [11, 12].

4.3 Applications of Artificial Intelligence in Agriculture

AI models can revolutionize the farming industry and streamline a number of manual operations, among other benefits, when they are integrated with traditional farming practices [13]. Intelligent agriculture systems can facilitate chores effectively and efficiently by enhancing current technologies see (Fig. 4.2). Here are expanded explanations of typical AI use cases in agriculture [14].

4.3.1 Optimizing Automated Irrigation Systems

Autonomous crop management is made possible by combining AI algorithms with Internet of Things sensors that track soil moisture levels and meteorological variables. Making decisions about the distribution of water in real time helps to save water supplies and encourages sustainable agricultural methods.

4.3.2 Leaks or Damage to Irrigation Systems

AI is essential in the early identification of leaks because it can analyze data and spot trends or abnormalities that could point to leaks. Real-time monitoring improves water efficiency and conservation efforts by preventing crop damage and water waste.

4.3.3 Crop and Soil Monitoring

AI-driven computer vision models accurately monitor soil conditions and crop health, allowing farms to maximize harvest production along with the condition of crops by making the appropriate modifications.

Figure 4.2 Applications of artificial intelligence in agriculture.

4.3.4 Distinguishing Infection and Bug Issues

In image recognition, the AI intelligent computer vision instantly relates bugs, sicknesses, and trim well-being issues in the images of people to ensure intervention to avoid trim harm quickly. Sophisticated mathematical techniques provide better comparability in determining sundry risks of adapt adjust to boost wellness.

4.3.5 Checking Animals Well-Being

Here are some examples of imaginative AI arrangements, including Cattle Eye, which applies rambles, cameras, and computer vision to screen the well-being and behavior of animals. This knowledge contributes to the continuity of the creature's advancement and enhanced operation in the area of well-being.

4.3.6 Pesticide Application

Control on rambles peruses pesticide use by hitting troubled areas influenced by natural effects while at the same time passing on efficiency to all performances of computer vision.

4.3.7 Surrender Projection and Expository Estimating

Big data business intelligence computations enable decentralized trim yield probability calculations at the same time as asset distribution. Integration of information from sensors, rambles, and 3D displaying helps in deterministic planning and decision-making for maximum profits with regard to investment.

4.3.8 Computerized Weeding and Collecting

Computer vision and machine learning assist to distinguish a weed from the crops and make mechanized automatism of the tasks such as weeding and picking possible. This innovative headway streamlines operations and decreases labor necessities.

4.3.9 Create Sorting

AI-driven computer vision frameworks precisely categorize gathered create based on quality properties, refining sorting forms and empowering proficient categorization for different markets.

4.3.10 Improved Reconnaissance

Integration of computer vision and ML supports cultivate security by instantly recognizing security breaches and recognizing between authorized staff and unauthorized guests, guaranteeing the security of resources and people.

By leveraging AI innovations over these regions, ranchers can improve proficiency, efficiency, and maintainability, eventually driving to progressed rural homes and favorable results for both ranchers and shoppers [15, 16].

4.4 Part of AI Within the Farming Data Administration Cycle

AI-powered rural information administration has numerous focal points for distinctive features of cultivating and counting:

4.4.1 Hazard Control

AI-powered prescient analytics encourages chance distinguishing proof and moderation for ranchers, cutting down on blunders and progressing decision-making. AI models can offer bits of knowledge into conceivable issues like climate swings, advertise patterns, and bug episodes by assessing past information and show circumstances. This permits agriculturists to require preventative activity to decrease misfortunes.

4.4.2 Plant Breeding

Counterfeit insights analyze plant development information to pinpoint characteristics connected to resistance against serious climate, ailments, or bothers. Ranchers can progress trim yields and flexibility to changing natural circumstances by utilizing this information to advise their trim choice and breeding techniques [17].

4.4.3 Investigation of Soil and Edit Wellbeing

Manufactured insights calculations look at soil tests to decide supplement shortages and assess their chemical cosmetics. Moreover, by looking at information from various sources, counting adj. imaging, climate designs, and past illness episodes, AI is able to distinguish and figure trim sicknesses. In arrange to protect the wellbeing of the soil and crops, this empowers ranchers to carry out particular mediations, such as changing supplement levels or spraying pesticides.

4.4.4 Edit Bolstering

AI-driven water system frameworks optimize the organization of water and supplements by utilizing information from real-time sensors, soil dampness substances, and editing needs. Manufactured insights (AI) help ranchers in maximizing edit yields while minimizing asset utilization and natural impact by accurately determining the perfect timing and dosage of agronomic inputs.

4.4.5 Gathering

Counterfeit insights (AI) makes strides trim yield figures and streamlines harvesting methods. Manufactured insights calculations are able to estimate the leading time to harvest crops in arrange to realize the leading conceivable abdicate and quality by analyzing factors counting trim development markers, soil dampness substance, and climate designs. Moreover, amid the gathering handle, AI-driven sorting and reviewing frameworks increment precision and effectiveness, permitting agriculturists to optimize incomes and diminish squandering.

All things considered, joining AI into agrarian information administration gives ranchers valuable bits of knowledge that offer assistance to them to make shrewd choices, maximize the utilization of assets, and make strides in cultivating methods' supportability and efficiency [18, 19].

Figure 4.3. shows that AI can transform agriculture by using technology to improve crop monitoring and management. Seedlings growing in soil represent a

Figure 4.3 Use of sensors and intelligent automation for plant growth.

typical crop, while the graphic shows advanced technology used in intelligence-based agriculture.

4.5 Optimizing AI for Farming and Agrarian Forms

AI and other advanced advances in farming must work together agreeably to realize their full potential and fathom the issues that modern ranchers stand up to. Here's a made strides duplicate:

4.5.1 Huge Information for Educated Decision-Making

Ranchers may presently utilize enormous information sets to urge bits of knowledge that are valuable and significant by combining AI and huge information analytics. AI calculations give exact suggestions for moving forward, cultivating procedures, and raising efficiency while cutting costs by collecting and analyzing information in real time. Agriculturists are able to create well-informed choices by utilizing exact and convenient data that is much obliged to this cooperative energy.

4.5.2 IoT Sensors for Capturing and Analyzing Information

IoT sensors are basic for following, measuring, and putting away real-time information on an assortment of agrarian factors. They work in conjunction with supporting innovations like GIS and AI rambles. Agricultural sending of these sensors permits for the quick collection of exact and intensive information. This data becomes amazingly vital when coordinated with AI, permitting agriculturists to memorize approximately edit well-being, soil conditions, climate designs, and other subjects. This data at that point makes a difference when they make well-informed choices and actualize proactive administration strategies.

4.5.3 Cleverly Mechanization and Mechanical Technology for Minimizing Manual Work

AI-driven computerization fathoms the issue of labor deficiencies in farming when combined with IoT and independent tractor innovation. AI-enabled independent tractors can carry out tedious occupations like planting, watering, and gathering with precision and productivity, nullifying the requirement for human labor. Furthermore, agricultural robots are being utilized more as often as possible for occupations like picking vegetables since they can utilize AI to perform assignments with more prominent exactness, longer workdays, and lower botch rates than human labor. Agriculturists can move forward operations, boost yield, and lessen the impact of labor deficiencies on cultivated yield by utilizing these advances.

Ranchers may overcome impediments, make way better choices, and execute feasible agrarian homes for expanded efficiency and benefit by utilizing AI-driven analytics, IoT sensors, and clever automation [20, 21].

4.6 AI's Limitations with Regard to Agriculture

Nowadays, in a changing world in various domains, AI is useful in [22–24] medical [25, 26] and educational fields [27–29] only. This perception is due to the absence or lack of understanding of AI and its methods. The general public, specifically those not from a computer background, is not well taught about AI, which is causing the problem in the agricultural industry to adopt AI more efficiently. Despite all advancements in the agriculture industry, many farmers are still using traditional techniques. It's unlikely that many farmers have probably worked on AI-related projects [30].

Furthermore, agriculture suppliers usually do a job of defining the advantages of novel technology and its applications from the ground level. Technological companies have a great job of work in educating the public in the area of AI and its application machines implementation in agriculture business. Using AI in the agricultural business will be a profitable for all farmers, due to its advantages for upcoming farming future, still there are many obstacles to be addressed, even fulfilling the growth.

4.6.1 Large Upfront Costs

There is clear visibility that the preliminary investment in agriculture with AI can be extremely costly. Furthermore, it may be economical over time. It will be difficult for farmers and agribusinesses to implement AI due to expenses and cost difficulties, especially in small-scale farming. But, as technology advances, the cost of inculcating AI into agriculture will come down. Various companies will also find other financing options, including government and private firms' grants [31].

4.6.2 Reluctance to Adopt New Procedures and Technologies

Despite AI's many advantages, farmers are still finding it difficult to learn and embrace the technology as they are unfamiliar with new ideas. The growth of farming with AI techniques and its overall profitability is hindered by aversion to innovation and unwillingness to learn new procedures and techniques. Its high time that farmers must understand that AI for growth and innovation in the field of agriculture is must. For the government and private sectors, AI adoption in farming should be encouraged by providing training resources and incentives. Governments must create rules to ensure that employees learn that technology for their growth.

4.6.3 Insufficient Hands-on Familiarity with Emerging Innovation

In the agricultural sector, there are various technological developments around the world. Some areas are still able to implement AI. Countries are still lacking in learning and implementing various AI-based agricultural innovations. However, they want to work in areas where the agricultural economy is uplifting. In addition to providing resources, they should also provide support and education for farmers and agricultural professionals who are willing to take on new challenges for upcoming innovations [32].

4.6.4 An Extended Process of Adopting New Technology

Apart from inadequate learning and knowledge, the agriculture industry still lacks the infrastructure and resources for AI to function. Software developers are facing challenges in adopting new technology. One of the best strategies to get around this is to engage farmers slowly and slowly. It is better to give them the use of less complex technologies in the starting, such as an agricultural trade platform and other less complex resources to use. Further provide them with some manuals too to learn theoretically and practically.

4.6.5 Technological Limitations

AI has limitations because this technology is still in its evolving face. Various, high-quality data are required for the implementation of accurate models, but that might be hard to implement in agriculture. To overcome these limitations, more studies and data analytics are done. Learn additionally that farmers should continue to participate in the process of technology growth. Also, it will be helpful to manually monitor various AI judgments.

4.6.6 Privacy and Security Issues

Regulations governing the application of AI in all areas are still being processed. In particular, there are certain legal concerns that need to be worked on when applying AI for smart farming and precision agriculture. Cyberattacks, securities, and data breaches. Likely that hackers may attack AI-based farming systems in an attempt to destroy food supplies and affect its future outcomes [33, 34].

4.7 Future of AI in Agriculture

In the coming years, AI will play an important part in agriculture and food sustainability. The agricultural industry has always been at the forefront, moving from early tools like tractors to irrigation technology and to AI. Farming is becoming more efficient with every advancement while minimizing its difficulties. More significantly, AI's advantages in agriculture are highlighted as intelligent farming equipment, intelligent automation, and AI-powered products that really free up workers' time so that this time can be used for more strategic activities that require human judgment. AI in farming and technology tools could grow more quickly because of agricultural robotics and enhanced computer vision.

AI is becoming more and more advanced to tackle the various problems brought up by rising food prices environmental issues, and rapid climate changes by enhancing productivity, resource allocation, sustainability, and real-time monitoring for healthier and good-quality products. This will completely transform agriculture in the future. But we can't just buy AI and start using it right away. AI is not a physical object rather, it is a collection of programming-automated technologies. An AI algorithm learns from training data and then applies that knowledge to solve problems, just like a human thinks. The agricultural sector has to adapt AI to drive the industry's revolution in today's world. It is necessary to educate and train farmers on the use of AI-powered technologies [35]. Farmers' jobs as manual laborers will probably transform into planners and administrators of intelligent agricultural systems as a result of AI. It's possible that knowledge of IT solutions and agro intelligence will be more valuable than proficiency with traditional instruments or manual work.

Indeed, whereas Machine Learning operation administrations, machine learning, and AI have the potential to totally alter cultivation, they depend on other innovations to work together. Agriculturists must have an innovative foundation put in some time recently so they can completely benefit from AI. Building that framework may take a long time, but the payoff seems to be a solid, long-lasting, innovative biological system. Maximizing AI's benefits requires an understanding of how it works and the best ways to join specialized information into regular methods. Because of this, working with a learned program advancement group is an awesome put to begin. Ag Tech arrangement suppliers have a noteworthy part to play. Everyone needs to think about how they can make their instruments way better, bargain with issues, and communicate the quantifiable preferences of AI and machine learning. AI in agribusiness is beyond any doubt to have a shining future if this can be finished [36]. The success of human society is primarily determined by how well-functioning its agricultural systems are. When conventional farming methods become outdated, there is an increasing need for better technical alternatives. Across all industries, automation has always had a significant impact. Agriculture will surely be greatly impacted by AI, as the industry is currently undergoing tremendous change due to digital technologies [37].

4.8 The Future Research of AI in Small-Scale Farming

The future of AI in small-scale farming looks very exciting. With new technology, small farmers will be able to improve their crops more easily and efficiently. They can use affordable drones and sensors to keep an eye on the health of their plants and the weather. This means they can quickly spot any problems and take action to keep their crops in great shape. By analyzing data on humidity, temperature, and moisture, intelligence can optimize water and equipment, increase crop yields, and save resources.

Machine learning models will predict crop yields, identify potential pests, and recommend preventive measures, allowing farmers to make informed decisions that will increase productivity and profitability. AI will also support culture by analyzing environmental data to recommend environmentally friendly technologies, reduce chemical waste, and promote healthy drinking on clean soils. This technology freedom will give small farmers around the world access to smart tools, increasing productivity and improving food security. A better decision support system that combines satellite imagery, weather forecasts, and economic models will increase efficiency. Additionally, education and training tools will help farmers use these innovations effectively and ensure better results. Overall, AI research will transform small-scale farming into a viable, profitable, and profitable business that will benefit farmers around the world. Agriculture will surely be greatly impacted by AI, as the industry is currently undergoing tremendous changes due to digital technologies.

References

1 Minwoo Ryu, Jaeseok Yun, Ting Miao, Il-Yeup Ahn (2015). Design and implementation of a connected farm for smart farming system. *2015 IEEE SENSORS*, pp. 1–4.November 2015, Busan, Korea (South).

2 Nayyar, A. and Puri, E.V. (November 2016). IoT based smart sensors agriculture stick for live temprature and moisture monitoring using Arduino cloud computing & solar technology. *International Conference on Communication and Computing Systems (ICCCS-2016).*

3 Navulur, S., Sastry, A.S.C.S., and Giri Prasad, M.N. (2017). Agricultural management through wireless sensors and internet of things. *International Journal of Electrical and Computer Engineering* 7: 3492–3499.

4 Rekha, P., Rangan, V.P., Ramesh, M.V., and Nibi, K.V. (2017). High yield groundnut agronomy: An IoT based precision farming framework. *2017 IEEE Global Humanitarian Technology Conference (GHTC)*, pp. 1–5., 19–22 October 2017, San Jose, CA, USA.

5 Mekala, M.S. and Viswanathan, P. (2017). A Survey: Smart agriculture IoT with cloud computing. *Proceedings of the 2017 International Conference on Microelectronic Devices, Circuits and Systems (ICMDCS)*, Vellore, India (10–12 August 2017), pp. 1–7. IEEE: Vellore, India.

6 Chandgude, A., Harpale, N., Jadhav, D. et al. (2018). A review on machine learning algorithm used for crop monitoring system in agriculture. *International Research Journal of Engineering and Technology (IRJET)* 5 (4): 1470.

7 Mishra, S., Paygude, P., Chaudhary, S., and Idate, S. (2018). Use of data mining in crop yield prediction. *2018 2nd International Conference on Inventive Systems and Control (ICISC)*, pp. 796–802. January 2018,

8 Medar, R., Rajpurohit, V.S., and Shweta, S. (2019). Crop yield prediction using machine learning techniques. *2019 IEEE 5th International Conference for Convergence in Technology (I2CT)*, pp. 1–5. 29–31 March 2019, Bombay, India.

9 Anusha, A., Guptha, A., Sivanageswar Rao, G., and Tenali, R.K. (2019). A model for smart agriculture using IoT. *International Journal of Innovative Technology and Exploring Engineering (IJITEE)* 8 (6): 2278–3075.

10 Bu, F. and Wang, X. (2019). A smart agriculture IoT system based on deep reinforcement learning. *Future Generation Computer Systems* 99: 500–507.

11 Shadrin, D., Menshchikov, A., Ermilov, D., and Somov, A. (2019). Designing future precision agriculture: detection of seeds germination using artificial intelligence on a low-power embedded system. *IEEE Sensors Journal* 19 (23): 11573–11582.

12 Triantafyllou, A., Sarigiannidis, P., and Bibi, S. (2019). Precision agriculture: a remote sensing monitoring system architecture. *Information* 10: 1–25. https://doi.org/10.3390/info10110348.

13 Singh, D. and Kumar, A. (2024). A deep recurrent neural network for plant disease classification. *SN Computer Science* 5 (8): 1053.

14 Rodzalan, S.A., Yin, O.G., and Noor, N.N.M. (2020). A foresight study of artificial intelligence in the agriculture sector in Malaysia. *Journal of Critical Reviews* 7 (8): 1339–1346.

15 Shepherd, M., Turner, J.A., Small, B., and Wheeler, D. (2020). Priorities for science to overcome hurdles thwarting the full promise of the 'digital agriculture'revolution. *Journal of the Science of Food and Agriculture* 100 (14): 5083–5092.

16 Erakis, K., Lampathaki, F., Nikas, K. et al. (2020). Cybele–fostering precision agriculture & livestock farming through secure access to large-scale HPC enabled virtual industrial experimentation environments fostering scalable big data analytics. *Computer Networks* 168: 107035.

17 Talaviya, T., Shah, D., Patel, N. et al. (2020). Implementation of artificial intelligence in agriculture for optimisation of irrigation and application of pesticides and herbicides. *Artificial Intelligence in Agriculture* 4: 58–73.

18 Afzaal, H., Farooque, A.A., Abbas, F. et al. (2020). Computation of evapotranspiration with artificial intelligence for precision water resource management. *Applied Sciences* 10 (5): 1621.

19 Anubha Pearline, S. and Sathiesh Kumar, V. High performance ensembled convolutional neural network for plant species recognition. *International Conference on Computer Vision and Image Processing*, pp. 526–538. March 2021, Singapore.

20 Bertoglio, R., Corbo, C., Renga, F.M., and Matteucci, M. (2021). The digital agricultural revolution: a bibliometric analysis literature review. *IEEE Access* 9: 134762–134782.

21 Bhardwaj, H.H., Tomar, P., Sakalle, A., and Sharma, U. (2021). Artificial intelligence and its applications in agriculture with the future of smart agriculture techniques. In: *Artificial Intelligence and IoT-Based Technologies for Sustainable Farming and Smart Agriculture*, 25–39. IGI Global.

22 Kumar, A., Vohra, R., Jain, R. et al. (2024). Correlation filter based single object tracking: a review. *Information Fusion* 112: 102562.

23 Kumar, A. (2023). *Visual Object Tracking Using Deep Learning*. CRC Press.

24 Kumar, A., Walia, G.S., and Sharma, K. (2020). Robust object tracking based on adaptive multicue feature fusion. *Journal of Electronic Imaging* 29 (6): 063001–063001.

25 Kumar, A. and Bhalla, S. (2025). Exploring brain tumor detection through artificial intelligence. In: *Revolutionizing Medical Systems using Artificial Intelligence* (ed. Ashish Kumar, Divya Singh), 93–120. Academic Press.

26 Kumar, A., Raja, and Muhammad Fazalljaz, A. (2023). Augmented reality as a boon to disability. In: *Smart Distributed Embedded Systems for Healthcare Applications*, 77–92. CRC Press.

27 Kumar, A., Singh, D., and Vohra, R. (2023). Improving learning abilities using ai-based education systems. In: *AI-Assisted Special Education for Students With Exceptional Needs*, 137–155. IGI Global.

28 Kumar, A., Nayyar, A., Sachan, R.K., and Jain, R. (2023). *AI-assisted Special Education for Students with Exceptional Needs*. IGI Global.

29 Kumar, A. and Rubeena, V. (2023). Impact of deep learning models for technology sustainability in tourism using big data analytics. In: *Deep Learning Technologies for the Sustainable Development Goals: Issues and Solutions in the Post-COVID Era*, 83–96. Singapore: Springer Nature Singapore.

30 Ben Ayed, R. and Hanana, M. (2021). Artificial intelligence to improve the food and agriculture sector. *Journal of Food Quality* 2021: 2021.

31 Jung, J., Maeda, M., Chang, A. et al. (2021). The potential of remote sensing and artificial intelligence as tools to improve the resilience of agriculture production systems. *Current Opinion in Biotechnology* 70: 15–22.

32 Spanaki, K., Karafili, E., Sivarajah, U. et al. (2021). Artificial intelligence and food security: swarm intelligence of AgriTech drones for smart AgriFood operations. *Production Planning and Control* 33: 1–19.

33 Olipa, N.L., Lydia, M.C., Chabala, S., and Chizumba, S. (2021). Satellite-based crop monitoring and yield estimation – a review. *The Journal of Agricultural Science* 13: 180–194.

34 Sood, A., Sharma, R.K., and Bhardwaj, A.K. (2022). Artificial intelligence research in agriculture: a review. *Online Information Review* 46 (6): 1054–1075.

35 Amr, M.E., Al-Awamry, A.A., Elmenyawi, M.A., and Tag Eldien, A.S. (2022). Design and implementation of a low-cost IoT node for data processing case study: smart agriculture. *Journal of Communication* 17 (2): 99–109.

36 Qazi, S., Khawaja, B.A., and Farooq, Q.U. (2022). IoT-equipped and AIenabled next generation smart agriculture: a critical review current challenges and future trends. *IEEE Access* 10: 21219–21235.

5

Foundations of Agricultural AI

Divya Singh, Naman Agrawal, Jaya Saini, and Manoranjan Kumar

School of Computer Science Engineering and Technology, Bennett University, Greater Noida, Uttar Pradesh, India

5.1 Introduction

Agriculture is facing grand challenges, including increased demand, reduced land, water, and labor resources, degradation of the environment, and the effects of climate change. Information technology, automation, and artificial intelligence (AI) offer opportunities to address many of these issues. This course will introduce computer scientists and others to core concepts in agriculture and researchers in agriculture to the research problems that are currently being addressed by AI researchers. We will cover phenomena at many spatial scales, from the growth of individual plants to weather and ecosystem-scale processes [1]. We will also consider problems related to data analysis and acquisition, including robot scouting and genome-based predictions. At the end of this course, students will be positioned to address important challenges in agriculture.

When faced with the unsustainability of modern agriculture, many proposals suggest a return to traditional agricultural methods [2]. But, by embracing scientific and technological advances, we can achieve the increased productivity and scalability required by a growing population while reducing negative environmental impacts. AI offers powerful tools to model, analyze, and manage agricultural systems. Although great progress has been made in developing the foundations of AI, including pattern recognition, computer vision, machine learning (ML), and planning for control, this progress has largely taken place in a narrow context that does not capture agricultural systems and their problems. By

Emerging Smart Agricultural Practices Using Artificial Intelligence, First Edition.
Edited by Ashish Kumar, Jai Prakash Verma, and Rachna Jain.

Published 2025 by John Wiley & Sons, Inc.

focusing on crop phenotyping, i.e. measuring a crop's growth and performance, and intelligent scouting for pests, diseases, and stress using robots, we have made significant progress in understanding and modeling vital aspects of agricultural systems. This understanding enables estimation, prediction, and control, providing information for enhanced decision-making [3].

5.1.1 Definition and Scope

There is not a single universally accepted definition of AI. Among these different views, there are differences between introspective and external conceptions. However, despite these differences, most definitions take into consideration the idea that intelligence is largely determined by its environment as a result of the adaptation of an entity in a certain niche.

According to Nill and Rushton, "Artificial intelligence is the development and study of computer systems capable of performing tasks that would require intelligence if done by a human being." These authors quote Herbert Simon: "The study of artificial intelligence seeks to develop a computer that can be integrated into a wide range of problem domains, a computer that can carry out those abstract functions we attribute to human minds: reasoning, problem solving, learning, planning, perception, understanding natural language, and manipulation of objects, including the navigation, locomotion, and manipulation of the body." McCorduck states, finally, that the idea currently in the popular mind of AI – of providing machines with the ability to understand, respond to, and replicate human abilities, amounts to having them take the Turing test.

Agricultural AI is not long-standing and is not as precisely defined as some other fields if compared with crop production, animal production, or agricultural machines. For instance, at an international conference on Agricultural Engineering in 1989, it was difficult to find a paper within this field. In the case of agricultural meteorological applications with AI, published papers can be found as far back as 1982. More recently, AI has been used in grain crop storage, grading, and test analytics, providing reasonably good results [4]. The scope of AI for agriculture may include both software and hardware. Examples of the first are what we can do with expert systems, knowledge-based systems, object-oriented problem solving, distributed computing, robotics, machine vision, speech processing, or neural networks. Hardware developments have also been growing quite fast, involving mechanical manipulation, control applications, instrumentation, communication, sensory measurements, and robotics. Despite the fact that AI software that is living together with some agricultural activities has been greatly increasing over the last few years, future actions may come from hardware developments [5].

5.1.2 Importance and Benefits

Highly efficient agricultural systems that maximize yield and quality at a reduced production cost, minimal environmental impact, and minimal use of nonrenewable resources are key to supporting healthy and prosperous economies located in increasingly large and densely populated urban centers. What is more, the output that these systems generate is needed in enormous volume to support the growing demand for organic fuels, industrial products, and, not least, high-quality, nutritious food for a family of 9 billion people by the year 2050. To satisfy all these requirements, we need wise management of the world's limited agricultural land, as well as savvy and sustainable use of water, nutrients, and other inputs [6]. At the same time, we need to maintain ecosystem services such as biodiversity and natural pest and disease control. Unfortunately, the sustainability and efficient use of vital agricultural resources such as water, land, nutrients, and energy have not been a high priority for the world's agriculture industry in the past. That situation must change [7].

Renewed lessons from more than 10,000 years of settlements of agricultural and civil society underscore the strong relationship between food, culture, health, and prosperity. The critical role that agriculture has played in the development dynamics of the global economy over the last 200 years is also very apparent. Agricultural markets determine the price signals that influence how people use land, water, nutrients, energy, technologies, and knowledge. It is the use of these resources "beyond the farm gate" and the resulting effects on waste and pollution that are becoming of major public concern and depend on the development of new policies and sometimes new regulations [8]. We believe that AI technologies have an important role to play in assisting these new developments and facilitating the transition of our current agricultural systems to vibrant and increasingly productive Agro-Ecosystems (AEs) designed around multifunctional principles that benefit society, the economy, and the environment. This paper will have achieved its goal if it identifies the new research directions and policy options that are needed when the employment of AI technologies in agriculture is considered as an endoscopy of the AE concept [9].

5.1.3 Fundamental Concepts in AI

The goal of AI problems is to make computers capable of intelligent behavior. One measure of whether a computer is intelligent is if it can perform tasks that are difficult for humans to perform employing our training set. In other words, the power of AI comes from representing complex, irregular problems in a regular form upon which a general algorithm may be employed. A general approach for understanding intelligent processes is the following by looking at many

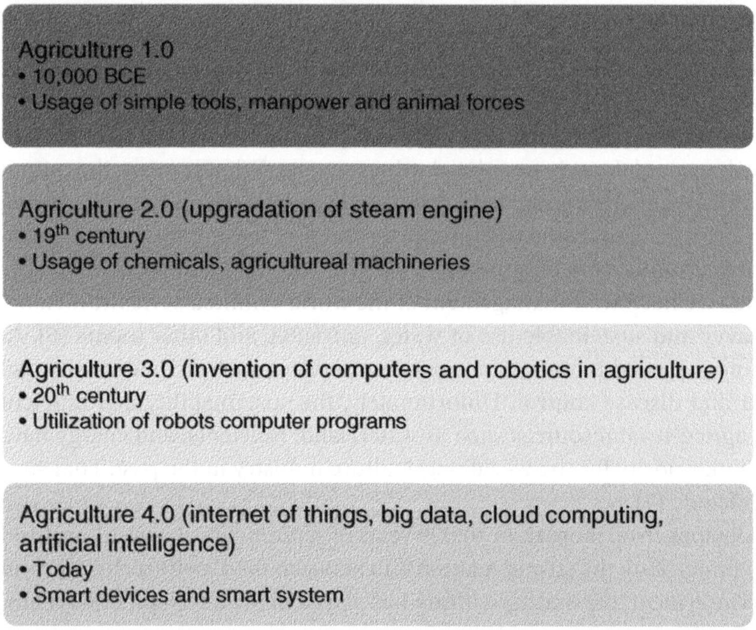

Figure 5.1 Growth of technology for the agriculture sector. *Source:* Pugun & Photo Studio/Adobe Stock Photos.

different examples of real-world intelligent action to abstract out the simple principles and common forms behind intelligent action. Organize these principles into a computer-implemented organization, a form of computer program that is capable of producing the intelligent actions we observe. Part of the content of AI, then, is simply the recognition that such a model may be possible and the clarification and specification of these simple expressed principles and forms [10]. The growth of technology for the smart farming is shown in Figure 5.1.

Of special importance to AI are the principles of search, pattern recognition, abstraction processes, reasons and planning, representation and model-based reasoning, and problem-solving methods. Because psychology is the study of the principles and forms of intelligence underlying animal and then human behavior, much AI research is based upon psychological research into these topics. However, AI includes studies beyond psychology, for example studies of performance characterizations of knowledge processes in organizations' management science or economics. AI is not restricted to the study of only human-like intelligence but also the study of types of intelligence differing from the human form, such as the learning behavior found in ants or the highly abstract reasoning found in expert chemists [11, 12].

5.1.4 Highlights of the Chapter

- In recent years, farming has made substantial use of AI.
- AI helps farmers choose the right seed for a given set of weather conditions.
- Important applications of AI in agriculture are emphasized and investigated.
- AI-powered solutions will improve crop quality and allow farmers to produce more with fewer resources.

5.2 Machine Learning

Machine learning (ML) is a rapidly expanding discipline and has demonstrated tremendous success in a variety of fields, including computer vision, speech recognition, medicine, finance, and adversarial cyber. ML consists of several key components. The first is the training data, which is used to teach the machine, usually at scale, to learn from past experiences. The design of the feature space, where the data resides, is another critical aspect of ML [13]. Selecting the right architectural type, such as a convolutional neural network (CNN) for image data or a recurrent neural network (RNN) for time series data, is also fundamental to success. Next, the optimal model hyperparameters need to be chosen, such as the learning rate and the dropout rate, which are unlikely to be abstractly optimal for different applications. Then, the model needs to be trained, with the appropriate form of data augmentation and regularization, to avoid overfitting. Finally, the work of ML isn't done; the model must be verified and continually updated to ensure that it continues to stay trained on the most recent data, among other issues [14].

In agriculture, ML has shown great success in both weather crop yield prediction and disease identification in many crops, e.g. sorghum, spinach, and strawberries. While ML has shown great success in predicting plant phenotypes, especially when large sample sizes and color imagery are used, ML has shown less success with respect to important traits such as nitrogen and water content [15]. Additionally, ML models do not provide actionable decisions that timely address the broad range of agricultural problems. Such answers can only come by broadening the current computer vision-dominated AI models with the inclusion of additional data sources – including spectral and time series data – and structuring models to directly address key questions.

5.3 Deep Learning

Deep learning is based on artificial neural network theory and involves using many layers of both linear and nonlinear processing units to extract relevant information from raw data. The deep learning algorithm can model complex patterns directly from data. Deep learning can model data with increasing levels of abstraction [16]. In computer vision, deep learning is often modeled using the CNN. In natural language processing, another important subarea with deep learning, the RNN is used in sequence data and the transformer is applied to sequence-to-sequence problems. Deep learning breaks through the era of AI, brings new opportunities for AI, and also advances the development of AI [17].

Deep learning has many advantages. It is powerful in capturing complex structures. It can build deep hierarchies of representations that ultimately can represent very abstract concepts, going far beyond just digital sequences, although deep learning also has some disadvantages – such as requiring many more parameters and large amounts of data, suffering from overfitting, and one inability to completely understand the learning process. Because deep learning is powerful and, has stronger generalization ability and is more suitable for extracting useful information from complex data, it has received more and more attention. To describe the data generation process through neural network and then use training data for learning, this results in better predictive accuracy. This is the driving force that leads to deep learning being the dominant technology used today because we live in an era characterized by the creation and availability of large amounts of data [18, 19].

5.4 Applications of AI in Agriculture

Many AI applications have been developed in recent decades for agricultural tasks, in which the AI paradigm has been successful with many commercialized systems. Specifically, many ML applications have been implemented in various agricultural applications. The importance of in-field operations is evident in the report of the U.S. Department of Energy (DOE) in 2009, which states that it is believed that a potential saving of $12 billion could be captured by networking and automating agricultural vehicles. Whether in-field or in-post-field management or operations, several commercial robots, robotic subsystems, or AI software focused on specific market needs in agricultural sectors have been developed. Some common commercial robots in agriculture are precision seeding, robotic transplanters, thinning and shoot removal robots, cultivation robots, weeding robots, and harvesting robots [20]. The uses of modern technology in advanced farming is shown in Figure 5.2.

Large farm equipment has attracted much attention among researchers, companies, farmers, soil scientists, and agronomists in recent decades. Agricultural field operation assessments are often conducted based on standard tractor performance conditions or labor-related regulation directives of agricultural enterprises. The main job of big farm equipment is to carry out various in-field operations, which

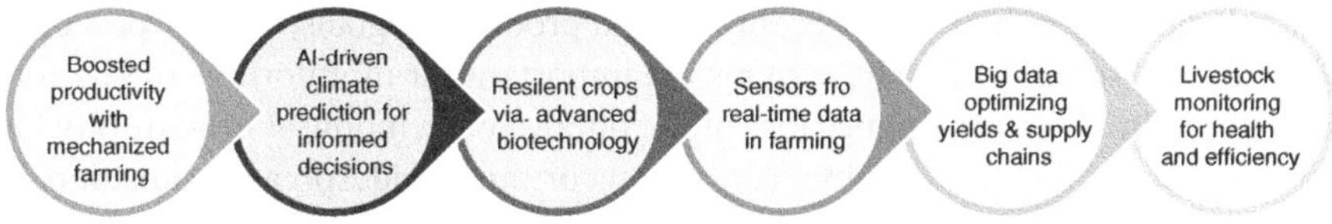

Figure 5.2 Uses of modern technology in agriculture.

are generally composed of a series of repetitive operations, often in the same operation category [20]. These operations include soil preparation (such as plowing, subsoiling, plowing, and stone removal operations), seeding and transplantation operations (such as seeding, fertilizing, and transplanting operations), plant care operations (such as moldboard operations, row spacing operations, and protection operations), and plant harvest and postharvest operations (such as mowing, collection, and dividing operations). These operations can be highly automated and address various problems common to several agricultural products [21].

5.4.1 Precision Farming

The increased use of technology in farming is often referred to as "Precision Farming," a concept that can take many forms. One is the use of a global positioning system (GPS) installed on combines, which then coordinates with switches on an irrigation system to maximize water usage. In effect, the irrigation system can turn the water on and off as the combine moves through the field. In the United States, the use of precision farming technologies, whether through GPS mapping of fields to only irrigate specific areas or the more advanced use of machines to directly plant and harvest specific plots separately, is not widespread. However, farmers are pushing to stretch resources as much as possible because margins are already thin, and the predictions are for even tighter profit margins in the future [22].

On the positive side, many operators advocate the use of these technologies as useful tools to keep resources available. By avoiding damaging soils, over-fertilizing, or creating other harms, it is possible to keep agriculture in business for the long term without having to pay expensive clean-up bills. However, the increasing use of these technologies to tailor farming practices to the "needs" of the crop has implications for humans who eat the crop, for soil health, and for the health of the surrounding ecosystem [23].

5.4.2 Crop Monitoring and Management

Surveillance, optimization, and machine intelligence processes are also applied in crop environments. "Crops" can include any kind of plants important to humans. This chapter provides crop management and supports decision-making systems that help growers achieve their economic and production goals, such as pest control and yield monitoring. The systems are important for monitoring crop performance and timely treatment because of the often short window of opportunity for the necessary action. In addition, continuous or more widespread acquisition of accurate characterization data by decreased cost of the new AI application contributes data for fundamental research and development purposes [24].

A variety of strategies are developed on the basis of remote and local surveillance data. They also refer to the method by which the monitoring data is obtained. Some strategies are generally appropriate for all sorts of diseases, pests, or nutritional problems, such as the percentage of plants within a field that have visible symptoms of these problems. Other tools have been developed specifically for a disease, pest, or nutritional problem. For example, Visual Sponge Trap Count methods have been developed for the use of pheromone traps to monitor pests, such as the acarine varroa in honeybees. There is also the development of applications that monitor the type of pests in a specific location as they move between environmental conditions that cause distinct variations in the spectrum of electromagnetic radiation reflected cameras [25].

5.5 Challenges and Opportunities

The motivation for integrating AI with agriculture is clear: the field is facing an aging global population and concurrent shortages of skilled labor. The problem sets, such as high cost of crop production, lifestyle-related health problems, and inability of visually impaired farmers to independently manage the crops, seedlings, and do the post-harvest sorting and reaching the markets, further justify the need for making available personalized and dependable robotic assistants. This is even more important on a global scale. Trustworthy agricultural AI systems present the potential of not only mitigating these problems but doing so in a way that can improve the efficiency, longevity, and base of knowledge of the agricultural community as well as the quality of life of its members [26].

One of the primary challenges to agriculture is addressing Threat Level Zero. The Food and Agricultural Organization's (FAO) declaration that famine on the scale of the Ituri crisis is not possible, vis-a-vis its certification of a free country from a given threat, is predicated on, among other things, easy access to markets. From a computer systems perspective, this can be analyzed as all Mode C actors have committed to creating a Rule of Law at their expense and opportunity cost (and therefore, everyone would be within easy access to a market). This assumes that the non-intentional physical systems are responsible for gates, security clearances, certifications, and logistics work. Satisfying all the open conditions of the activities, Mode C lies outside the abilities of concrete physical systems, as that would entail having it work in the wild without power or operator intervention for a week and have exponential adversarial robustness. Therefore, the role and mission assurance of agricultural AI agents is not solving an optimization problem, its spirit is to provide assistance and support for agents that are doing that. They should support mission assurance of agents who lie out of Mode C and do not conform to open conditional fidelity [26, 27].

5.5.1 Data Quality and Availability

In general, ML models benefit from both large and diverse data. Many small-scale and experimental agricultural datasets share certain biases. While five main biases were well-documented in the literature – location, institution, cropping systems, crop coverage, and type of data – other industry-related biases were also identified according to our plant ontology knowledge and were treated in the context of abiotic and biotic stress. Along another axis of data availability, every ML application has a target accuracy, and on the one hand, no dataset is complete. On the other hand, realistic individuals who are labeling or integrating these datasets cannot invest more resources than the potential increase in accuracy they would observe [28].

The cost of obtaining a high-quality human label motivated the exploration of novel data sources, selective sampling heuristics, and semi-supervised methods that leverage background knowledge and invariants of the observed system. Ecologists recently developed a benchmark with higher diversity in species, which considers more frequent scenarios. On the other hand, the Traffic4cast challenge sets the spotlight on street view images in more cities to predict and nowcast city-wide traffic in highly urbanized areas. Finally, there was a consistent public demand for that. Dubbed LILA BC Label, this benchmark is an extension of a previous work, and the dataset features more high-concept diversity [29].

5.5.2 Interdisciplinary Collaboration

In addition to our own expertise in agricultural science and AI, we interact with researchers in many other fields, including remote sensing, soil science, phenomics, ML, formal methods, computer vision, and reinforcement learning. This collaboration style has included both deeper collaborations with true biologists and agronomists, with us bringing knowledge of AI and learning and the scientists bringing the nuanced understanding of how their scientific disciplines work, and more shallow collaborations in which we use AI to automatically learn from background scientific knowledge without necessarily working closely with any biologists or agronomists [30].

We need both depth and breadth at once. Depth is needed to avoid the hype cycle with real users in real settings, with user interfaces that address real needs with established technologies, and with support for deployment at scale. Breadth is needed to enable deeper models by bootstrapping on the tractable scientific models. In traditional collaboration, both these requirements are achieved by having many members of both collaborating teams. AI makes this approach scalable, both by enabling high-throughput small-scale interactions to avoid the need for a group to have people with a deep understanding of seemingly unrelated fields and by learning automatically from knowledge acquired from these interactions.

5.6 Ethical and Social Implications

In contrast to the recommendations focused on technical and policy issues, ethical and social implications of AI in a broader context are proposed here. The ethical implications include value judgments, such as privacy. The social implications encompass considerations of behavioral studies that reflect the direction of human innovation and learn the characteristics of traditional Japanese intellectual curiosity.

With the focus placed on ethics and security in the technology development phase, the attention naturally tends to become focused primarily on the development of the underlying technology alone. But when that technology is left to commercial production, should ethical considerations not arise at the societal level? The social consequences of innovation are debated in this study, such as the degradation of labor or the harms of closing the gap, since with AI, we all face fundamental questions about being human. A unique Japan-focused debate focusing on questions regarding whether the future will separate people from food can lead to a discussion of what desirable rural and agricultural societies should look like.

5.6.1 Privacy and Data Security

Privacy and data security have always been challenging issues for data-intensive industries such as healthcare and finance. Today, the vast explosion of sensing and measuring devices in both public and private environments is extending the boundary of data generation to every corner of the physical world. This offers great potential to improve our quality of life and optimize resource utilization in many of the world's most fundamental industries. Agriculture is a prime example [31]. Unfortunately, the expansion and deployment of sensing and measuring devices have also amplified privacy and data security concerns to a point where the very people, communities, and companies that might benefit the most from today's information technology advancements are hesitant to embrace them for fear of loss of data control and potential harm [32].

How much data is enough for good farming decisions? The answer is the same for all industries: it depends on the specific questions being asked. As the data problem becomes more complex, the number of sensors, wireless networks, and data analysis and display software and hardware expands. As a result, the seed of privacy concern becomes an invasive weed with tendrils that can potentially affect every aspect of agricultural data flow. Examples of highly visible sensors include the GPS units used for sky and earth signal propagation and timing. HD spectral imaging evaluates reflected light against a spectrum that is useful in determining plant stress of interest. In some agribusiness applications, DNA gene chips probe plant health and function by detecting the expression of thousands of genes simultaneously. As these plants grow, they can also be observed with leaf and canopy movement sensors that use microwave or acoustic backscatter [33, 34].

For crop irrigation applications, farmers use ground-based, piezoelectric moisture reticulation systems to adjust and control water usage. In response to the complexity of crop growth and nutrient management, researchers are experimenting with a root sensor nozzle-in-mixer cultivator system to provide accurate feed-to-root-zone delivery and control. A comprehensive summary of plant health sensors is found in Heun et al. A recent industry growth study by Piper Jaffray predicts a considerable increase in commercial demand and the vendor community. This broad universe of plant O.P.S. sensors and the steady increase of options created by innovative firms bring several privacy issues to the fore. Furthermore, one characteristic of many plant-based sensing systems is that the data they generate rolls up to specific, and often valuable, metadata.

5.6.2 Job Displacement

The problem of job displacement is often raised in discussions of agricultural automation. However, several developments complicate this issue. First, the high degree of rural underemployment and the large numbers of children needed for African household farm labor suggest that there is substantial slack in the African farm labor markets. Second, the demographic megatrend of large numbers of African farmers leaving farming and taking up other professions will resolve this dilemma. The ultimate outcome is presumably similar to what happens in developed economies undergoing industrialization. From the twilight of the Roman Empire to the present day, services and manufacturing have steadily gained relative importance over agriculture. Today, less than 1% of the US workforce is employed in agriculture, and those people have steadily growing salaries in real terms and are utilized for relatively high-value-added work [35].

While it was technologically possible to exceed real salaries and net employment in the preindustrial agricultural economy, other institutions allowed the expansion of the numbers of educated people and the levels of capital accumulation that allowed the industry to flourish and pull people from agriculture. Four forces were the public sector provision of educational infrastructure, the creation of a system of private property rights in land and labor, free trade, and dedicated public and private investment. Although it might not imply more for the average rural worker, in developed economies, widespread automation contributes to increased incomes in the agricultural industry as a whole. In our prior work, we report that on average, the share of value added in agriculture increases by about 20% as a result of the introduction of known agricultural robots and AI technologies. Friedman et al. estimate total social benefits from the introduction of agricultural robots in the United States in the range of 10–15% of the total value of US agricultural production [36].

5.7 Current Trends and Future Directions

In this section, we identify trends in the industry, noting the impressive amount of AI innovation taking place in both established companies and newly created startups. We then take a look at research, and in particular deep learning and reinforcement learning, which are two areas of AI research that are currently showing huge growth. Understanding these areas provides important insights into the uses of AI for agriculture in the future. Finally, we review models for prioritizing research in AI for agriculture.

1) **Industry Trends**: At present, the most dominant uses of AI in agriculture use standard ML approaches to solve specific tasks, including image recognition, robotics, and control algorithms used in 3D mapping [37, 38]. The majority of the investments are in Berlin, London, and Paris, and in the Far East, the agriculture AI investments are mainly in India and across China [39]. A number of established companies (e.g. Monsanto, John Deere) use AI in their products and services, and a large number of startups have been recently created that focus only on agricultural uses of AI [40].
2) **Research in AI**: AI draws from computer science, psychology, and many sub-areas, including statistics for learning, optimization for search and planning, control theory for robotics, and symbolic logic for planning and general problem-solving. Since the early days of AI, there has always been tension between symbolic approaches based on logic, abstraction, and deduction and subsymbolic approaches based on probability and statistics, analogy, learning, perception, and induction. After a series of rapid successes and even more

rapid failures, and by rebranding themselves as "big data," the current kind of AI work focuses on deep learning and reinforcement learning [41]. Accordingly, a worldwide community of theorists, engineers, and domain experts has taken a renewed interest in AI for agriculture. With the advent of deep learning in 2012, the community has focused on "big data," and there have been numerous applications of deep learning techniques in diverse agriculture problems such as plant disease detection, weed detection, threatened animal species management, climate sensing, yield prediction, complex planning, and multi-robot coordination for precision agriculture, and even AI aesthetics. AI offers tools crucial to maintaining agricultural production to meet the future demands under these potential adversarial effects of climate changes [1].

5.7.1 Advancements in AI Technologies

In this section, we identify the five key attributes of advanced AI technology, which are detailed below.

1) **High Accuracy**: It is axiomatic that the most advanced precursor technology of all time is the best possible specimen of its kind. So, it should be no surprise that highly accurate AI technologies are better than less accurate AI technologies for a great many applications, for example, better early warning systems that use weather and climate data to predict future temperature, humidity, and weather patterns [42]. Of course, practical people pay attention to cost as well, and the best forecasting technology that is cheap may substitute for today's best and most costly forecasting technologies, such as the supercomputers that figure prominently in the world's major forecasting technologies. Practical people also understand costs and benefits and may find that the likely benefits to be derived from predicting future weather patterns may not be large enough to warrant the investment of substantial resources in next-generation forecasting technologies [43].
2) **High Software and Hardware Efficiency**: Economic considerations of scale and scope are as basic in AI as they are in other activities, and it is natural that a more efficient AI technology is better than a less efficient AI technology whenever all other aspects are held constant, where efficiency in practical applications means the lowest cost of using all of the relevant inputs. These inputs include, of course, the cost of AI hardware and the time needed to complete an AI-funded task [44].
3) **Causes in Conclusion**: High-quality technologies are capable of deductive reasoning, an activity critical to many decision-making activities. Inductive processes, too, can be more accurate the more signals and predictors that are

identified [45]. The more of these signal AI technologies that assist with decision-making, the better the decision quality that may thereby be achieved [46].

4) **Consistent Outcomes**: An achievement of great importance is the predictable consistency of the AI system's output. Even the best faculties and provide inconsistent grading of the same written paper, misinformation and alternative truths that surface more and more frequently in the world's information infrastructures provide consistent support for political perspectives on the left and the right side of the ideology spectrum [47].
5) **Clearly Labeled and Precisely Defined Inputs**: It is one thing to be able to predict the likely impact of economic policies or technological change on employment, income, and community development, and another to know who lives now in areas that are particularly vulnerable to the policy and technological changes, and how these people and the communities in mainland experience or perceived the impacts of policy and technological transformations [48].

5.7.2 Emerging Applications

Emerging applications in agriculture implement ML techniques for various tasks: in machine vision, most frequently for plant disease detection, species discrimination, and phenotyping; in ML for plant stress detection, identification, and classification, including model-free applications [47]. In biosystems engineering and horticulture, we find most frequently ML applications, then within precision crop production. Also, other applications can be found in agriculture, such as those that are related to predicting market commodity pricing, finding optimal agricultural weather conditions for crop insurance, early crop yield prediction, irrigation decision-making, greenhouse environmental control, throughputs, poultry weight prediction, methane clamp prediction, as well as other tasks [49, 50]. In most cases, the quality rate of the applications is benchmarked against ground truth data, except where an analysis of historical datasets is performed [51].

5.8 Conclusion

AI in agriculture appears to have a bright and promising future despite being neglected, and the upcoming outcomes may be revolutionary. AI holds great promise for the future of agriculture. AI will help farmers find better occupations, enhancing their standard of living in the process. Compared to conventional techniques, it can lower the labor required in agriculture. Overall, in order to solve the issues of food security, sustainability, and environmental stewardship in a world that is changing quickly, agricultural science and technology must be applied. Smart agriculture is a cutting-edge tactic that leverages technology to enhance farming methods and address issues facing the sector. By merging several technologies, such as IoT, AI, and data analytics, smart agriculture offers many benefits and opportunities for farmers, the environment, and food production. One of the main advantages of smart agriculture is an increase in productivity and efficiency in the field. Farmers may monitor and enhance several attributes with sensors and connected devices, including temperature, nutrient levels, and soil moisture. This real-time data can be used to make informed decisions that lead to focused fertilizer distribution, timely insect control, and precise watering. Crop yields could rise, and resource use could decrease as a result, leading to greater profitability and less of an adverse effect on the environment.

References

1. Javaid, M., Haleem, A., Khan, I.H., and Suman, R. (2023). Understanding the potential applications of artificial intelligence in agriculture sector. *Advanced Agrochem* 2 (1): 15–30.
2. Tzachor, A., Devare, M., King, B., Avin, S., and Ó hÉigeartaigh, S. (2022). Responsible artificial intelligence in agriculture requires systemic understanding of risks and externalities. *Nature Machine Intelligence* 4 (2): 104–109.
3. Jung, J., Maeda, M., Chang, A., Bhandari, M., Ashapure, A., and Landivar-Bowles, J. (2021). The potential of remote sensing and artificial intelligence as tools to improve the resilience of agriculture production systems. *Current Opinion in Biotechnology* 70: 15–22.
4. Subeesh, A. and Mehta, C.R. (2021). Automation and digitization of agriculture using artificial intelligence and internet of things. *Artificial Intelligence in Agriculture* 5: 278–291.
5. Elbasi, E., Mostafa, N., AlArnaout, Z. et al. (2022). Artificial intelligence technology in the agricultural sector: a systematic literature review. *IEEE Access* 11: 171–202.
6. Akkem, Y., Biswas, S.K., and Varanasi, A. (2023). Smart farming using artificial intelligence: a review. *Engineering Applications of Artificial Intelligence* 120: 105899.
7. Ganeshkumar, C., Jena, S.K., Sivakumar, A., and Nambirajan, T. (2023). Artificial intelligence in agricultural value chain: review and future directions. *Journal of Agribusiness in Developing and Emerging Economies* 13 (3): 379–398.
8. Yuan, Y., Chen, L., Wu, H., and Li, L. (2022). Advanced agricultural disease image recognition technologies: a review. *Information Processing in Agriculture* 9 (1): 48–59.
9. Wang, T., Chen, B., Zhang, Z., Li, H., and Zhang, M. (2022). Applications of machine vision in agricultural robot navigation: a review. *Computers and Electronics in Agriculture* 198: 107085.

10 Sharma, S., Verma, K., and Hardaha, P. (2023). Implementation of artificial intelligence in agriculture. *Journal of Computational and Cognitive Engineering* 2 (2): 155–162.

11 Li, X., Zhou, Y., Hejazi, M., Wise, M., Vernon, C., Iyer, G., and Chen, W. (2021). Global urban growth between 1870 and 2100 from integrated high resolution mapped data and urban dynamic modeling. *Communications Earth & Environment* 2 (1): 201.

12 Pise, A.A., Almuzaini, K.K., Ahanger, T.A., Farouk, A., Pant, K., Pareek, P.K., and Nuagah, S.J. (2022). Enabling artificial intelligence of things (AIoT) healthcare architectures and listing security issues. *Computational Intelligence and Neuroscience* 2022 (1): 8421434.

13 Obaigbena, A. et al. (2024). AI and human-robot interaction: a review of recent advances and challenges. *GSC Advanced Research and Reviews* 18 (2): 321–330.

14 Chen, J., Sun, J., and Wang, G. (2022). From unmanned systems to autonomous intelligent systems. *Engineering* 12: 16–19.

15 Gupta, S. et al. (2022). Machine learning-and feature selection-enabled framework for accurate crop yield prediction. *Journal of Food Quality* 2022: 1–7.

16 Shook, J., Gangopadhyay, T., Wu, L., Ganapathysubramanian, B., Sarkar, S., and Singh, A.K. (2021). Crop yield prediction integrating genotype and weather variables using deep learning. *PLoS One* 16 (6): e0252402.

17 Ju, S., Lim, H., Ma, J.W. et al. (2021). Optimal county-level crop yield prediction using MODIS-based variables and weather data: a comparative study on machine learning models. *Agricultural and Forest Meteorology* 307: 108530.

18 Morales, A. and Villalobos, F.J. (2023). Using machine learning for crop yield prediction in the past or the future. *Frontiers in Plant Science* 14: 1128388.

19 Elbasi, E., Zaki, C., Topcu, A.E. et al. (2023). Crop prediction model using machine learning algorithms. *Applied Sciences* 13 (16): 9288.

20 Bali, N. and Singla, A. (2022). Emerging trends in machine learning to predict crop yield and study its influential factors: a survey. *Archives of Computational Methods in Engineering* 29 (1): 95–112.

21 Staie, B., Kinzer, A., Macknick, J. et al. (2024). Pathways for Agricultural Decarbonization in the United States (No. NREL/TP-5100-86071). *National Renewable Energy Laboratory (NREL)*, Golden, CO (United States).

22 Kim, Y.S., Lee, S.D., Baek, S.M. et al. (2022). Analysis of the effect of tillage depth on the working performance of tractor-moldboard plow system under various field environments. *Sensors* 22 (7): 2750. https://doi.org/10.3390/s22072750.

23 D'Antonio, P. et al. (2023). Operating performance of manual, semi-automatic, and automatic tractor guidance systems for precision farming. *Research in Agricultural Engineering* 69 (4): 179.

24 Jalajamony, H.M., Nair, M., Mead, P.F., and Fernandez, R.E. (2023). Drone aided thermal mapping for selective irrigation of localized dry spots. *IEEE Access* 11: 7320–7335.

25 Jing, Y. et al. (2023). Development of a GNSS/INS-based automatic navigation land levelling system. *Computers and Electronics in Agriculture* 213: 108187.

26 Obi Reddy, G., Dwivedi, B., and Chary, G.R. Kandiah Pakeerathan (2023). Applications of geospatial and big data technologies in smart farming. In: *Smart Agriculture for Developing Nations: Status, Perspectives and Challenges*, 15–31. Springer.

27 Attri, I., Awasthi, L.K., and Sharma, T.P. (2024). Machine learning in agriculture: a review of crop management applications. *Multimedia Tools and Applications* 83 (5): 12875–12915.

28 Nie, J. and Yang, B. (2021). A detailed study on GPS and GIS enabled agricultural equipment field position monitoring system for smart farming. *Scalable Computing: Practice and Experience* 22 (2): 171–181.

29 Sáiz-Rubio, V. and Rovira-Más, F. (2020). From smart farming towards agriculture 5.0: a review on crop data management. *Agronomy* 10 (2): 207.

30 Cravero, A., Pardo, S., Sepúlveda, S., and Muñoz, L. (2022). Challenges to use machine learning in agricultural big data: a systematic literature review. *Agronomy* 12 (3): 748.

31 Soussi, A., Zero, E., Sacile, R., Trinchero, D., and Fossa, M. (2024). Smart sensors and smart data for precision agriculture: a review. *Sensors* 24 (8): 2647.

32 Huo, D., Malik, A.W., Ravana, S.D., Rahman, A.U., and Ahmedy, I. (2024). Mapping smart farming: addressing agricultural challenges in data-driven era. *Renewable and Sustainable Energy Reviews* 189: 113858.

33 Vangala, A. et al. (2023). Security in IoT-enabled smart agriculture: architecture, security solutions and challenges. *Cluster Computing* 26 (2): 879–902.

34 Mohsan, S.A.H., Othman, N.Q.H., Li, Y., Alsharif, M.H., and Khan, M.A. (2023). Unmanned aerial vehicles (UAVs): practical aspects, applications, open challenges, security issues, and future trends. *Intelligent Service Robotics* 16 (1): 109–137.

35 Kumar, N., Chaudhary, A., Ahlawat, O.P. et al. (2023). Crop residue management challenges, opportunities and way forward for sustainable food-energy security in India: a review. *Soil and Tillage Research* 228: 105641.

36 Gyamfi, E.K., et al. (2024). Agricultural 4.0 leveraging on technological solutions: study for smart farming sector. arXiv preprint arXiv:2401.00814.

37 Ashish Kumar, Rachna Jain, Ajantha Devi Vairamani, Anand Nayyar (2023). Deep learning-based multi-object tracking. In: *Object Tracking Technology: Trends, Challenges and Applications*, 183–199. Springer Nature Singapore: Singapore.

38 Kumar, A., Vohra, R., Jain, R., Li, M., Gan, C., and Jain, D.K. (2024). Correlation filter based single object tracking: a review. *Information Fusion* 112: 102562.

39 Ha, L.T. (2024). Employment of the DCC-GARCH Copula Model to explore a link between robotics and artificial intelligence and green crypto investments. *Environ Sci Pollut Res Int*. 2024 May 30. doi: 10.1007/s11356-024-33765-1. Epub ahead of print. PMID: 38814559.

40 Uddin, M., Chowdhury, A., and Kabir, M.A. (2024). Legal and ethical aspects of deploying artificial intelligence in climate-smart agriculture. *AI & Society* 39 (1): 221–234.

41 Attri, I., Awasthi, L.K., Sharma, T.P., and Rathee, P. (2023). A review of deep learning techniques used in agriculture. *Ecological Informatics* 77: 102217.

42 Kumar, A., Aelgani, V., Vohra, R. et al. (2024). Artificial intelligence bias in medical system designs: a systematic review. *Multimedia Tools and Applications* 83 (6): 18005–18057.

43 Preeti Nagrath, Jafar A. Alzubi, Bhawna Singla, Joel J. P. C. Rodrigues, A. K. Verma (2023). Healthcare engineering using AI and distributed technologies. In: *Smart Distributed Embedded Systems for Healthcare Applications*, 1–14. CRC Press.

44 Shang, Y. et al. (2024). The impact of artificial intelligence application on enterprise environmental performance: evidence from microenterprises. *Gondwana Research* 131: 181–195.

45 Malherbe, W., Biggs, R., and Sitas, N. (2024). Comparing apples and pears: linking capitals and capacities to assess the resilience of commercial farming operations. *Agricultural Systems* 217: 103934.

46 Leonori, S. et al. (2024). Cable-driven agribot prototype: enabling precision agriculture through innovative design. *Smart Agricultural Technology* 7: 100426.

47 Wani, J.A., Sharma, S., Muzamil, M., Ahmed, S., Sharma, S., and Singh, S. (2022). Machine learning and deep learning based computational techniques in automatic agricultural diseases detection: methodologies, applications, and challenges. *Archives of Computational Methods in Engineering* 29 (1): 641–677.

48 Shaikh, T.A., Rasool, T., and Lone, F.R. (2022). Towards leveraging the role of machine learning and artificial intelligence in precision agriculture and smart farming. *Computers and Electronics in Agriculture* 198 (C): 29.

49 Kumar, A. and Jha, R. (2022). IoT in smart farming: a boon to farmers. In: *Internet of Everything: Smart Sensing Technologies* (ed. T. Kavitha, V.A. Devi, S.N. Pari, and S. Ramanathan), 185–206. Nova Publisher.

50 Singh, D. and Kumar, A. (2024). A deep recurrent neural network for plant disease classification. *SN Computer Science* 5 (8): 1053.

51 Ahmad, A., Saraswat, D., and El Gamal, A. (2023). A survey on using deep learning techniques for plant disease diagnosis and recommendations for development of appropriate tools. *Smart Agricultural Technology* 3: 100083.

6

AI in Agriculture: A Comprehensive Exploration of Technological Transformation

Manya Gupta[1], Gargi Mishra[1], Supriya Bajpai[2], Abhinav Bhardwaj[1], and Milind Gautam[1]

[1] *Department of Computer Science and Engineering, Bharati Vidyapeeth's College of Engineering, New Delhi, India*
[2] *IIT Bombay-Monash Research Academy, IIT Bombay, Mumbai, Maharashtra, India*

6.1 Introduction

With the rapid changes brought about by industrialization, globalization, and societal pressures, technology emerges as a catalyst for transformative shifts. Automation, artificial intelligence (AI), and data-driven approaches are revolutionizing agriculture, promising improved productivity, profitability, and sustainability. AI's integration into agriculture signifies a profound change as it reshapes how crops are grown, harvested, and managed for future food production. Traditional farming practices face hurdles due to limited technology integration, leading to inefficiencies in resource management and processes.

Manual labor-intensive tasks, coupled with reliance on outdated methods, hinder scalability and increase operating costs. Moreover, inadequate decision-making stemming from incomplete information on weather, soil, and crop health further impacts output and quality. The labor-intensive nature of tasks like weed control and pesticide spraying adds to the challenges, affecting productivity and farmer health. Addressing these obstacles is crucial for ensuring the economic viability and sustainability of agriculture, necessitating a shift toward more sophisticated AI-based farming systems. Figure 6.1 represents the implementation of AI techniques in different phases of agriculture.

Emerging Smart Agricultural Practices Using Artificial Intelligence, First Edition.
Edited by Ashish Kumar, Jai Prakash Verma, and Rachna Jain.

Published 2025 by John Wiley & Sons, Inc.

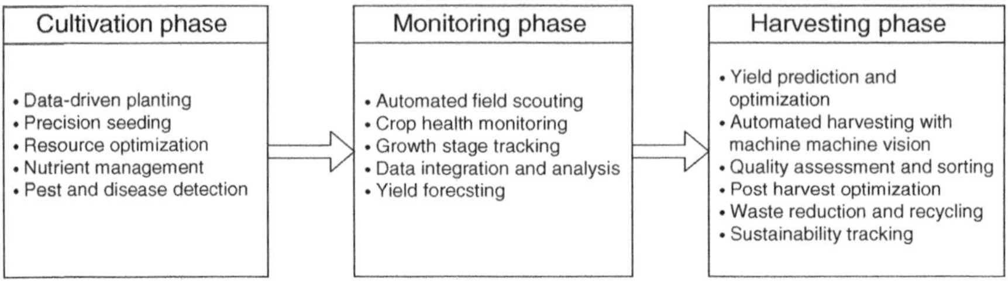

Figure 6.1 Implementation of AI techniques in different phases of agriculture.

Demand for agriculture, which has historically required a lot of work, is rising as a result of population growth and increased production requirements. AI-driven automation turns out to be essential for many facets of farming. Crop quality is guaranteed by predictive analytics and improved farm management, which uses satellite data for in-the-moment monitoring [1–3]. AI helps with yield estimates, price forecasts, and pest detection, advising farmers on the best decisions and application of pesticides. It addresses the lack of personnel and resources and calls on big businesses to invest. Its implementation overcomes obstacles in a variety of industries, including finance, transportation, healthcare, and agriculture [4–6]. Figure 6.2b mentions some benefits of using AI in agriculture. Farmers are under pressure to increase yield due to urbanization and the world's population growth. Because there is a limited supply of fertile soil, innovation is essential,

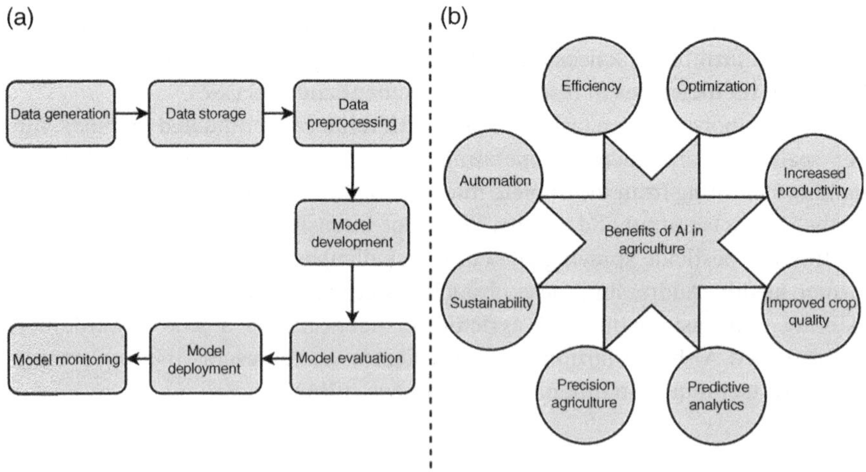

Figure 6.2 Figure showing (a) Process of AI implementation in agriculture (b) Benefits of using AI in agriculture.

and risk management plans are needed. With its ability to handle issues including pests, illnesses, climate change, monoculture, and pesticide use, AI has a bright future in agriculture [7]. Unpredictable weather, a manpower shortage, and the yearly requirement for higher yields all contribute to this stress. AI shows up as a crucial component that can automate agriculture and solve impending industrial problems [8]. In this chapter, the vast potential of AI in revolutionizing various aspects of agriculture has been explored by delving into the extensive literature available on this topic. Table 6.1 represents the data collection indicator, and Table 6.2 lists some top institutions pursuing extensive research in the field of AI in agriculture. Figure 6.3 lists some top journals publishing articles in the field of AI and agriculture, along with the number of articles published by them [9].

The objectives of this chapter are summarized as follows:

1) Explore how data-driven farming methods and precision agriculture are being adopted by farmers as a result of technology improvements.

Table 6.1 Data collection indicator.

Indicator	Description
Search interval	2016–2024
Databases	ScienceDirect; IEEE Xplore; Springer, MDPI, Scopus
Screening and original article	Title, abstract, DOI and year
Language	English

Table 6.2 Top institutions encouraging research in this field.

Institution	Country
University of Florida	USA
ICAR-Indian Agricultural Statistics Research Institute	India
Wageningen University and Research	Netherlands
China Agricultural University	China
Vellore Institute of Technology	India
University Putra Malaysia	Malaysia
Shihezi University	China
Dalian University of Technology	China
Universidad Catolica del Maule	Chile

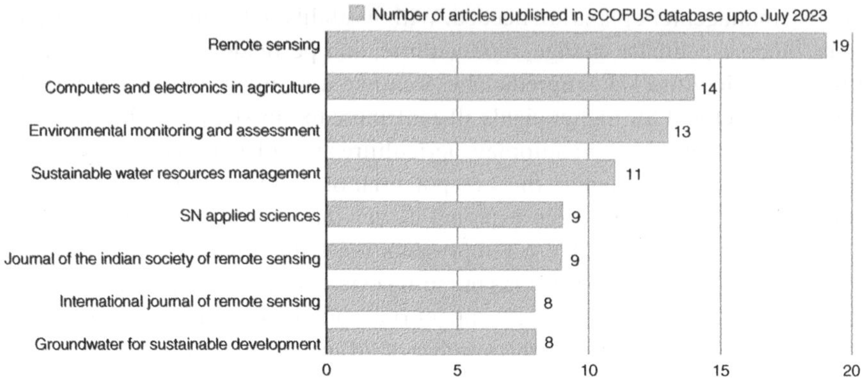

Figure 6.3 Top SCOPUS indexed journals publishing research on AI in agriculture.

2) Emphasize the critical role AI plays in improving crop quality, overcoming obstacles faced by conventional approaches, and transforming decision-making.
3) Investigate AI-monitored agricultural parameters and the technologies used to gather and analyze data from diverse sources, including wireless sensor networks (WSNs) and Internet of Things (IoT) devices.
4) Explore major application areas of AI in agriculture, i.e. crop management, disease control, weed control, and soil management.
5) Discuss issues with data protection, technical obstacles, and farmer adoption when implementing AI in agriculture.

In Section 6.2, the focus is on how agricultural processes are being changed by AI methods, such as robotics and machine learning (ML). Specifically, attention is given to duties like crop monitoring and disease identification. In Section 6.3, the monitoring of variables like soil moisture and weather by AI systems to assist farmers in making more informed decisions is discussed. In Section 6.4, examples of AI applications in agriculture, ranging from weed control to soil management, disease control, and crop monitoring, are presented. Finally, in Section 6.5, the restrictions and difficulties surrounding the application of AI in agriculture are addressed.

6.2 AI Integration in Agricultural Practices

AI's ability to solve problems in agriculture is largely dependent on the quality of data that is available. AI has the potential to transform agricultural practices, but obtaining relevant data at the farmer level is still a major barrier. Utilizing remote and local sensing data in conjunction with image classification techniques can revolutionize the efficiency and use of farm machinery, especially when it comes to weed management, early disease identification, soil management, irrigation management, crop harvesting, and grading [10, 11]. AI makes it possible to continuously monitor agricultural products by utilizing data from sensors, drones, cellphones, precision software, and other sources. Comparing soil data to past yields optimizes nutrient management and evaluates the environmental effects of fertilizer, encouraging environmentally friendly farming [12–14]. Agribusiness may boost productivity and efficiency by implementing smart irrigation systems that make use of IoT, ML, and the Internet. Additionally, it improves the sustainability and environmental suitability of farming practices [15]. Farmers have increasing challenges in determining the best times to plant seeds in light of climate change due to rising pollution levels and unpredictable weather patterns. AI presents a crucial solution by providing insights into how weather variables affect planting cycles. AI makes remote sensing technologies for recognizing and controlling plants, weeds, pests, and diseases possible by utilizing advances in computer vision, mechatronics, and ML. AI also makes it easier to use accurate seeding techniques, which improves fertilization accuracy and reduces waste, ultimately leading to higher-quality products and more efficient market access. Drones collect vital data for well-informed decision-making while remotely monitoring farm and crop health. Automation using GPS-enabled tractors streamlines agricultural operations [16–18]. AI recognizes key plants, helps with pest management, and maximizes pesticide application with drones. While ML models provide information on soil moisture, temperature, and condition, predictive analytics estimate precipitation and evapotranspiration. Farmers may irrigate crops more profitably and more sustainably by using this data-driven technique [19–21] Automated farm monitoring systems optimize crop care for higher yields and more profitability while streamlining agriculture. Environmental sustainability and resource efficiency are guaranteed by precision management. AI technologies are improving agricultural productivity worldwide [22, 23]. As civilization moves from rural to urban areas, there is a labor deficit in the agricultural sector. This problem can be solved through automation, thanks to AI, since businesses are creating self-driving bots to complete labor-intensive activities. These automated workers augment human labor, improving efficiency, velocity, affordability, and quality of output. AI is being used in agriculture to help farmers make more informed decisions by doing everything from modeling crop development to predicting diseases. The sector's development and prospects for future expansion are shown by the effective application of AI tools and trustworthy data collection [24]. AI has been a useful tool for enhancing agricultural decision-making in recent years. Farmers can increase livestock and agricultural productivity by using AI technology to make decisions more effectively. AI-based analytics are essential to data mining, which advances human intelligence in the agricultural industry and research [25, 26]. Standard data-processing systems find it difficult to handle the massive and complex big data sets that contain significant insights. AI uses ML algorithms to help. This facilitates the study of crop development patterns, the detection of possible diseases, and the use of certain pesticides or fertilizers in accordance with disease patterns and leaf growth. It uses a range of sensors, such as cameras, feed, weight, soil, temperature, and intensity sensors, to navigate fields efficiently. These sensors can be installed on apparatuses such as miniature robots or low-altitude drones [27, 28].

The first step in the lengthy process of integrating AI into agriculture is obtaining crucial information on crop development and environmental conditions from a range of sources, including satellite imaging and sensor data. After that, the collected data is preprocessed to ensure that it is suitable for analysis and cleaned before being put into structured databases. By building ML models, past data is used to train algorithms for a variety of tasks, such as disease detection, crop yield prediction, and fertilizer optimization. After being trained, these AI models are integrated into functional systems, facilitating instantaneous decision-making and response. AI enables preemptive interventions and necessary modifications by facilitating the monitoring of farm conditions through integration with agricultural machinery, drones, or mobile applications. Furthermore, AI models can offer insightful analysis and prognostications that help farmers make decisions that will increase sustainability and productivity. The significance of AI in promoting efficiency and innovation in agriculture is further supported by the constant monitoring and assessment of model performance, which guarantees continuous improvement and refinement [29]. Figure 6.2a shows the process of AI adoption in agriculture.

6.3 AI-Monitored Agricultural Parameters

Automation in the agricultural sector, such as intelligent irrigation systems and AI-based harvesting robots, can assist in addressing the labor deficit in the sector. Outdated forecasting techniques are being replaced by ML and AI, which are adapting to the industry. Using biosensors to track soil fertility and moisture will improve crop development. AI finds deficiencies, manages pests, and harvests fields. Crucial crops like rice, wheat, and maize can be produced by timing the sowing of seeds according to weather patterns. Software firms might, however, have trouble explaining the AI system to farmers in its entirety [30, 31]. In order to make educated decisions about planting seeds, choosing crops, and using hybrid seeds for improved yields, farmers employ AI and ML models to collect data on temperature, soil, water use, and weather. Early illness and pest identification are also aided by AI technology. The utilization of AI sensors can help reduce pesticide usage by specifically targeting weeds. AI is being used to advance several fields, and cutting-edge technologies like IoT, ML, and AI are being used to secure, fertilize, and irrigate crops [32, 33]. ML models and IoT sensors are being used in intelligent irrigation systems to effectively water crops. In order to complete labor-intensive jobs, autonomous robots are also being created. Early identification of pollutants in food and agriculture is being investigated using smart sensors and sensing systems based on nanotechnology [34–36]. Farmers may now make educated judgments instantly by using AI technology to collect and analyze millions of data points from agricultural land. Additionally, it has been aided with weather forecasting, water efficiency, hybrid seed evaluation, and labor scarcity management for farmers [37–39]. New agricultural technologies, such as AI, support sustainable goals by assisting farmers in crop monitoring and optimization. To help with agricultural production, data is gathered via sensors, drones, and satellites. By analyzing this data, AI helps farmers make wise decisions. By identifying pests, contaminants, and illnesses in crops, AI can increase total production. Weather prediction and water management are two areas where ML-based applications are helpful [40–42]. Agriculture could undergo a transformation thanks to AI, which can solve a variety of problems and help farmers by making forecasts. It makes it possible to grow new and established crops under controlled conditions, helps combat problems like pests and climate change, and replaces human labor with robots [43, 44]. By 2050, there will be around 10 billion people on the planet, according to the World Resources Institute. It will be necessary to raise food production by 53% in order to accommodate the overall expansion of agricultural lands and reduce emissions by 67% in order to feed this many people sustainably [42]. Smart farming is one method of satisfying these goals. We can track and enhance farming outputs by utilizing IoT devices, cloud computing, AI, wireless and wired networks, and software management systems. There are two primary aspects of farming that can be observed: crops and livestock. Every category is unique and has requirements of its own. Crop monitoring considers one or more of the elements listed below:

Environmental factors, including temperature, humidity, solar radiation, fertilizer, and pesticide application, can be monitored using WSNs and IoT sensors to gather data [45]. Crop diseases, including visual data that can be gathered via unmanned aerial vehicles (UAVs) or fixed, high-resolution cameras that are mobile [46].

For anomaly categorization, prediction, and risk assessment, the data gathered by these devices in both situations must be further processed [43]. Bauer and Aschenbruck introduced a cutting-edge agricultural monitoring system based on the IoT [44]. Their analysis was centered on the leaf area index, which offers details on the photosynthetic processes and physiological state of plants. The photosynthetically active radiation range was determined by measuring solar radiation using WSN clusters of sensors, along with temperature, humidity, and light. Every cluster node had a Raspberry Pi, which used an LTE modem to communicate with the central base unit to exchange data. After that, the information was processed in a farm management information system to produce reports and make choices. In order to enhance the processes for monitoring temporal changes in agriculture using a UAV, Bagheri [47] created a remote sensor system with high spatial and temporal resolution. The primary onboard system, 182 Volume 11, 2023, made up the system architecture: ground station subsystems equipped with multispectral cameras for high-precision capture, a GPS tracking system, and a telemetry system to transport data among the sub-systems. The accuracy of crop classification could be improved, and monitoring procedures could be accelerated using this method. A maximum likelihood model was used to create multispectral imaging classification maps once images were taken. With a kappa coefficient of 0.9 and an accuracy of 94%, the results were extremely encouraging. To enhance the UAV-based temporal change monitoring procedures in agriculture, Bagheri [48] created a remote sensor system with high spatial and temporal resolution. With multispectral cameras for high-precision capture, a GPS tracking system, and a telemetry system to transport data between the subsystems, the system architecture comprised a core onboard system and ground station subsystems. The accuracy of crop classification could be improved, and monitoring procedures accelerated using this method. Following the acquisition of images, a maximum likelihood model was used to create multi-spectral imaging classification maps. With a 94% accuracy rate and a 0.9 kappa value, the results were extremely encouraging.

6.4 Application Areas of AI in Agriculture

In this section, applications of AI in agriculture are discussed. Table 6.3 mentions some priority areas for applications of AI research. Four major applications are discovered and discussed here.

6.4.1 Soil Management

An essential component of agricultural operations is soil management. Crop production can be increased and soil resources preserved with a solid understanding of the different types and conditions of soil. It is the application of procedures, methods, and treatments to raise soil performance. A conventional soil survey approach can be used to examine the presence of contaminants in urban soils [49].

Table 6.3 Priority areas for application of research in AI and agriculture.

Broad theme	Sub-theme	Field crops priority	Horticulture priority	Livestock priority	Fisheries priority
AI and farming efficiency	Seed quality and traceability	High	Moderate	—	Low
	Soil health monitoring	High	High	—	—
	Water conservation techniques	Moderate	Low	—	—
	Precision irrigation systems	High	High	—	—
	Crop yield prediction	High	Moderate	—	—
	Pest and disease detection	High	High	—	—
Climate adaptation	Climate-resilient crop varieties	High	High	—	—
	Extreme weather forecasting	High	High	—	—
	Drought and flood mitigation	High	Moderate	—	—
AI in supply chains	Inventory management with AI	Moderate	Low	—	—
	Market demand prediction	High	Moderate	—	—
	Price volatility analysis	High	Low	—	—
	Sustainable packaging solutions	Low	Low	—	—
Livestock management	Animal health monitoring	—	—	High	—
	Feed optimization	—	—	High	—
	Breeding selection	—	—	High	—
	Livestock tracking and traceability	—	—	High	—
Fisheries management	Aquaculture monitoring	—	—	—	High
	Stock assessment	—	—	—	High
	Fishing zone optimization	—	—	—	High

Manure and compost applications increase the aggregation and porosity of the soil. An improved aggregation suggests the addition of organic components, which are crucial in avoiding the formation of soil crusts. Alternative tillage techniques can be used to stop the physical deterioration of the soil.

A summary of AI soil management techniques is shown in Table 6.4. The IoT, predictive analytics, precision farming, ML algorithms (including SVM, DT, NN, and Random Forest), and other technical advances in agriculture are highlighted in the table. While ML and predictive analytics have great potential for streamlining agricultural operations, they demand high-quality data and significant upfront expenditures. Although it requires money for infrastructure and data collecting, Random Forest excels at predicting soil conditions. Though they come with challenges like initial investment, data reliability, and network access, precision farming, and the IoT allow for advanced monitoring and optimization. All things considered, these technologies have a lot of promise to improve agricultural productivity, but their successful implementation and broad adoption depend on resolving issues and encouraging cooperation

6.4.2 Crop Management

Table 6.5 provides an overview of crop management strategies. The first step in crop management is sowing, and it continues with keeping an eye on crop distribution, storage, and harvesting. It can be summed up as the endeavors to enhance the productivity and expansion of agricultural products. Crop yield will undoubtedly rise with a thorough comprehension of each crop class in relation to its timing and type of thriving soil. In order to maximize profitability and safeguard the environment, precision crop management, or PCM, is an agricultural management system that targets crop and soil inputs based on field conditions. The paucity of accurate and dispersed information regarding crop and soil conditions has hindered PCM [61]. Water deficits caused by poor soil, unfavorable weather, or insufficient irrigation require farmers to use a variety of crop management techniques. It is best to use adaptable crop management systems based on decision-making guidelines. When selecting between cropping options, drought prediction is a crucial factor to consider [60]. Making decisions that will lead to a high-quality crop output is aided by having a thorough awareness of weather trends. For assessing the operational behavior of a farm system, PROLOG makes use of meteorological data, machinery capacity, labor availability, and details on approved and prioritized operators, tractors, and implements. Additionally, it projects crop yield, total farm revenue, and net profit for each field and the farm as a whole [63]. Using sensors for a variety of soil and atmospheric characteristics, crop prediction methods are used to determine the best crop.

Table 6.4 AI in soil management.

Reference	Techniques used	Strength	Limitations
Sarkar et al. [50]	Predictive analytics	Enhances agricultural efficiency through predictive modeling, reducing labor and resource requirements.	Requires substantial initial investment in technology and infrastructure; may face challenges in adoption and implementation.
Awais et al. [51]	ML	Facilitates development of predictive models based on available soil data and auxiliary environmental variables.	Requires substantial amounts of high-quality data for training; expertise in model development and tuning may be necessary.
Chen et al. [52]	SVM, DT, NN	Extensively studied and proven to be highly efficient and accurate in data-driven modeling for modern farming development.	Data preparation challenges and weather/climate change can impact temperature, soil moisture, photosynthesis rates, and ecological equilibrium.
Folorunso et al. [53]	Random forest (RF)	Excels in soil prediction, outperforming traditional methods, enhancing fertilizer, pest/disease management, and irrigation optimization.	Tackling challenges involves investing in data collection, network upgrades, computing resources, education, and training. Partnerships and collaborations are key for success.
Ranjan et al. [54]	Precision farming	Optimizes practices with advanced tech.	Initial investment, data reliability, expertise needed, privacy concerns.
Alqadhi et al. [55]	Artificial neural network (ANN)	High accuracy in predicting soil erodibility. Ability to handle complex data patterns effectively.	Requires large amounts of training data. Computationally intensive for model training.

(*Continued*)

Table 6.4 (Continued)

Reference	Techniques used	Strength	Limitations
Nguyen et al. [56]	Extreme gradient boosting regression with genetic algorithm (XGBR with GA)	Achieved high performance ($R^2 = 0.891$; RMSE = 0.875%) in predicting soil moisture.	Utilized remote sensing data from Sentinel-1 (S1), Sentinel-2 (S2), and ALOS DSM, enabling precise predictions at a 10 m spatial resolution.
Ou et al. [57]	Semi-supervised deep neural network regression (semi-DNNR)	Successfully addressed inconsistent features and low accuracy issues in SOM, As, and Cr detection using hyperspectral imagery.	Utilized HyMap sensor's airborne hyperspectral imagery and mitigated overfitting and weak expression problems with Semi-DNNR.

Table 6.5 AI in crop management.

Reference	Techniques used	Strength	Limitations
Arashpour et al. [58]	Context-based augmentation, shapely additive explanations (SHAP)	Proper learning of essential image features	Specialized input datasets for different aspects
Gautron et al. [59]	Reinforcement learning (RL)	Interactive and action-oriented approach	Underutilized in crop management decision support systems (DSS)
Navinkumar et al. [60]	CNN	Analyzes a diverse range of data related to natural calamities, crop growth stages	Data availability and quality, complexity of environmental factors, local variability and specificity
Ghosh et al. [61]	ANN	It predicts comprehensive information on carbon (C), nitrogen (N), phosphorus (P) and sulphur (S)	Lack of comprehensive information, absence of discriminant biological factors
Ryan et al. [62]	Smart information systems (SIS)	Provides a detailed overview of how SIS, powered by big data and artificial intelligence, are utilized in various aspects of agriculture	Addressing ethical concerns
Parihar et al. [63]	Permanent bed (PB), zero tillage (ZT), and conventional tillage (CT)	Providing valuable insights into the effects of different tillage practices and cropping systems over an extended period	Potential lack of generalizability of the findings beyond the specific context of northwestern India

Variables such as soil type, depth, temperature, rainfall, humidity, PH, nitrogen, phosphate, potassium, organic carbon, calcium, magnesium, sulfur, manganese, copper, iron, and so forth [61]. With two video cameras and a global positioning sensor for navigation, Demeter is a computer-controlled speed-rowing device. It can plan and carry out the harvesting process for a full field, chopping crop rows, turning to cut the next row, shifting positions in the field, and spotting unforeseen obstructions. AI is being used to harvest cucumbers by utilizing the various hardware and software components of the robot, such as the autonomous vehicle, manipulator, end-effector, two computer vision systems for 3D imaging and fruit detection, and a control scheme.

6.4.3 Disease Management

Disease control is essential for agricultural harvests to have the best possible yield. Effective disease control is crucial for optimizing agricultural yields, as both plant and animal diseases pose significant obstacles to growth. Various factors, including genetics, soil composition, moisture, wind, dry weather, and temperature, contribute to the development of illnesses in agriculture. Managing the consequences is challenging due to these variables and the unpredictable nature of some disease-causing influences, especially in large-scale farming.

Table 6.6 gives an overview of the many methods and tools used in agricultural applications, with an emphasis on disease assessment, pest identification, and monitoring systems. Various methods are investigated, including convolutional neural network (CNN), InceptionV3, YoloV5, transfer learning, ML, IoT, and AI with CNNs. Although certain techniques demonstrate high-accuracy rates, real-time precision, and applicability to agricultural problems, they may be limited in their range of crop varieties, reliant on image quality, devoid of implementation specifics, and lacking sufficient data on system improvements. Moreover, other factors that should be taken into account are the reliance on singular datasets, the lack of information regarding the performance of hybrid models, and the scope limitation of current advances in AI. The table also discusses feature selection techniques and the application of unmanned aerial systems (UASs) in agriculture for remote sensing, along with the related limitations, including weather, flying restrictions, and expensive operating expenses.

6.4.4 Weed Management

Farmers' earnings and yields are greatly impacted by weeds; reports suggest that unchecked infestations might cut dry bean and maize crops by 50% [73]. Weed competition causes wheat production losses of around 48% [74, 75], with the potential to increase to 60% [76]. Sesame crops lose 50–75% of their production, while soybean yield declines range from 8% to 55%. Crop exposure time and weed geographic heterogeneity cause variable yield losses. According to the WSSA report, weeds also have both beneficial and harmful consequences on ecosystems, including competition for resources, liver damage, floods, and fire resistance. Despite decades of intensive pesticide treatment to reduce the impact of weeds on crops, it is estimated that western Canadian fields lose more than $500 million in crops per year. This emphasizes the requirement for cutting-edge weed control strategies to counteract these losses [73]. With the help of UAV imagery, a system may more precisely identify weed and crop pixels by dividing, computing, and converting vegetation indexes into binary data, detecting crop rows, optimizing parameters, and learning a classification model. This strategy helps overcome the spectral resemblance between weeds and crops, which is especially advantageous for crops arranged in rows.

Table 6.6 AI in disease management.

Reference	Techniques used	Strength	Limitations
Tirkey et al. [64]	CNN, inceptionV3, yoloV5, transfer learning	High accuracy rates, real-time precision, comprehensive dataset	Limited scope of crop varieties, generalization to diverse insect species, dependence on image quality
Ngongoma et al. [65]	ML, IoT	Relevance to agricultural challenges, evaluation of plant disease detection models	Lack of implementation details for improvements, insufficient detail on monitoring system enhancement
Ibrahim et al. [66]	AI, CNN, deep learning (DL)	Real-world relevance, effective use of CNNs, utilizes a range of performance metrics	Dependency on single dataset, sparse detail on hybrid model performance, no discussion on explainability of the proposed models
Sujawat et al. [67]	IoT, CNN	High performance	May not account for the most recent advancements in AI technologies, scope limitation
Hesham et al. [68]	CNN, RNN	Provides a comprehensive overview of the research domain, covering the use of AI to process diverse data types	Limited availability or quality of data
Afifi et al. [69]	CNN	Robust evaluation and high accuracy of 99%, conducts a comparative analysis with metric learning methods	Dependency on quality of baseline model, long-term model stability
Esgario et al. [70]	CNN	High accuracy metrics, use of ResNet50 Architecture	Lack of comparative analysis, relies on a new database
Whitmire et al. [71]	Feature selection	Simplified models and reduced computational time	Feature selection method choice depends on dataset characteristics and ML algorithm requirements
Neupane et al. [72]	Unmanned aerial system (UAS)	Enables access to remote areas through remote sensing technology	Subject to weather constraints, flight regulations, and high operational costs due to skilled operators and limited payload capacity

Table 6.7 gives an overview of the many methods and tools used in agricultural applications, with an emphasis on weed management systems. These include Faster R-CNN with Inception-V2, YOLOv5's CNN, and SVM, RF, XGB, and KNN. This initiative is in line with the urgent demand for weed control techniques that work to improve crop productivity and farming practices that are sustainable. Among these algorithms, Faster R-CNN with Inception-V2 addresses the problem of weed proliferation among a variety of plant species in mixed agriculture fields with resilience and enhanced accuracy. However, if weed infestation continues over time, algorithms such as YOLOv5s CNN demonstrate sustained accuracy advances even after reaching a plateau, highlighting the constant need for effective weed detection techniques. Additionally, the effectiveness of RF, XGB, and KNN algorithms in detecting certain weed species, such as mouse-ear hawkweed, emphasizes the significance of precision agriculture in focused weed management strategies. Novel techniques like PSPNet, semi-supervised learning, and deep transfer learning present promising paths toward automating weed detection procedures, hence simplifying farmers' weed management tasks as technology advances.

Besides the applications discussed above, there are some other applications emerging in this field. These applications aim to improve the productivity and sustainability of agricultural practices. Table 6.8 represents these applications. Open-source datasets for use in developing AI applications for precision agriculture are listed in Table 6.9. Table 6.9 contains the name of the dataset, platform, url for accessing the dataset and the applications for which it can be used.

6.5 Limitations

In agriculture, AI has significantly advanced, offering tools for crop management and enhancing profitability, sustainability, and productivity. Recent years have seen notable progress in the field, with AI-driven systems providing integrated guidance to farmers [87]. However, for AI in agriculture to realize its full potential, it must overcome several challenges.

Data Challenges: The ability of AI systems to process vast amounts of input data is essential to their efficacy in agriculture. Real-time systems have to weed out irrelevant data and respond quickly to important events [88]. To guarantee the use of pertinent data, field experts must collaborate, yet standardizing and integrating data from many sources might be difficult [89]. In addition to creating accessible and affordable AI tools, researchers are looking into solutions such as open data platforms and standardized formats [88].

Accuracy and Response Time: It's imperative that AI systems be able to do jobs quickly and precisely. Many systems have issues with response time or accuracy,

Table 6.7 AI in weed management.

Reference	Techniques used	Strength	Limitations
Ajayi et al. [77]	Faster region-based convolutional neural network (faster R-CNN) with inception-V2.	In mixed crop fields, Faster R-CNN shows resilience in automatic weed detection and classification; accuracy increases dramatically with the number of training epochs requirements.	After 200,000 training epochs, the model reaches a saturation point in performance, suggesting diminishing returns.
Ajayi et al. [78]	You only look once (YOLO) v5s, CNN	With more training epochs, YOLOv5s shows considerable accuracy gains in automatic crop and weed categorization, indicating its resilience.	Higher spatial resolution and more cannabis imaging data are required for future studies.
Amarasingam et al. [79]	SVM, RF, XGB, KNN	High precision in the identification of mouse-ear hawkweed flowers and foliage using multispectral data captured by a UAV.	The accuracy of the SVM model was poorer, especially when it came to detecting foliage.
Cai et al. [80]	Incorporating ECA module in SPP layer for improved occlusion handling and enhanced global context in PSPnet	Enhanced feature extraction and performance in complex field environments.	Use of limited data.
Jiu et al. [81]	Semi-supervised learning with attention mechanism	Reduces time and labor by using fewer labelled training photos to achieve high weed detection accuracy.	Restricted to finding weeds in wheat fields.
Jiang et al. [82]	YOLOv5 detection model	Better rates of weed removal, less harm to crops, and less need for herbicides.	May not work well in some environments.
Albraikan et al. [83]	Modified barnacles mating optimization with deep learning (MBMODL-WD)	Compared to existing deep learning models, the MBMODL-WD approach outperforms them in automated weed detection in agricultural fields, achieving up to 98.99% accuracy.	Future study must examine the computational difficulty.

(*Continued*)

Table 6.7 (Continued)

Reference	Techniques used	Strength	Limitations
Chen et al. [84]	Deep transfer learning (DTL) with 35 state-of-the-art deep learning models.	Achieved high classification accuracy (>95%) with reasonably short training time (<2.5 h).	Using traditional loss functions, the performance on minority weed groups was less satisfactory.
Khan et al. [85]	DL system	Identifies weeds and crops with great accuracy, surpassing current methods.	Restricted analysis focusing on just two crop varieties (strawberry and pea).
Dasgupta et al. [86]	Naïve Bayes for crop recommendation and deep learning for weed detection.	Combining IoT, WSN, and AI to achieve high accuracy	Dependency on precise data, possible deployment and maintenance difficulties, need for high-quality training data and photos, and financial consequences for small-scale farmers.

Table 6.8 Applications of AI in agriculture.

Application of AI in Agriculture	
Solid management	Crop yield management
Weed management	Crop management
Crop management	Weather forecasting
Disease management	Smart irrigation

Table 6.9 Open-source datasets.

Name	Platform	Url	Application
Ladybird Cobbitty Brassica	Ground vehicle	https://ses.library.usyd.edu.au/handle/2123/20187	Weed control
weedNet	UAV	https://github.com/inkyusa/weedNet	Weed control
Maize disease	Multiple platforms	https://osf.io/p67rz/	Disease detection
Oil radish growth	Ground vehicle	https://vision.eng.au.dk/oil-radish/	Yield estimation
Capsicum Annuum	No imaging platform	https://data.4tu.nl/repository/uuid%3a884958f5-b868-46e1-b3d8-a0b5d91b02c0	Semantic plant segmentation
Open Plant Phenotype Database (OPPD)	Ground fixed platform	https://gitlab.au.dk/AUENG-Vision/OPPD/-/tree/master/	Weed control
DeepWeeds	Ground vehicle	https://www.nature.com/articles/s41598-018-38343-3	Weed detection
Sugarcane billets	Ground based	https://github.com/The77Lab/SugarcaneBilletsDataset	Quality inspection

which affects how users make decisions [90]. To help farmers make wise decisions, researchers are refining task techniques and creating AI technologies that are more accurate and responsive [90].

Ethical Concerns: As AI becomes more integrated into agriculture, ethical questions like data ownership, algorithm openness, and responsible AI use start to surface. It is imperative to have data privacy legislation, transparent AI development processes, and impartial algorithms [91, 92]. Legislators need to support ethical development methods and provide various agricultural groups fair access to AI [93, 94]. AI should support environmentally and socially responsible farming practices and be regularly evaluated [91].

AI Equitability: Efforts must be made to lower the cost of AI solutions and increase their accessibility for underprivileged farming communities. To close the adoption gap for AI, governments should provide access through free or inexpensive web services and training initiatives [95, 96].

Data Security and Privacy: Robust rules pertaining to data security and privacy are essential in order to avert the misappropriation of agricultural data that AI systems gather. Ensuring data security and privacy helps to maintain a just and sustainable agricultural ecosystem by protecting against threats like intellectual property theft and price manipulation [95].

6.6 Conclusion and Future Scope

AI in agriculture has made significant advances by providing farmers with new tools to solve problems and increase yields. AI robots can make farming easier by completing tasks such as crop selection faster and more accurately. AI is changing agriculture in many ways, including helping farmers detect and manage diseases, weeds, and soil problems early. It also improves crop yield estimates. AI has the potential to improve global agriculture by making it more sustainable and efficient. However, results so far have been limited to large farms with large numbers of machines. This world shows how AI is changing agriculture and leading business to the information path. AI should improve productivity and communication through its automated processes, data analysis, and smart devices. Recent developments have made AI superior to humans in important areas such as communication and business. It uses automation and predictive analytics to solve problems such as labor dependency and resource inefficiency in agriculture. As demand for agricultural products continues to grow, AI-powered automation becomes increasingly important in solving problems such as labor shortages and unpredictable weather conditions. Researchers are developing the best treatments against diseases, pests, and climate change. Despite the many promises of AI, data privacy and equity issues remain. These challenges need to be overcome to make agriculture sustainable in the future.

References

1 Gryshova, I., Balian, A., Antonik, I. et al. (2024). Artificial intelligence in climate smart in agricultural: toward a sustainable farming future. *Access to Science, Business, Innovation in the Digital Economy, ACCESS Press* 5 (1): 125–140.

2 Subeesh, A. and Mehta, C.R. (2021). Automation and digitization of agriculture using artificial intelligence and internet of things. *Artificial Intelligence in Agriculture* 5: 278–291.

3 Kumar, A., Mamgai, R., and Jain, R. (2023). Application of IoT-Enabled CNN for natural language processing. In: *IoT-enabled Convolutional Neural Networks: Techniques and Applications* (ed. M. Naved, V.A. Devi, L. Gaur, and A.A. Elngar), 149–177. River Publishers.

4 Awasthi, Y. (2020). Press "A" for artificial intelligence in agriculture: a review. *JOIV: International Journal on Informatics Visualization* 4 (3): 112–116.

5 Skvortsov, E.A. (2020). Prospects of applying artificial intelligence technologies in the regional agriculture. *Ekonomika Regiona= Economy of Regions* 2: 563.

6 Kumar, A., Walia, G.S., and Sharma, K. (2020). Recent trends in multicue based visual tracking: a review. *Expert Systems with Applications* 162: 113711.

7 R. Sharma, "Artificial Intelligence in Agriculture: A Review," *2021 5th International Conference on Intelligent Computing and Control Systems (ICICCS), Madurai, India*, 2021, pp. 937–942, https://doi.org/10.1109/ICICCS51141.2021.9432187.

8 J. Angelin Blessy and A. Kumar, "Smart Irrigation System Techniques using Artificial Intelligence and IoT," *2021 Third International Conference on Intelligent Communication Technologies and Virtual Mobile Networks (ICICV), Tirunelveli, India*, 2021, pp. 1355–1359, https://doi.org/10.1109/ICICV50876.2021.9388444.

9 Saxena, R., Pant, D.K., Pant, S.C. et al. (2024). Artificial intelligence-led innovations for agricultural transformation: a scoping study. *Agricultural Economics Research Review* 36: 1–18.

10 Elbasi, E., Zaki, C., Topcu, A.E. et al. (2023). Crop prediction model using machine learning algorithms. *Applied Sciences* 13 (16): 9288.

11 J. Mahale, S. Degadwala and D. Vyas, "Crop Prediction System based on Soil and Weather Characteristics," *2022 Sixth International Conference on I-SMAC (IoT in Social, Mobile, Analytics and Cloud) (I-SMAC), Dharan, Nepal*, 2022, pp. 340–345, https://doi.org/10.1109/I-SMAC55078.2022.9987366.

12 Mkrttchian, V. (2021). Artificial and natural intelligence techniques as IoP-and IoT-based technologies for sustainable farming and smart agriculture. In: *Artificial Intelligence and IoT-Based Technologies for Sustainable Farming and Smart Agriculture* (ed. P. Tomar and G. Kaur), 40–53. IGI Global.

13 Kumar, A. and Jha, R. (2022). IoT in smart farming: a boon to farmers. In: *Internet of Everything: Smart Sensing Technologies* (ed. T. Kavitha, V.A. Devi, S.N. Pari, and S. Ramanathan). Nova Publisher.

14 Gupta, R., Sharma, A.K., Garg, O. et al. (2021). WB-CPI: weather based crop prediction in India using big data analytics. *IEEE Access* 9: 137869–137885.

15 Janaki, G., Deepika Sree, D., Sheela, A., Kalaiselvi, V.K.G., Ranjani, J. (2021). Crop Yield Prediction Using Machine Learning Algorithm. *2021 4th International Conference on Computing and Communications Technologies (ICCCT), Chennai, India*, 2021, pp. 611–616. https://doi.org/10.1109/ICCCT53315.2021.9711853.

16 S. Jain and D. Ramesh, "Machine Learning convergence for weather based crop selection," *2020 IEEE International Students' Conference on Electrical, Electronics and Computer Science (SCEECS), Bhopal, India*, 2020, pp. 1–6, https://doi.org/10.1109/SCEECS48394.2020.75.

17 A. Nigam, S. Garg, A. Agrawal and P. Agrawal, "Crop Yield Prediction Using Machine Learning Algorithms," *2019 Fifth International Conference on Image Information Processing (ICIIP), Shimla, India*, 2019, pp. 125–130, https://doi.org/10.1109/ICIIP47207.2019.8985951.

18 Kumar, A. (2023). *Visual Object Tracking Using Deep Learning*. CRC Press.

19 Y. Gandge and Sandhya, "A study on various data mining techniques for crop yield prediction," *2017 International Conference on Electrical, Electronics, Communication, Computer, and Optimization Techniques (ICEEECCOT), Mysuru, India*, 2017, pp. 420–423, https://doi.org/10.1109/ICEECCOT.2017.8284541.

20 R. Sujatha and P. Isakki, "A study on crop yield forecasting using classification techniques," *2016 International Conference on Computing Technologies and Intelligent Data Engineering (ICCTIDE'16), Kovilpatti, India*, 2016, pp. 1–4, https://doi.org/10.1109/ICCTIDE.2016.7725357.

21 Kumar, A., Jain, R., Vairamani, A. D., & Nayyar, A. (Eds.). (2023). Object Tracking Technology: Trends, Challenges and Applications.

22 Singh, D. and Kumar, A. (2024). A deep recurrent neural network for plant disease classification. *SN Computer Science* 5 (8): 1053.

23 Dozono, K., Amalathas, S., and Saravanan, R. (2022). The impact of cloud computing and artificial intelligence in digital agriculture. In: *Proceedings of Sixth International Congress on Information and Communication Technology. Lecture Notes in Networks and Systems*, vol. 235 (ed. X.S. Yang, S. Sherratt, N. Dey, and A. Joshi). Singapore: Springer https://doi.org/10.1007/978-981-16-2377-6_52.

24 Vazquez, J.P.G., Torres, R.S., Perez, D.B.P. et al. (2021). Scientometric analysis of the application of artificial intelligence in agriculture. *Journal of Scientometric Research* 10 (1): 55–62.

25 Sivarethinamohan, R., Yuvaraj, D., Shanmuga Priya, S., and Sujatha, S. (2021). Captivating profitable applications of artificial intelligence in agriculture management. *Intelligent Computing and Optimization: Proceedings of the 3rd*

International Conference on Intelligent Computing and Optimization 2020 (ICO 2020), pp. 848–861. Springer International Publishing, 8 February 2021.

26 Tarik, H. and Mohammed, O.J. (2020). Big data analytics and artificial intelligence serving agriculture. In: *Advanced Intelligent Systems for Sustainable Development (AI2SD'2019) Volume 2-Advanced Intelligent Systems for Sustainable Development Applied to Agriculture and Health*, 57–65. Springer International Publishing.

27 Afzaal, H., Farooque, A.A., Abbas, F. et al. (2020). Computation of evapotranspiration with artificial intelligence for precision water resource management. *Applied Sciences* 10 (5): 1621.

28 Khan, R., Dhingra, N., and Bhati, N. (2022). Role of artificial intelligence in agriculture: a comparative study. In: *Transforming Management with AI, Big-Data, and IoT* (ed. F. Al-Turjman, S.P. Yadav, M. Kumar, et al.), 73–83. Cham: Springer International Publishing.

29 Javaid, M., Haleem, A., Khan, I.H., and Suman, R. (2023). Understanding the potential applications of artificial intelligence in agriculture sector. *Advanced Agrochem* 2 (1): 15–30.

30 Wongchai, A., Shukla, S.K., Ahmed, M.A. et al. (2022). Artificial intelligence-enabled soft sensor and internet of things for sustainable agriculture using ensemble deep learning architecture. *Computers and Electrical Engineering* 102: 108128.

31 Bu, F. and Wang, X. (2019). A smart agriculture IoT system based on deep reinforcement learning. *Future Generation Computer Systems* 99: 500–507.

32 Singh, A., Mehrotra, R., Rajput, V.D. et al. (2022). Geoinformatics, artificial intelligence, sensor technology, big data. In: *Sustainable Agriculture Systems and Technologies* (ed. P. Kumar, A.K. Pandey, S.K. Singh, et al.). https://doi.org/10.1002/9781119808565.ch14.

33 Sabrina, F., Sohail, S., Farid, F. et al. (2022). An interpretable artificial intelligence based smart agriculture system. *Computers, Materials & Continua* 72: 3777–3797.

34 Malhotra, C. and Anand, R. (September 2020). Accelerating public service delivery in India: application of internet of things and artificial intelligence in agriculture. In *Proceedings of the 13th International Conference on Theory and Practice of Electronic Governance (ICEGOV '20). Association for Computing Machinery, New York, NY, USA*, 62–69. https://doi.org/10.1145/3428502.3428510.

35 Sankaranarayanan, S. (2021). Applications of artificial intelligence for smart agriculture. In: *AI-Based Services for Smart Cities and Urban Infrastructure* (ed. K. Lyu, M. Hu, J. Du, and V. Sugumaran), 277–288. IGI Global.

36 Gupta, N., Khosravy, M., Gupta, S. et al. (2022). Lightweight artificial intelligence technology for health diagnosis of agriculture vehicles: parallel evolving artificial neural networks by genetic algorithm. *International Journal of Parallel Programming* 50: 1–26.

37 Shaikh, F.K., Memon, M.A., Mahoto, N.A. et al. (2021). Artificial intelligence best practices in smart agriculture. *IEEE Micro* 42 (1): 17–24.

38 Panpatte, S. and Ganeshkumar, C. (2021). Artificial intelligence in agriculture sector: case study of blue river technology. *Proceedings of the Second International Conference on Information Management and Machine Intelligence: ICIMMI 2020*, pp. 147–153. Springer Singapore, 23 January 2021.

39 Mokaya, V. (2019). Future of precision agriculture in India using machine learning and artificial intelligence. *International Journal of Computer Sciences and Engineering* 7 (2): 1020–1023.

40 Nawaz, A.N., Nadaf, H.A., Kareem, A.M., and Nagaraja, H. (2020). Application of artificial intelligence in agriculture-pros and cons. *Vigyan Varta* 1 (8): 22–25.

41 Tangwannawit, S. and Tangwannawit, P. (2022). An optimization clustering and classification based on artificial intelligence approach for internet of things in agriculture. *IAES International Journal of Artificial Intelligence* 11 (1): 201.

42 Chen, G., Li, J., Chen, H., and An, Y. (2018). A survey of researches on artificial intelligence in the field of agriculture in big data era. *Journal of Jilin Agricultural University* 40 (4): 502–510.

43 Singh, G., Singh, A., and Kaur, G. (2021). Role of artificial intelligence and the internet of things in agriculture. In: *Artificial Intelligence to Solve Pervasive Internet of Things Issues* (ed. A. Srivastava, M.S. Gupta, and G. Kaur), 317–330. Academic Press.

44 Zhou, Y., Xia, Q., Zhang, Z. et al. (2022). Artificial intelligence and machine learning for the green development of agriculture in the emerging manufacturing industry in the IoT platform. *Acta Agriculturae Scandinavica Section B Soil and Plant Science* 72 (1): 284–299.

45 Pathan, M., Patel, N., Yagnik, H., and Shah, M. (2020). Artificial cognition for applications in smart agriculture: a comprehensive review. *Artificial Intelligence in Agriculture* 4: 81–95.

46 Liakos, K.G., Busato, P., Moshou, D. et al. (2018). Machine learning in agriculture: a review. *Sensors* 18 (8): 2674.

47 Shaikh, T.A., Rasool, T., and Lone, F.R. (2022). Towards leveraging the role of machine learning and artificial intelligence in precision agriculture and smart farming. *Computers and Electronics in Agriculture* 198: 107119.

48 Petrakieva, S., Garasym, O., and Taralova, I. (2014). http://ieeexplore. ieee. org/s- tamp/stamp. jsp? tp= &arnumber= 7038771.

49 De Kimpe, C.R. and Morel, J.L. (2000). Urban soil management: a growing concern. *Soil Science* 165 (1): 31–40.

50 Sarkar, A., Maity, P.P., and Mukherjee, A. (2021). Application of AI in soil science. *Technology* 62: 131–136.

51 Awais, M., Naqvi, S.M.Z.A., Zhang, H. et al. (2023). AI and machine learning for soil analysis: an assessment of sustainable agricultural practices. *Bioresources and Bioprocessing* 10 (1): 90.

52 Chen, Q., Li, L., Chong, C., and Wang, X. (2022). AI-enhanced soil management and smart farming. *Soil Use and Management* 38 (1): 7–13.

53 Folorunso, O., Ojo, O., Busari, M. et al. (2023). Exploring machine learning models for soil nutrient properties prediction: a systematic review. *Big Data and Cognitive Computing* 7 (2): 113.

54 P. Ranjan, R. Garg and J. K. Rai, "Artificial Intelligence Applications in Soil & Crop Management," *2022 IEEE Conference on Interdisciplinary Approaches in Technology and Management for Social Innovation (IATMSI), Gwalior, India*, 2022, pp. 1–5, https://doi.org/10.1109/IATMSI56455.2022.10119362.

55 Alqadhi, S., Mallick, J., Talukdar, S., and Alkahtani, M. (2023). An artificial intelligence-based assessment of soil erosion probability indices and contributing factors in the Abha-Khamis watershed, Saudi Arabia. *Frontiers in Ecology and Evolution* 11: 1189184.

56 Nguyen, T.T., Ngo, H.H., Guo, W. et al. (2022). A low-cost approach for soil moisture prediction using multi-sensor data and machine learning algorithm. *Science of The Total Environment* 833: 155066.

57 Ou, D., Tan, K., Lai, J. et al. (2021). Semi- supervised DNN regression on airborne hyperspectral imagery for improved spatial soil properties prediction. *Geoderma* 385: 114875.

58 Arashpour, M. (2023). AI explainability framework for environmental management research. *Journal of Environmental Management* 342: 118149.

59 Gautron, R., Maillard, O.A., Preux, P. et al. (2022). Reinforcement learning for crop management support: review, prospects and challenges. *Computers and Electronics in Agriculture* 200: 107182.

60 Navinkumar, T.M., Kumar, R.R., and Gokila, P.V. (2021). Application of artificial intelligence techniques in irrigation and crop health management for crop yield enhancement. *Materials Today Proceedings* 45: 2248–2253.

61 Ghosh, A., Singh, A.B., Kumar, R.V. et al. (2020). Soil enzymes and microbial elemental stoichiometry as bio-indicators of soil quality in diverse cropping systems and nutrient management practices of Indian Vertisols. *Applied Soil Ecology* 145: 103304.

62 Ryan, M. (2019). Ethics of using AI and big data in agriculture: the case of a large agriculture multinational. *Orbit Journal* 2 (2): 1–27.

63 Parihar, C.M., Yadav, M.R., Singh, A.K. et al. (2018). Long-term conservation agriculture and intensified cropping systems: effects on growth, yield, water, and energy-use efficiency of maize in northwestern India. *Pedosphere* 28 (6): 952–963.

64 Tirkey, D., Singh, K.K., and Tripathi, S. (2023). Performance analysis of AI-based solutions for crop disease identification, detection, and classification. *Smart Agricultural Technology* 5: 100238.

65 Ngongoma, M.S., Kabeya, M., and Moloi, K. (2023). A review of plant disease detection systems for farming applications. *Applied Sciences* 13 (10): 5982.

66 A. O. Ibrahim and S. Elghamrawy, "Plant Disease Detection using Artificial Intelligence techniques for Agricultural productivity enhancement in Egypt," *2023 International Telecommunications Conference (ITC-Egypt), Alexandria, Egypt*, 2023, pp. 490–496, https://doi.org/10.1109/ITC-Egypt58155.2023.10206149.

67 Sujawat, G.S. (2021). Application of Artificial Intelligence in detection of diseasesin plants: A Survey. *Turkish Journal of Computer and Mathematics Education (TURCOMAT)* 12 (3): 3301–3305.

68 Hesham, M. (2023). Artificial intelligence in plant disease management (no.11206). EasyChair.

69 Afifi, A., Alhumam, A., and Abdelwahab, A. (2020). Convolutional neural network for automatic identification of plant diseases with limited data. *Plants* 10 (1): 28.

70 Esgario, J.G., Krohling, R.A., and Ventura, J.A. (2020). Deep learning for classification and severity estimation of coffee leaf biotic stress. *Computers and Electronics in Agriculture* 169: 105162.

71 Whitmire, C.D., Vance, J.M., Rasheed, H.K. et al. (2021). Using machine learning and feature selection for alfalfa yield prediction. *AI* 2 (1): 71–88.

72 Neupane, K. and Baysal-Gurel, F. (2021). Automatic identification and monitoring of plant diseases using unmanned aerial vehicles: A review. *Remote Sensing* 13 (19): 3841.

73 Neil Harker, K. (2001). Survey of yield losses due to weeds in central Alberta. *Canadian Journal of Plant Science* 81 (2): 339–342.

74 Khan, M. and Haq, N. (2002). Wheat crop yield loss assessment due to weeds. *Sarhad Journal of Agriculture (Pakistan)* 18 (4): 449–453.

75 Fahad, S., Hussain, S., Chauhan, B.S. et al. (2015). Weed growth and crop yield loss in wheat as influenced by row spacing and weed emergence times. *Crop Protection* 71: 101–108.

76 Rao, A.N., Wani, S.P., and Ladha, J.K. (2014). Weed management research in India-an analysis of past and outlook for future.

77 Ajayi, O.G. and Ashi, J. (2023). Effect of varying training epochs of a faster region-based convolutional neural network on the accuracy of an automatic weed classification scheme. *Smart Agricultural Technology* 3: 100128.

78 Ajayi, O.G., Ashi, J., and Guda, B. (2023). Performance evaluation of YOLO v5 model for automatic crop and weed classification on UAV images. *Smart Agricultural Technology* 5: 100231.

79 Amarasingam, N., Hamilton, M., Kelly, J.E. et al. (2023). Autonomous detection of mouse-ear hawkweed using drones, multispectral imagery and supervised machine learning. *Remote Sensing* 15 (6): 1633.

80 Cai, Y., Zeng, F., Xiao, J. et al. (2023). Attention-aided semantic segmentation network for weed identification in pineapple field. *Computers and Electronics in Agriculture* 210: 107881.

81 Liu, T., Jin, X., Zhang, L. et al. (2023). Semi-supervised learning and attention mechanism for weed detection in wheat. *Crop Protection* 174: 106389.

82 Jiang, W., Quan, L., Wei, G. et al. (2023). A conceptual evaluation of a weed control method with post-damage application of herbicides: A composite intelligent intra-row weeding robot. *Soil and Tillage Research* 234: 105837.

83 Albraikan, A.A., Aljebreen, M., Alzahrani, J.S. et al. (2022). Modified barnacles mating optimization with deep learning based weed detection model for smart agriculture. *Applied Sciences* 12 (24): 12828.

84 Chen, D., Lu, Y., Li, Z., and Young, S. (2022). Performance evaluation of deep transfer learning on multi-class identification of common weed species in cotton production systems. *Computers and Electronics in Agriculture* 198: 107091.

85 Khan, S., Tufail, M., Khan, M.T. et al. (2021). Deep learning- based identification system of weeds and crops in strawberry and pea fields for a precision agriculture sprayer. *Precision Agriculture* 22 (6): 1711–1727.

86 Dasgupta, I., Saha, J., Venkatasubbu, P., and Ramasubramanian, P. (2020). AI crop predictor and weed detector using wireless technologies: a smart application for farmers. *Arabian Journal for Science and Engineering* 45: 11115–11127.

87 Rajotte, E., Bowser, T., Travis, J. et al. (1992). Implementation and adoption of an agricultural expert system: the penn state apple orchard consultant. *Acta Horticulturae* 313: 227–232. https://doi.org/10.17660/actahortic.1992.313.28.

88 Washington, R. and Hayes-Roth, B. (August 1989). Input data management in real-time ai systems. *IJCAI*, Vol. 89, No. 11.

89 Ishak, N.A., Idrus, S.Z.S., Saraih, U.N. et al. (2021). Exploring digital parenting awareness during covid-19 pandemic through online teaching and learning from home. *International Journal of Business and Technopreneurship* 11 (3): 37–48.

90 Steven L. Teal and Alexander I. Rudnicky. 1992. A performance model of system delay and user strategy selection. In *Proceedings of the SIGCHI Conference on Human Factors in Computing Systems (CHI '92). Association for Computing Machinery, New York, NY, USA*, 295–305. https://doi.org/10.1145/142750.142818.

91 Al-Awfi, A.H.H. and Ayasrah, F.T.M. (2022). The effectiveness of digital game activities in developing cognitive achievement and cooperative learning skills in the science course among female primary school students in Medina. *Arab Journal of Specific Education* 6 (21): 17–58.

92 Al-Harbi, A.A. and Ayasrah, F.T.M. (2021). The effectiveness of using augmented reality technology in developing spatial thinking and scientific concepts in the chemistry course among female secondary school students in Medina. *Arab Journal of Specific Education* 5 (20): 1–38.

93 Ayasrah, F.T., Abu-Bakar, H., and Ali, A. (2015) Exploring the fakes within online communication: a grounded theory approach (phase two: study sample and procedures).

94 Ayasrah, F.T.M., Alarabi, K., Al Mansouri, M. et al. (2024). Enhancing secondary school students' attitudes toward physics by using computer simulationsInternational Journal of Data and Network. *Science* 8: 369–380.

95 Jack, W. and Bagh, S. (2024). Revolutionizing agriculture: AI-powered crop yield forecasting and precision farming for optimal harvests (no. 11942). EasyChair.

96 Eli-Chukwu, N.C. (2019). Applications of artificial intelligence in agriculture: a review. *Engineering, Technology & Applied Science Research* 9 (4): 4377.

7

Integrating AI and Climate-Smart Agricultural Mechanization: Strategies for Enhancing Productivity and Sustainability in a Changing Climate

Anil Kumar

Bharati Vidyapeeth's College of Engineering, New Delhi, India

7.1 Introduction

Natural disasters and climate change pose a threat to agricultural production. The agricultural production system and farmers' revenues are both at risk by climate change, according to Ali et al. [1]. Limited natural resources are a challenge for agricultural production systems. Climate change has led to inefficient management techniques, which in turn have an impact on the amount and quality of these resources. In order for agriculture to adjust to the new reality, it has to drastically modify in order to be more resilient to the effects of climate change without compromising food security. There is a connection between all of these problems, and the concept of climate-smart agriculture was to solve them all at once. Farm mechanization is an excellent choice for these weather-related problems. As a result, we can ensure a steady supply of food even as the world continues to warm. Agricultural automation that employs climate-smart technology to boost agricultural output has the potential to make agriculture more resistant to the effects of climate change and increase agricultural profitability. Additionally, carbon sequestration might be enhanced if conservation agriculture or climate-smart agriculture proves to be more feasible. Although mechanization in agriculture is not new, using technology to climate-smart farming offers a fresh approach to boosting crop yields while reducing environmental impacts. This course of action would better equip agricultural machines to cope with climate change. Taken together, agronomic methods, field operations, and the evaluation and use of climate-smart technology make up climate-smart agriculture. Important components

Emerging Smart Agricultural Practices Using Artificial Intelligence, First Edition.
Edited by Ashish Kumar, Jai Prakash Verma, and Rachna Jain.

Published 2025 by John Wiley & Sons, Inc.

include formulating investment plans, creating a research base, and enacting government regulations that are helpful.

7.1.1 Mechanization Status, Demands, and Opportunities

Improved input usage efficiency, higher output, and higher productivity can only be achieved via the application of farm mechanization. According to Grant Thornton [2] and Fang [3], the "*United States of America, Western Europe, Russia, Brazil, and China*" have reached different levels of mechanization: 95.5%, 80.5%, 75.5%, and 59.5%, respectively. In contrast, it is found that India's mechanization level was a meager 40%. Figure 7.1 shows the degree of mechanization for several of India's most important crops. The states of Punjab, Haryana, and Uttar Pradesh have the highest availability of agricultural power in India, whereas the states in the northeast are severely short (Figure 7.2). While other agricultural tools are falling behind, tractor penetration has been on the rise.

One limitation of agricultural mechanization is the fragmentation of farm size. Small and "marginal agricultural land holdings (<2.0 ha) account for 86.21% of total operational land holdings and cover 47.35% of farms in India," while the average farm size is less than 1.08 ha [4]. Climate change poses a significant threat to agricultural systems, even those with modest holdings of land. In order to achieve food security, it may also be subjected to less agricultural automation in the operations of the farm. Land fragmentation, which has led to

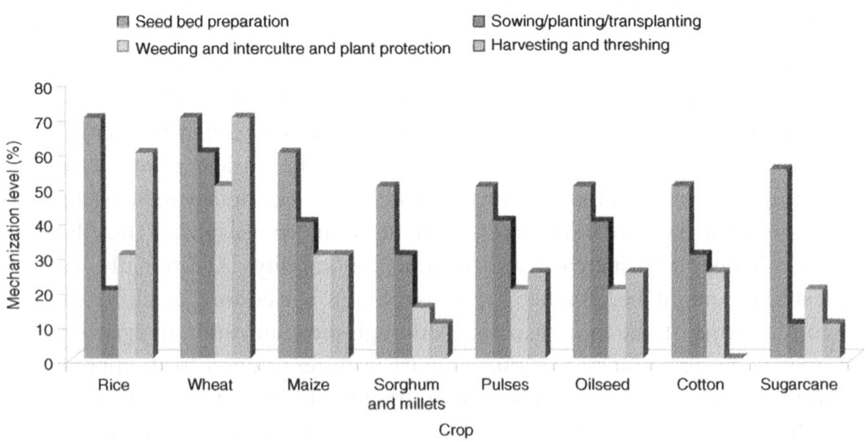

Figure 7.1 Mechanization level for major crops in India. *Source:* Fang [3]/Ama-agricultural Mechanization in Asia Africa and Latin America.

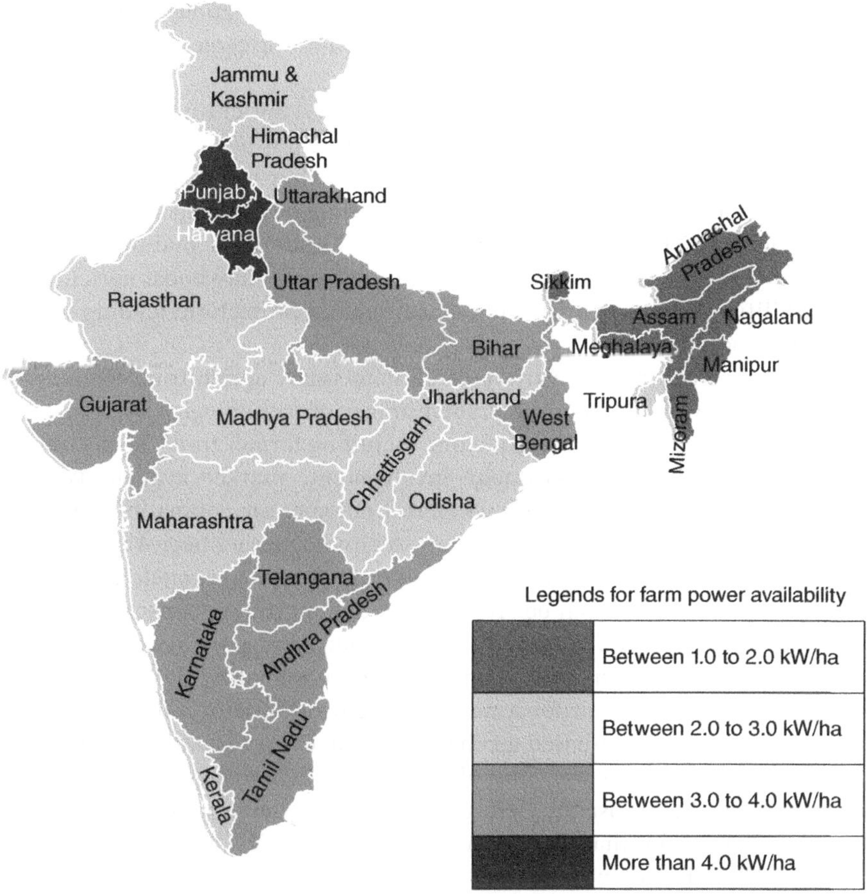

Figure 7.2 Farm power availability in India. *Source:* DACFW [4]/Ministry of Agriculture and Farmers Welfare, Government of India.

tiny agricultural holdings, is a serious problem. Another issue is the predicted 26% decline in the availability of agricultural laborers by the year 2050 [5]. Opportunities to improve India's current agricultural situation include custom hiring centers, cooperative societies, and farmer-producer organizations. As part of its Make in India program, the Indian government is helping domestic producers create agricultural machinery and equipment that meets domestic demand rather than relying on imports from wealthy nations. One practical option for small-scale farmers is the rise of custom-hiring centers and small-scale automation.

In many Indian states, the number of developed agricultural technologies is still low, ranging from 2 to 3 kW/ha, and there has been little to no front-line demonstration or reach to the end user. One targeted way to affect Indian agriculture is to promote the adoption of existing climate-smart technology via climate-smart solutions. In addition to maintaining and improving natural resources, managing ecosystems may boost agricultural profitability via food security and environmental protection, leading to sustainable production.

Because there is a wide variety of climate-smart agricultural techniques that could be available to farmers, the number of technologies that are actually adopted depends on a lot of factors, including farmer traits, farm circumstances, affordability, government involvement, market availability, and information access. In situations when technologies are compatible with one another, farmers might implement many technologies at once. Water conservation measures can be achieved through various means, such as a happy seeder, and soil erosion can be reduced through minimum conservation, tillage, and mulching. On the other hand, a farmer who raises livestock can use the manure from that business to improve the soil's health [6]. The developed world has long advocated for a new paradigm in agricultural development known as climate-smart-based mechanization in order to adapt farming practices to the effects of a warming planet. Technical, managerial, and policy initiatives may help close the yield gap and boost agricultural output in India. Research has shown that some automated technologies, when implemented properly, have the potential to

- Raised output and efficiency on farms by 10–15%.
- Increase of 5–20% in the degree of cropping.
- You may save up to 15–20% on seeds.
- You may save time and effort by 20–30% and save fertilizer and chemicals by 15–20%.

7.1.2 Efforts by the Government to Automate Farming

There is an immediate need for climate-smart agriculture in light of the current situation of increased input usage efficiency and better use of agricultural inputs.

The Indian government has launched a program called Sub-Mission on Agricultural Mechanization (SMAM) to increase the availability of farm electricity up to 2.5 kW/ha and encourage the use of farm mechanization. The Government of India provides 100% to SMAM, a Central Sector Scheme that has many components aimed at teaching, testing, promoting, demonstrating, and enhancing agricultural mechanization and management makes a contribution. With the exception of the states in the Northeast and Himalayan areas, where the percentages are 90 and 10, respectively, the amount allocated to centrally sponsored schemes by the Indian government and the states is 60% and 40%. According to SMAM [7], in the case of Union Territories, the center share is 100%. Another big problem that arose was residue burning, which prompted efforts to develop climate-smart solutions, such as investing Rs. 591.62 crore to promote agricultural mechanization for in situ management of crop residue [4].

7.1.3 Research Gap and Chapter Contribution

Despite the considerable strides in farm mechanization and climate-smart technologies, a significant gap remains in their integration and application, particularly in regions like India with fragmented landholdings and labor shortages. This chapter aims to bridge this gap by providing a comprehensive analysis of how AI and climate-smart mechanization can be effectively integrated to enhance productivity and sustainability [8]. By focusing on the practical challenges and opportunities in diverse global regions, particularly India, this research offers a roadmap for adopting innovative farming techniques that can withstand the impacts of climate change while optimizing resource use and minimizing environmental damage.

7.2 Literature Review

Authors	Title	Focus	Findings	Research contribution	Research gap
Ali [1]	Advances in Climate-Smart Mechanization for Sustainable Agriculture	Climate-smart mechanization and sustainability	Integrating AI and climate-smart mechanization enhances sustainability	Provides a comprehensive review of AI integration in agriculture, offering a roadmap for future studies and practical applications	Need for more regional studies on AI integration
Bhambota et al. [9]	Field Evaluation of Automated Vegetable Transplanters	Automated vegetable transplanters	Automated transplanters improve efficiency and reduce labor	Reviews the effectiveness and efficiency gains of automated transplanters in vegetable farming	Long-term impact studies are needed
Fang [3]	Impact of AI on Precision Agriculture: A Global Perspective	AI in precision agriculture	AI significantly impacts precision agriculture globally	Provides a global perspective on the impact of AI on precision agriculture, highlighting various applications and benefits	Regional impacts of AI not fully explored
Gautam et al. [10]	Mechatronics Application in Precision Sowing: A Review	Mechatronics in precision sowing	Mechatronics enhances precision sowing and efficiency	Offers a detailed review of mechatronics applications in precision sowing, emphasizing their benefits and potential	Practical applications in diverse settings are needed
Government of India [11]	Sub-mission on Agricultural Mechanization (SMAM) Annual Report	Agricultural mechanization policies in India	SMAM policies are effective in increasing mechanization	Reviews the effectiveness of SMAM policies in promoting mechanization in Indian agriculture	Long-term impact of policies not assessed
Grant Thornton [2]	Mechanization Key to Higher Productivity to Double Farmers' Income	Mechanization and productivity in India	Mechanization is key to productivity and income doubling	Highlights the link between mechanization and productivity, offering insights for policymakers	Scalability of mechanization solutions

National Academy of Agricultural Sciences (NAAS) [12]	Policy Paper No. 84: Practical and Affordable Approaches for Precision in Farm Equipment and Machinery	Precision in farm equipment and machinery	Precision in equipment enhances efficiency and sustainability	Reviews practical and affordable approaches to precision in farm equipment, providing policy recommendations	Field-level implementation studies
Onwude et al. [13]	Mechanisation of Large-Scale Agricultural Fields in Developing Countries – A Review	Mechanization in developing countries	Mechanization improves large-scale agriculture in developing countries	Provides a comprehensive review of mechanization in large-scale agriculture in developing countries, highlighting benefits and challenges	Scalability in different contexts
Parmar [14]	Development of an Agricultural Spraying System for Unmanned Aerial Vehicles	Agricultural spraying systems with UAVs	UAVs improve efficiency in agricultural spraying	Reviews the development and effectiveness of UAV-based spraying systems in agriculture	Field trials and farmer adoption rates
Press Information Bureau (PIB) [15]	Union Ministry for Agriculture and Farmers Welfare Adopts Several Steps to Tackle Stubble Burning	Government steps to tackle stubble burning	Steps effective in reducing stubble burning	Summarizes government interventions to tackle stubble burning, providing an overview of policies and their impact	Long-term efficacy of interventions
Reddy et al. [16]	NICRA-Compendium of Climate Resilient Technologies	Climate resilient technologies in agriculture	Technologies enhance climate resilience in agriculture	Reviews various climate-resilient technologies, compiling best practices for enhancing resilience in agriculture	Adoption barriers in different regions
Sidhu et al. [17]	Development and Evaluation of the Turbo Happy Seeder for Sowing Wheat into Heavy Rice Residues in NW India	Turbo Happy Seeder for wheat sowing in heavy rice residues	Turbo Happy Seeder effective for sowing in heavy residues	Provides a detailed review of the development and evaluation of the Turbo Happy Seeder, highlighting its benefits and potential for wider adoption	Broader applicability in varied conditions

7.3 Methodology

This research employs a multifaceted methodology to explore the integration of AI and climate-smart agricultural mechanization, focusing on enhancing productivity and sustainability across diverse global regions, with a particular emphasis on India. The approach includes a combination of literature review, data analysis, field surveys, and expert consultations.

7.3.1 Literature Review

The study begins with an extensive review of existing literature to understand the current state of agricultural mechanization and AI applications in climate-smart agriculture. Sources include academic journals, government reports, industry publications, and case studies from the United States, Western Europe, Russia, Brazil, China, and India. This review helps identify key technologies, practices, and challenges in different regions.

7.3.2 Data Collection

Data is collected from various sources, including government databases, agricultural surveys, and industry reports. Key metrics such as mechanization levels, crop yields, input usage, and environmental impact are gathered. Specific attention is given to the mechanization status in India, where data on farm size, labor availability, and technology adoption rates are critical.

7.3.3 Field Surveys and Interviews

Field surveys are conducted with farmers, agricultural experts, and machinery manufacturers in India. These surveys aim to gather firsthand information on the practical challenges faced in adopting mechanization and AI technologies. Interviews with stakeholders provide insights into the effectiveness of current policies and the need for future interventions.

7.3.4 Case Studies

Detailed case studies are developed for regions where climate-smart mechanization has been successfully implemented. These case studies highlight best practices, successful technologies, and the impact of mechanization on productivity and sustainability. Comparative analysis of Indian conditions helps to identify transferable strategies.

Specific case studies include:

- **United States**: Implementation of AI-driven precision farming techniques to optimize resource use and enhance crop yields. Examples include the use of predictive analytics for seedbed preparation and nutrient management.
- **Western Europe**: Adoption of advanced soil management practices and climate-smart technologies such as automated irrigation systems and drones for crop monitoring.
- **Russia**: Large-scale mechanization efforts and integration of AI for efficient soil and water management, addressing challenges of vast agricultural landscapes.
- **Brazil**: Use of climate-smart mechanization to tackle deforestation and improve the sustainability of soybean and sugarcane production.
- **China**: Integration of AI and mechanization in smallholder farms to boost productivity and ensure food security, with a focus on precision agriculture and smart irrigation.
- **India**: Case studies from Punjab, Haryana, and Uttar Pradesh, where high levels of mechanization exist, contrasted with the northeastern states with low mechanization levels. Emphasis is placed on custom hiring centers and cooperative societies to address land fragmentation and labor shortages.

7.3.4.1 Data Analysis

The collected data is analyzed using statistical tools to assess the impact of mechanization and AI on agricultural productivity and sustainability. Predictive analytics models are employed to forecast the potential benefits of adopting climate-smart mechanization in different scenarios. The analysis also includes evaluating the cost-effectiveness and environmental impact of these technologies.

7.3.4.2 Development of Recommendations

Based on the analysis and expert feedback, the study develops a set of recommendations for integrating AI and climate-smart mechanization in agriculture. These recommendations focus on policy measures, investment strategies, and technological innovations tailored to the needs of different regions, particularly small-scale farmers in India.

7.3.4.3 Validation and Dissemination

The recommendations are validated through workshops and feedback sessions with stakeholders, including farmers, policymakers, and industry representatives. The final findings and recommendations are disseminated through academic publications, policy briefs, and industry conferences to ensure broad reach and impact.

7.4 Analysis

7.4.1 Climate-Smart Technology-Based Farm Mechanization

7.4.1.1 Mechanization in Seedbed Preparation

The first and most important step in achieving optimal crop growth, seedbed placement greatly affects yield potential. The first step in preparing a seedbed is tillage. To prevent weeds, prepare the soil for planting, and create an ideal root zone, farmers use a process known as tillage. For mouldboard and chisel plows, it decreases organic matter in the soil, adds air to the soil, and increases biological activity.

The conservation tillage (CT) approach is designed to reduce the intensity of plowing and retain or mulch crop waste, which helps to save natural resources, including soil, water, and energy. Using CT, the soil is disturbed very little throughout the planting, growing, and harvesting processes. Planting using crop residue keeps at least 30% of the soil surface covered [11].

The four main techniques of CT are zero-till, strip-till, ridge-till, and mulch-till. When compared to traditional methods, using mulch on top of the soil may reduce evapotranspiration by one million liters per hectare (100 mm) of water. CT practices, such as no-tillage, result in a greater increase in soil organic carbon (SOC) sequestration compared to conventional tillage. The use of zero tillage in rice–wheat cropping systems was found by Fang et al. to result in a reduction of 0.25 mg of carbon dioxide emissions, one million liters of irrigation water, and around 98 L of diesel. A rise in SOC near the surface might be offset by a fall in SOC at lower depths; hence, CT does not always enhance SOC storage [7]. According to Lenka and Lenka, no-tillage is an alternative to traditional seedbed preparation methods that result in a rise in soil carbon. The breakdown of organic materials is accelerated by over-tilling, which results in the release of carbon dioxide gas into the air [18].

Removing and combining weeds with soil, as well as mixing manure or fertilizer, are both made easier by using a rotary tiller. It saves around 30–35% of the time and 20–25% of the money [9]. The term "combined tillage" refers to the practice of using two or more tillage tools in tandem to improve soil quality while decreasing labor requirements and operating times in the field. Because fewer trips are needed, soil compaction, labor, and gasoline costs are reduced [19].

Additional production and decreased variable input costs are only two of the many advantages of a controlled wheeling system. Although it improves the soil's structure by around 15%, it decreases the compaction yield. In addition, it uses 35% less gasoline while operating. According to Reddy et al. [16], a 30% reduction in power consumption is possible as a result of structural improvements that result in less draft. For this reason, reducing tilling improves soil carbon sequestration.

We may improve the use efficiency of all inputs by reducing losses by reducing the number of passes of equipment in the field. This would lower the cost of fossil fuels and the accompanying carbon emissions to the atmosphere.

7.4.2 Mechanization in Sowing/Transplanting

The most important part of growing food is planting seeds or transplanting seedlings. Doing it within the allotted time frame will provide the best possible harvest. There are a lot of fields that need to be planted with crops simultaneously. Due to the labor crisis and human limits, it is not feasible to seed or transplant a certain crop within the time range using the manual approach. This may cause weather fluctuations and, if done manually, delay the procedure. Timely operation, accuracy, cost-effectiveness, resource and input efficiency, and resource conservation (including water and soil) are all possible outcomes of mechanization. Some climate-smart solutions for sowing/transplanting.

7.4.3 Mechanization in Weeding and Intercultural Operations

Weeding is an essential activity in agricultural crop supervision, as it aids in averting weed rivalry for resources like moisture, nourishment, and sunshine, ultimately influencing crop production and excellence. Numerous elements impact the magnitude of yield reduction, such as crop and weed efficiency, plant density, timing of weed appearance, weed duration, and crop-weed interactions. Research

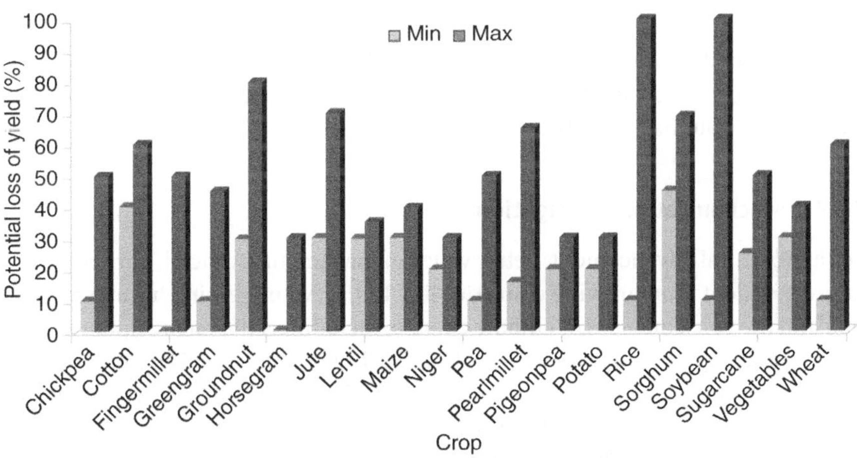

Figure 7.3 Potential loss of yield due to weeds in various major crops in India. *Source:* Reddy et al. [20]/(Rao and Chauhan, 2015).

has indicated that the competitiveness of weeds can result in noteworthy crop productivity reductions, varying from 10% to 100%.

Historically, farmers have employed mechanical weed eradication devices such as manual hoes, animal-drawn apparatuses, and tractor-drawn tillers to manage unwanted plants in agricultural furrows. Nevertheless, the accessibility of workforce and machinery frequently fails to coincide with the optimal period for removing unwanted plants, resulting in postponed and expensive weed management. To tackle these obstacles, scientists from across the globe have concentrated on mechanization in the process of removing weeds, creating smart weeders that utilize cutting-edge technologies like perception, automation, computer vision, and global positioning system (GPS) navigation systems. These innovations strive to diminish the laborious quality of removing weeds and enhance effectiveness [21].

Despite the challenges, automated herbicide management systems have a lot of potential for improvement. The topic of inter- and intra-line eradication for many agricultural plants needs more exploration. Mechanical weeding is a great alternative to chemical weed control, which is both expensive and harmful to the environment. Future weed control methods may advance by emphasizing mechanical and self-governing weeding principles, which are good for agriculture and the environment.

In addition, concerns about the potential damage to beneficial flora, wildlife, and the environment have been heightened by the increasing use of herbicides for the control of undesirable plants. There has to be a change in crop protection strategies due to the rising cost of weed herbicides and the risks they pose to the environment. Reduced chemical use means more money in your pocket and less harm to the environment thanks to better methods and innovations like mechanical weed control.

In a nutshell, weed management is a problem for farmers, but technology is making things better, and less reliance on chemicals means that we can grow better crops without harming the environment.

7.4.4 Mechanization in Irrigation

For agricultural advancement, better water management is crucial, particularly in rain-dependent crop-growing countries like India. An increasing human population and the consequences of climate change, including shifts in precipitation patterns and the loss of polar ice caps, both contribute to the already severe problem of water shortage. The agricultural sector of India is almost totally reliant on precipitation (around 65%), and there is a great deal of geographical variation in it.

Water efficiency is of the utmost importance in farming in order to reduce water waste. Nanotechnology and automated farming systems are two examples of new approaches that promise to cut down on water wastage. These approaches are being used by micro-scale farmers as a means to enhance agricultural productivity in areas with restricted water supply. Worldwide, regional governments are promoting these frameworks to increase the availability of micro-irrigation goods and services, as well as to lower their prices [5].

Irrigation mechanisms on agricultural lands are commonly categorized into groups like sprinkler mechanisms, superficial mechanisms, and trickle mechanisms. Sprinkler mechanisms distribute water in the shape of mists utilizing the above sprinklers. Subterranean trickle irrigation is also acquiring momentum for enhanced aqua administration, flaunting elevated utilization proportions and effectiveness levels ranging from 85% to 90%.

It may be difficult to measure and effectively control soil moisture despite its central function in irrigation. This issue may be resolved by using a mechanical watering system that incorporates a ground moisture sensor. These sensors monitor the soil moisture levels and activate the irrigation system when the levels drop below the optimal range for strong plant growth. You have the option to program the system to turn off automatically when a certain amount of time has passed or when the moisture level is restored. These systems are reliable and cost-effective since microprocessors are their brains.

In comparison to more conventional methods of watering plants, nano irrigation systems have shown a considerable improvement in water use efficiency (WUE). Water savings of 15–20% are possible with real-time irrigation scheduling that takes soil moisture data into account. In addition, moving toward nano-irrigation may improve WUE and harvest yields with less water used overall. It is equally important to recognize the diversity of disciplines. Irrigation water would normally be lost owing to uneven terrain, however laser terrain levelers may save around 20–25% of that water. In order to significantly reduce water waste, tensiometers are invaluable tools for predicting irrigation needs according to soil moisture and crop water demand. Also, with the help of system regulators, humidity detectors, thermometers, and up-to-the-minute chronometers, automated irrigation systems may improve water distribution for certain crops like cucumbers, leading to higher yields. Another factor in water conservation is crop choice; certain crops use less water than others. Particularly in regions facing water scarcity and climate change, sustainable agriculture requires efficient water management strategies, such as the use of nano-irrigation systems, automated irrigation, and careful agricultural practices [6].

Arid region cultivation plays a noteworthy role in the expansion of Indian farming. Therefore, the tractor-propelled former Phule inspection reservoir is advantageous for arid land agriculture in Maharashtra for on-site precipitation collection. It generates 6 m × 2 m dimensions inspection reservoirs and enables 30–40% punctuality and conservation of up to 50–60%. This precipitation preservation amplifies the yield of Rabi crops in the locality [22]. The agriculturists face challenges in attaining punctuality in different unit activities of peanut to capitalize on the advantageous circumstances in arid regions of Andhra Pradesh. Bollavathi et al. [23] devised an aqueous cultivator to address this predicament. Aqua sowing is simultaneous seed sowing and the provision of an equivalent amount of water, sufficiently ample for the seeds to sprout, irrespective of whether it rains or not during the initial few days. This is a contemporary technology crafted for the benefit of arid region cultivators and specifically for seed planting during the latter monsoon emergency period. The Pusa aqua-ferti seed drill is utilized for applying watery fertilizer alongside the seed, which has a tendency to enhance germination and initial crop yield, especially under rain-fed circumstances. It enhances sprouting and wheat harvest output by 53% and 35%, correspondingly, and possesses a field capability of 0.25 ha/h [13].

7.4.5 Mechanization in Nutrient Management

Live, noninvasive, rapid crop development information gathering is crucial for farming production supervision. Conventional identification methods result in harm to plants. The foliage hue diagram (FHD) is a simple method to identify crop nitrogen (N) deficiency, unlike indoor physiological and chemical assessment, which is laborious and sluggish. It fits rice, barley, and corn. Every domain's most recent mature foliage (second from the apex) of 10 arbitrarily chosen ailment-free organisms/mounds corresponds to the LCC hue band [1]. Spectroscopes have the capability to analyze 32% of wheat using the tractor-mounted N-detector. We created branching and sprouting algorithms to approximate wheat crop N-application rate.

For plots with different N levels, the algorithm and sensor value approximates the N-application rate similarly [24]. N evaluation and control are illustrated in Figure 7.4. Rasyid et al. [24] examined the portable chlorophyll content gauge (CCG) and discovered it valuable for assessing comparative chlorophyll content in chosen botanical varieties. Visual sensors rapidly and harmlessly evaluate cotton nitrogen status and plant mass during the growing season. Eco-friendly explorer in motion adaptable rate urea application machinery affixed to the operator's rear can top dress. It assessed nitrogen from 8.5 to 30 kg/ha N utilizing a grooved roller at 2.0 km/h. An accurate cartographic system might utilize a GPS-enabled variable-rate technology (VRT) fertilizer dispenser with a microchip, mechanism, and seed cum fertilizer bore for granulated fertilizer implementation. This diminishes fine fertilizer inconsistency by 15% and conserves 13–15% in comparison to traditional techniques [15]. These systems offer precise input by preserving input – input utilization effectiveness.

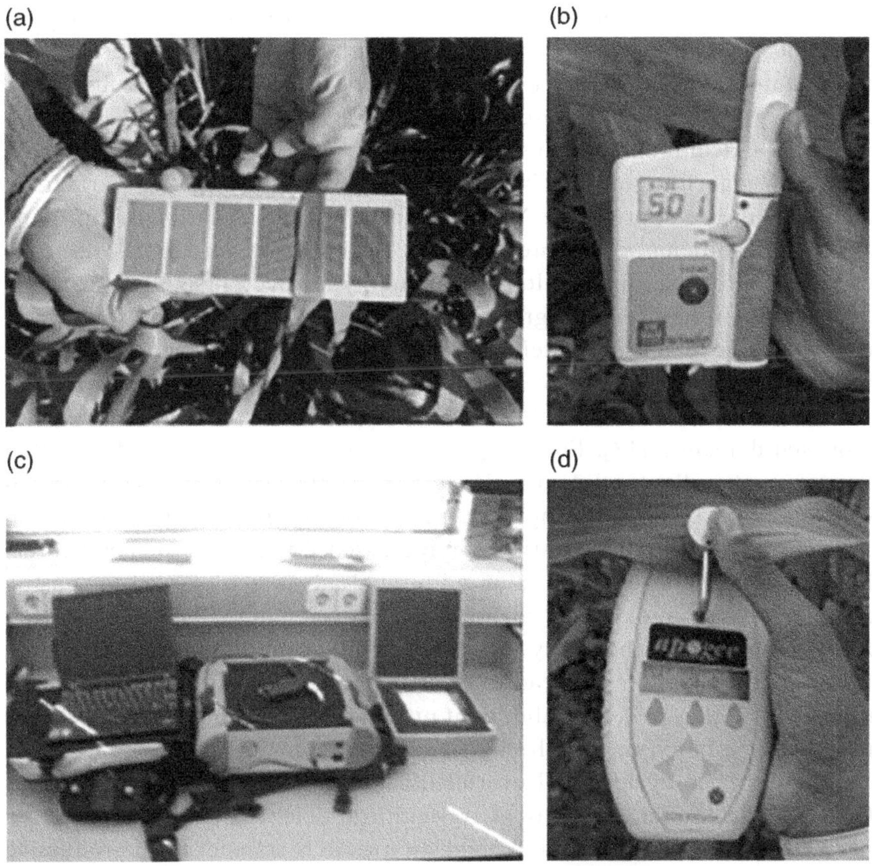

Figure 7.4 Measurement and management of N (a) LCC, (b) Spad meter and N-management device, (c) Spectroradiometer, and (d) CCI Meter. *Source:* Rao et al. 2020/ ICAR-National Academy of Agricultural Research Management/Public Domain.

N administration utilizing airborne and orbital imagery is effective in "fertilization. Multispectral, hyperspectral, and thermal aerial visuals from unmanned aerial vehicle (UAV) flights may detect N crop requirements. Various remote sensing imaging methods are employed in fertilization. RGB/CIR cameras, which incorporate infrared (CIR), crimson, emerald, and azure illumination (visible or RGB), evaluate verdant biomass and N status from NDVI. Multispectral cameras in the visible near-infrared regions may detect multiple spectral bands simultaneously and are employed to assess verdant biomass, fecundity, chromatic decay, and photosynthetic efficacy. To evaluate crop nutrient strain, infrared or thermal imaging cameras are beneficial.

Numerous establishments offer farmers aerial remote sensing facilities employing multispectral, hyperspectral, or thermal airborne imaging to identify crop nutrient deficiency in wheat, rice, cotton, horticulture, and other cultivations. Aerial mesh images are employed to generate nitrogen deposition blueprints for enhanced agricultural N administration."

These N algorithm models, built using N-sensor data, might aid farmers in determining the appropriate N dosage for crops and locations, improving agricultural efficiency and preventing ecological contamination caused by excessive N fertilization. Multispectral and hyperspectral satellite scanning additionally influences crop growth. Fertilization regulation and soil examination utilize extensive satellite imagery (occasionally complimentary). Satellite information for N field administration can be employed on catchments, landscapes, and fields, as opposed to portable or vehicle-mounted detectors [13]. Emerging affordable or widely accessible satellite systems such as Sentinel-2 with elevated temporal resolution and additional spectral bands could assess agricultural resources, enhancing nitrogen production and nutrient utilization for increased efficacy in food networks. Environmental conservation is an additional advantage of remote sensing and ground sensing systems for nitrogen management. Extended nitrogen fertilization harms the ecosystem by amassing contaminants in the soil. Excessive fertilization can lead to ecological harm and substantial costs in nutrient leaching [25]. Overabundant nitrogen fertilization of crops deteriorates water quality, one of the largest global ecological concerns. Nitrates originating from agricultural overfertilization harm groundwater and surface water. National Academy of Agricultural Sciences (NAASs) [18] connects freshwater eutrophication and estuaries to N transfer from productive fields to surface waters.

N damages from excessive fertilization rates impact water quality and greenhouse gas (GHG) levels. Surplus nitrogen fertilizer might be drained into the soil, surface watercourses, or the atmosphere as vapors, resulting in noteworthy ecosystem discharges. Harmonizing N formulation and crop standards lessen water nitrate contamination and gaseous discharges from inaccurate fertilization. All of these ecological concerns and inadequate nitrogen fertilizer require groundbreaking technology to enhance resource administration. Satellite imaging information to oversee nitrogen fertilizer amount and scheduling will protect the ecosystem and enhance crop administration for eco-friendly agriculture. Sustainable manuring in fertilizer utilization is environmentally conscious. Verdant manuring enhances soil vitality and diminishes chemical fertilizer expenses while safeguarding production. Intelligent mechanization and eco-friendly cover cropping render tractor-driven biomass integrators environmentally viable [12].

7.4.6 Mechanization in Plant Protection

The intricate interactions and climate fluctuations significantly impact agriculture by influencing the diversity and population of pests. These climate-related changes have led to increased pesticide usage, including higher quantities, frequencies, and various combinations of chemicals during the cultivation season. Technological advancements, such as location-specific pest management using sensors and GPS-equipped field sprayers, have revolutionized pest control in agriculture. Precision agriculture, with its targeted pesticide applications, has proven more efficient in managing pests compared to traditional blanket pesticide application methods. Innovative equipment like self-rotating model firearm sprayers and versatile elevated sprayers have further improved spraying efficiency, reducing costs, labor, and time requirements in crop protection [10].

Electrostatic spraying of pesticides or plant nutrients improves their efficacy and spatial dispersion throughout vegetation cover, especially under foliage where pests hide. For electrifying spray particles in the nozzle, a backpack-style pneumatic-assisted electrostatic sprayer with a compressor and spray gun was constructed. The outermost nozzle receives gases and fluids separately. Electrostatic sprayers deposit 80% and 85% more leaf top and bottom spray than knapsack sprayers. Electrostatic sprayers reduce erosion by 50% less than knapsack sprayers. Electrostatic sprayers kill 68% more insects than hand-operated knapsack sprayers [26]. Misting add-ons for UAVs are being developed for plant protection in varied cultivations [27]. Automated division regulation technology is auto-swath precision farming technology. Boom divisions automatically deactivate when they pass through processed areas, waterways, or uncultivated areas, reducing agricultural material use.

UAVs weigh under 20 kg without people. Technology is slowly taking over farming. A drone-based agricultural misting system was developed at PAU to improve crop spray deposition (Figure 7.5). Spray deposition bioeffectiveness was studied in cotton, rice, and moong crops. UAVs flying 0.5 m above vegetation at 2.0 m/s forward velocity provide the best spray dispersal results [14]. Farmers can ensure the efficacy of agricultural inputs like moisture or chemical fertilizers using this technique, improving productivity and product quality. Due to spray dispersion, drone-powered pesticide spraying on small Indian landholdings might harm nearby agricultural ponds, wells, and animal sanctuaries. Indian farmlands are often near rural communities, making UAV pesticide use dangerous if not properly done. Spray dispersion will carry sensitive, dangerous compounds beyond the implementation area, requiring dispersion management systems [8].

Figure 7.5 Drone spraying for the field crop. *Source:* Rao et al. 2020/ ICAR-National Academy of Agricultural Research Management/ Public Domain.

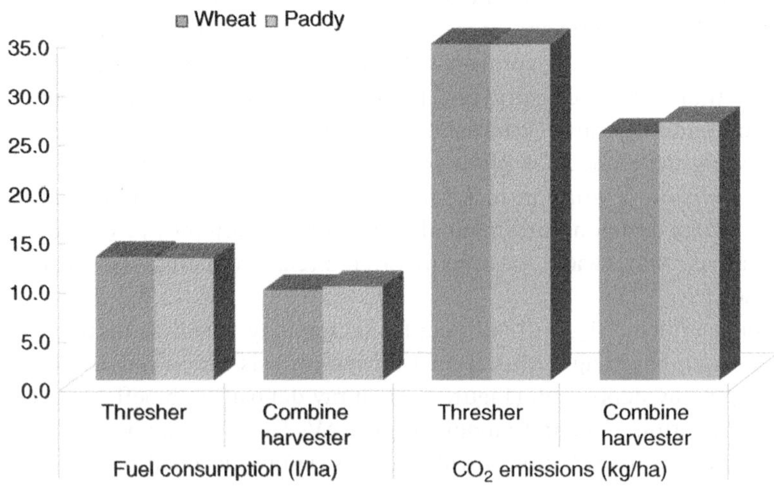

Figure 7.6 Fuel consumption and CO_2 emissions from thresher and combine harvesters.

7.4.7 Mechanization in Harvesting and Threshing

The degree of mechanization in India's agricultural sector differs considerably across crops. It was believed that between 60% and 70% of the harvesting and threshing of wheat and rice was done by machines. The use of equipment has greatly enhanced efficiency and production in the agricultural industry since it

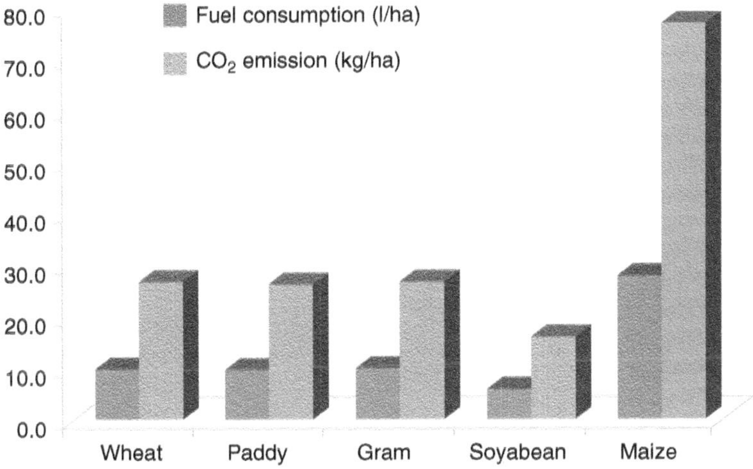

Figure 7.7 Emission footprints during combined harvesting (for Kartar 4000).

Figure 7.8 Burning of paddy crop residue and emission of GHGs. *Source:* Rao et al. 2020/ ICAR-National Academy of Agricultural Research Management/Public Domain.

now assists with a large amount of these activities. When it comes to other types of crops, however, things are quite different. According to the research, the amount of mechanization for these crops is far lower, at less than 30%. This points to the continued need for physical labor in many agricultural processes, which might lead to decreased productivity and higher labor demands. These results illuminate the existing level of mechanization in India's agriculture industry and draw attention to the need for more innovations and investments in equipment for a broader array of crops. Through the enhancement of mechanical paddy, or

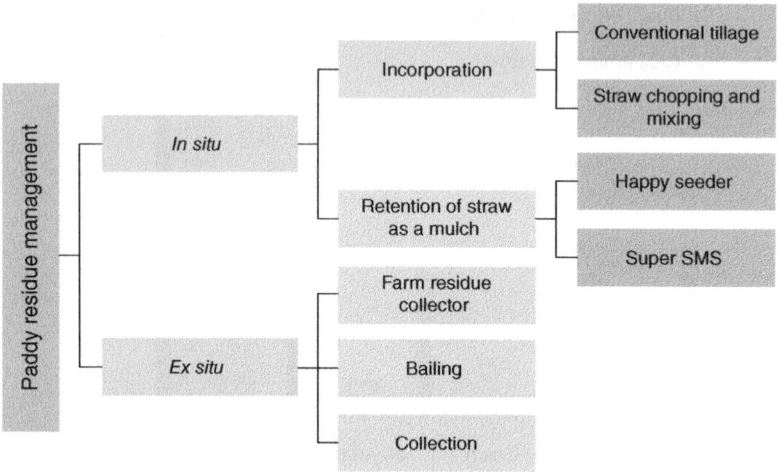

Figure 7.9 Paddy residue management techniques.

Figure 7.10 Happy seeder with press wheel operating in the combined harvested field. *Source:* Rao et al. 2020/ICAR-National Academy of Agricultural Research Management/ Public Domain.

rice produced in flooded fields, wheat, a staple grain eaten all over the globe, are the primary crops planted in this area. There would be no local economy or food supply without these two crops. It is crucial to assess the harvesting process's carbon footprint when thinking about the environmental effects of agricultural methods. When compared to other agricultural crops, this one leaves a somewhat large carbon footprint, so keep that in mind.

The fuel usage of self-propelled combine harvesters like Kartar 4000 varies depending on the harvested crops and emission coefficients. Combines are widely used, with around 5000 transactions per year. Threshers, on the other hand, are employed in undulating terrains or by marginal farmers who can't afford combines. Threshers for wheat, rice, and various crops are available, causing minimal grain damage.

Multi-harvest threshers handle wheat, sorghum, barley, soybean, millets, oats, grams, and legumes. Recent design changes have improved ergonomic features and straw output but led to slight carbon emissions increases. Combines consume less fuel than manual/reaper harvesting, followed by threshing in threshers.

For example, the Dasmesh MultiCrop 22×36 thresher consumes 12.5 L of fuel per hectare, while the Kartar 4000 combine can cover 1.10 hectares per hour for wheat and 0.92 hectares per hour for paddy, using 9.09 and 9.56 L of fuel per hectare, respectively. Corn crops show higher emissions compared to others.

In addition to petroleum fuel, sustainable biofuels from agricultural residue can be used in engines without compromising agricultural land, reducing emissions and aiding climate adaptability.

7.5 Future Mechanization Pathways Through Climate-Smart Technologies

Innovation plays a pivotal work in present-day agriculture, offering innovative methods for farm mechanization to enhance productivity. Leveraging information innovation within intelligent machinery can significantly enhance energy yields more capably than at any time in ongoing memory. While accuracy agriculture has demonstrated its capability, this present time is the ideal opportunity to transition to the following era of machinery and automation. Although current human operations are useful across large fields, there's as yet untapped potential for expanding autonomous framework administrations to additionally foster efficiencies additionally. Therefore, it's crucial to adopt practical accuracy methods in agricultural hardware, alongside various measures, to enhance farming viability, especially on limited cultivated land under evolving weather circumstances and diminishing ecological assets expected to take care of a growing population. This involves taking brief actions to help agro-specialist organizations through farmers' cooperatives, custom hiring centers, and machinery archives, ensuring that even small and marginal farmers can benefit from farm mechanization [28]. Innovations are expected in various aspects of agriculture, for example, improving wrinkle openers in sowing apparatus to diminish seed rate variations, mechanizing the planting of nursery mat sorts in fields for rice transplantation, and anticipating the transplantation of vegetables and horticultural produce. Additionally, enhancements in gear like VRT for flotation, ramble configuration, sprayer hardware, and computational control can be tailored to local circumstances and vegetation. Sugarcane, among all agricultural yields, requires substantial energy inputs and should benefit from mechanization in sowing, weed control, and harvesting through accurate agriculture tailored to regional attributes. The implementation of autonomous AI frameworks enhances the ability to cultivate another range of intelligent agricultural hardware capable of performing tasks with the most outrageous accuracy [32, 33]. These advances should be adaptable to the intricacy of real-world circumstances, ecological factors, and climate intelligence in agricultural creation scenarios.

7.5.1 Mechatronics

A vital part of contemporary farming is mechatronics, which combines mechanical engineering with electrical engineering, hardware, computation, and control. It has several reasons and can chip away at many parts of agricultural frameworks. Activities including planting, reaping, circulating compost, and checking plants depend upon parts like sensors and actuators. Concerning specialized necessities, mechatronics is continuously adjusting to help make cultivating more competent and naturally pleasing. The mixing of mechatronics and the IoT is growing, designing perspectives past the ordinary agrarian region. It is significant for computerized establishing systems, pesticide dosing, and cultivating cycles. Whether you're working in a no-till or revolving culturing environment, mechatronics courses of action can make seed drills and growers significantly more precise and suitable [25], which thus expands efficiency and yield rates. Beepers and divine pilots, both obliged by GPS, are likewise assisting farm vehicle drivers with augmenting efficiency in the field. These plans give a climate shrewd way to deal with overseeing climatic difficulties by decreasing the risk of regions being covered or missed. Further exactness and maintainability in horticultural systems are being presented by the continuous progression of farming by means of the joining of mechatronics and novel development.

7.5.2 Robotics

Drones have, as of late, seen dangerous expansion in the agricultural area, simplifying and removing obstacles to their usage. Researchers, for example,

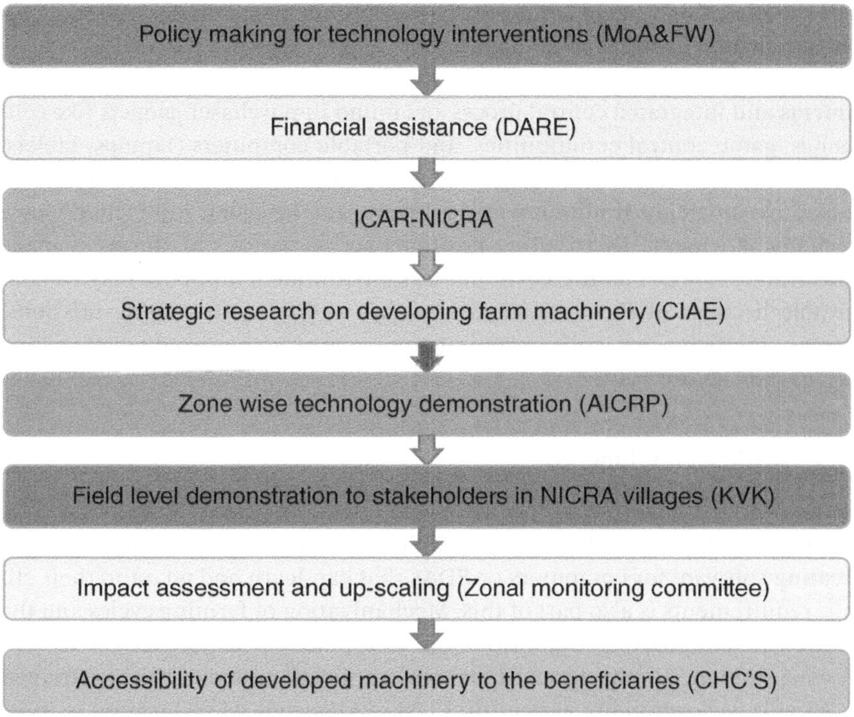

Figure 7.11 Flow chart of an action plan for climate-smart mechanization.

Prakash Arjun et al. [29] have investigated the utilization of robots in agriculture, highlighting the advantages and drawbacks of this innovation, particularly when used on a major scale. Drones have enabled a plethora of technical enhancements in agriculture, for example, crop mapping, soil sampling, and the utilization of automated hardware for planting and harvesting. In addition, research, for example, that of Raghavendra et al. [30] has illuminated the potential purposes and patterns of automated agricultural machinery later on. The aftereffects of this study demonstrate the increasing fascination with autonomous or semi-autonomous gadgets that may ease human effort in farming operations like planting, plowing, and harvesting.

Drones and other airborne robots have certain operational limitations, for example, how well they work in shady circumstances, how long it could take to transfer data, what kinds of grants are required, and which applications have the most to gain from them, all of which are reliant upon acceptance and regulation. On the other hand, farms may make great utilization of UAVs, which can transport several sensors for various tasks. They give an affordable decision that doesn't need exorbitantly stringent approvals. These tiny, integrated drones utilize of current automation by using smart, lightweight, and inexpensive parts [17].

Present-day automation has the advantage of using inexpensive, brilliant cameras and integrated central processors found in purchaser gadgets like cellphones, game control communities, and portable computers (laptops, tablets, and so on). Diminish soil degradation and potential injury to current yields caused by unwieldy traditional agricultural gear by using joint efforts in a mechanical armada. Distributing positions across various platforms is made conceivable. Lower manure costs and less environmental impact may be conceivable because of the automated armada's ability to streamline missions, decrease inefficiencies, and concentrate on areas of higher need by leveraging various data sources [14].

7.5.3 Artificial Intelligence

The goal of artificial intelligence (AI) is to create intelligent gadgets that can impersonate human intellect by mimicking human ideas and behavior. Creating software for computers or PDAs that can learn and adapt to their clients' requirements is also part of this. Mechanization of farming cycles and the creation of autonomous algorithms are two potential consequences of AI's growing influence in the agricultural area. Applications and cycles constrained by AI will perform many agricultural tasks. Although AI in farming is in its infancy, with additional new development and funding, all of these agricultural tasks will be mechanized, leading to more compelling utilization of assets and lower yield costs. It is critical to think about small-scale and unobtrusive farmers in solicitation to combat agricultural outcome misfortunes and encourage arrangements that can withstand climate change [23]. In addition, specialized business agencies could give their assistance to eco-accommodating agricultural machinery. The capable functioning of agricultural hardware is reliant upon the operators' level of expertise. Regardless of whether farming is underway, weather and environmental factors have a major influence [19]. One new instrument for dealing with climate-related catastrophes is the ascent of autonomous or unmanned agricultural tractors. Gear intended for farming may eliminate weeds, plant seeds, cultivate soil, spray pesticides, and run these tractors autonomously [2]. The utilization of machine learning has allowed some prominent tractor manufacturers to create totally autonomous tractors. The level of farm tractor mechanization in India may be redefined and tractor intelligence in agricultural operations can be elevated via this intervention. Work vehicles with man-made reasoning (simulated intelligence) abilities incorporate capacities including independent guiding, route inside foreordained limits, control of joined gear, evasion of deterrents, and security highlights like geo-fencing and remote beginning/stop. The necessities of an extending populace might be met with the help of man-made intelligence-driven work vehicle advancement, which offers additionally created ampleness, more pay, less well-being risks for ranchers, and more food yield [3].

7.6 Discussion

The integration of AI and climate-smart agricultural mechanization represents a transformative shift in how agriculture can adapt to and mitigate the impacts of climate change. By leveraging advanced technologies such as AI-driven precision farming, UAVs, and smart irrigation systems, farmers can enhance productivity while ensuring sustainability. This integration not only optimizes resource use but also minimizes environmental damage, thus contributing to more resilient agricultural practices [1]. One significant challenge highlighted in the literature is the disparity in mechanization levels across different regions. While countries like the United States and Western Europe have achieved high levels of mechanization, India lags with only about 40% mechanization [3]. This gap underscores the urgent need for tailored strategies that address specific local challenges, such as land fragmentation and labor shortages, particularly in regions with small and marginal farms [4]. For instance, the rise of custom hiring centers and cooperative societies in India could offer practical solutions to these issues by making advanced machinery more accessible to small-scale farmers [21].

The role of AI in precision agriculture is pivotal [31]. AI can provide predictive analytics for seedbed preparation, sowing, and nutrient management, leading to improved crop yields and reduced input costs [3]. The application of AI in agriculture is not just about replacing manual labor with machines but also about making informed decisions that enhance the overall efficiency of farming operations. For example, AI-powered tools can analyze soil health and predict weather patterns, allowing farmers to optimize their planting schedules and irrigation practices. This level of precision is crucial for maximizing productivity and sustainability in the face of unpredictable climate conditions [10]. Moreover, the use of UAVs and other robotics in agriculture has shown promising results. These technologies facilitate the precise application of fertilizers and pesticides, which not only improves crop health but also reduces environmental pollution [14]. UAVs can cover large areas quickly and efficiently, making them ideal for monitoring crop health and detecting issues such as pest infestations early. This early detection can significantly reduce crop losses and increase overall farm productivity [17]. Additionally, the integration of mechatronics and IoT in agricultural equipment further enhances the efficiency and effectiveness of farming operations [25].

Water management is another critical area where climate-smart mechanization can make a significant impact. The adoption of smart irrigation systems and nano irrigation technologies can lead to substantial water savings and improve WUE [22]. These technologies are particularly beneficial in regions facing water scarcity. For instance, automated irrigation systems that use soil moisture sensors can ensure that crops receive the optimal amount of water, reducing waste and improving yields. This approach not only conserves water but also supports sustainable agricultural practices by maintaining soil health and preventing over-irrigation [10]. Despite the advancements and benefits, there are still gaps in the widespread adoption of these technologies. One major barrier is the high initial cost and the lack of awareness and training among farmers, especially in developing countries [9]. Governments and policymakers need to play a crucial role in bridging this gap by providing subsidies, training programs, and supportive policies that encourage the adoption of AI and mechanization in agriculture. Initiatives like the SMAM by the Indian government are steps in the right direction, but continuous efforts are needed to ensure long-term impact and sustainability [11].

7.7 Conclusion

Environment-wise automation is radically changing the agribusiness business. To battle the effects of environmental change, industrialized countries are embracing robotization and naturally pleasing horticultural stuff. As well as eliminating transmissions of ozone-hurting substances, these advances help with carbon capacity in the soil, resource preservation, and working on farming activities' general sufficiency. Precision, ampleness, and resource improvement are the benefits that ranchers get from computerizing planting and relocating systems. Further examination on bury/intraline destruction for fluctuated crops and computerized herbicide control systems is additionally conceivable. The usage of sensors and robotized structures in water systems is one illustration of how contemporary development is improving water the board, decreasing waste, and ensuring economical agrarian practices. Also, crop-reaping-related carbon radiations might be diminished by the administration of additional rural items like the use of Straw Control Structures. In requesting to battle environmental change, it is basic to expand comprehension that ranchers might interpret the significance of present-day rural stuff and advancement and to furnish them with the resources they need to complete these developments.

References

1 Ali, M. (2024). Advances in climate-smart mechanization for sustainable agriculture. *Journal of Agricultural Science and Technology* 23 (1): 45–60. https://doi.org/10.1016/j.jast.2024.01.005.

2 Grant Thornton (2017). Mechanization key to higher productivity to double farmers' income. Grant Thornton India LLP.

3 Fang, H. (2023). Impact of AI on precision agriculture: a global perspective. *Computers and Electronics in Agriculture* 198: 105001. https://doi.org/10.1016/j.compag.2023.105001.

4 Department of Agriculture, Cooperation & Farmers Welfare (DACFW) (2019). Agricultural statistics at a glance 2018, 218. Directorate of Economics and Statistics, Ministry of Agriculture & Farmers Welfare, Government of India.

5 Kumar, A., Mamgai, R., and Jain, R. (2023). Application of loT-enabled CNN for natural language processing. In: *In lo T-enabled Convolutional Neural Networks: Techniques and Applications*, 149–177. River Publishers.

6 Kumar, A., Walia, G.S., and Sharma, K. (2020). Recent trends in multicue based visual tracking: a review. *Expert Systems with Applications* 162: 113711.

7 Grant Thornton (2022). Mechanization key to higher productivity to double farmers' income. Grant Thornton India LLP. https://www.grantthornton.in/en/news-centre/mechanization-key-to-higher-productivity/ (accessed May 2024).

8 Kumar, A. and Jha, R. (2022). IoT in smart farming: a boon to farmers. In: *Internet of Everything: Smart Sensing Technologies* (ed. T. Kavitha, V.A. Devi, S.N. Pari, and S. Ramanathan). Nova Publisher.

9 Bhambota, S., Dixit, A.K., Manes, G.S. et al. (2023). Field evaluation of automated vegetable transplanters. *Indian Journal of Agricultural Sciences* 89 (2): 112–124. https://doi.org/10.1007/s00484-023-00968-2.

10 Gautam, P.V., Kushwaha, H.L., Kumar, A., and Kushwaha, D.K. (2022). Mechatronics application in precision sowing: a review. *International Journal of Current Microbiology and Applied Sciences* 11 (5): 1793–1807. https://doi.org/10.20546/ijcmas.2022.1105.207.

11 Government of India (2022). Sub-mission on agricultural mechanization (SMAM) annual report. Ministry of Agriculture & Farmers Welfare. https://agricoop.nic.in/schemes/sub-mission-on-agricultural-mechanization.

12 National Academy of Agricultural Sciences (NAAS) (2021). Policy paper no. 84: practical and affordable approaches for precision in farm equipment and machinery. New Delhi. https://naas.org.in/document/policy-paper-no-84 (accessed May 2024).

13 Onwude, D.I., Abdulstter, R., Gomes, C., and Hashim, N. (2016). Mechanisation of large-scale agricultural fields in developing countries – a review. *Journal of the Science of Food and Agriculture* 96: 3969–3976.

14 Parmar, R.P. (2020). Development of an agricultural spraying system for unmanned aerial vehicles. Doctoral dissertation, Punjab Agricultural University, Ludhiana. https://www.pau.edu/dissertation/agri_spraying_UAV.pdf.

15 Press Information Bureau (PIB) (2020). Union ministry for agriculture & farmers welfare adopts several steps to tackle stubble burning. https://pib.gov.in/PressReleseDetailm.aspx?PRID=1590402 (accessed May 2024).

16 Reddy, R.G., Sudhakar, N., Srinivasarao, C.H. et al. (2021). NICRA-compendium of climate resilient technologies. ICAR ATARI Hyderabad, 11 Dec 2015, Malaysia Pahang. https://icar.nic.in/compendium/nicra.

17 Sidhu, H.S., Singh, M., Singh, Y. et al. (2021). Development and evaluation of the turbo happy seeder for sowing wheat into heavy rice residues in NW India. *Field Crops Research* 260: 107981. https://doi.org/10.1016/j.fcr.2021.107981.

18 Central Institute of Agricultural Engineering (CIAE) (2016). Annual Report 2015–16.

19 Kumar, A. (2023). *Visual Object Tracking Using Deep Learning*. CRC Press.

20 Reddy, R.G., Sudhakar, N., Srinivasarao, C.H. et al. (2015). NICRA-compendium of climate resilient technologies-ICAR ATARI Hyderabad.

21 Kumar, A., Jain, R., Vairamani, A.D., and Nayyar, A. (2023). *Object Tracking Technology: Trends, Challenges and Applications*. Springer Nature.

22 Prem, M., Swarnkar, R., Kantilal, V.D.K. et al. (2016). Combined tillage tools – a review. *Current Agriculture Research Journal* 4: 179–185.

23 Bollavathi, L., Joseph Reddy, S., Madhusudhana Reddy, K., and Veera Prasad, G. (2018). Performance evaluation of aqua planter for groundnut crop. *International Journal of Current Microbiology and Applied Sciences* 7: 810–818.

24 Rasyid, A.M., Shahidan, N., Omar, M.O. et al. (2015). Design and development of irrigation system for planting – part 1. *2nd Integrated Design Project Conference (IDPC)*, pp. 1–22.

25 Namdev, S.K., Pateriya, R., Modi, R.U., and Dash, B.S. (2019). Estimation of operational cost of various intercultural implements. *Pantnagar Journal of Research* 17: 250–255.

26 Onwude, D.I., Abdulstter, R., Gomes, C., and Hashim, N. (2021). Mechanisation of large-scale agricultural fields in developing countries – a review. *Journal of the Science of Food and Agriculture* 101 (15): 3969–3976. https://doi.org/10.1002/jsfa.11023.

27 Dashmesh (2020). Dashmesh multicrop 22-36 thresher model. http://dashmeshagro.com/multi-crop22x36.html (accessed May 2024).

28 Department of Agriculture, Cooperation & Farmers Welfare (DACFW) (2018). Final report on monitoring, concurrent evaluation, and impact assessment of

Sub-Mission on Agricultural Mechanization, 5–36. Mechanization & Technology Division, Ministry of Agriculture & Farmers Welfare, Government of India.

29 Prakash Arjun, K.V., Sajeena, S., and Lakshminarayana, S.V. (2017). Field level investigation of automated drip irrigation system. *International Journal of Current Microbiology and Applied Sciences* 6: 1888–1898.

30 Raghavendra, R., Veerangouda, M., Prakash, K.V. et al. (2013). Development and evaluation of ridge planter for cotton. *Karnataka Journal of Agricultural Sciences* 26: 88–91.

31 Singh, D. and Kumar, A. (2024). A deep recurrent neural network for plant disease classification. *SN Computer Science* 5 (8): 1053.

32 Vohra, R., Kumar, A., and Rongali, G. (2024). Multi-scale extraction and spatial analysis of growth pattern changes in urban water bodies using sentinel-2 MSI imagery: a study in the central part of India. *Stochastic Environmental Research and Risk Assessment* 38 (6): 2397–2423.

33 Vohra, R. and Kumar, A. (Eds.). (2024). Advanced Geospatial Practices in Natural Environment Resource Management. IGI Global.

8

Harvesting Tomorrow: Exploring Real-World Applications of AI in Agriculture

Priya and Neha Gupta

School of Computer Applications, Manav Rachna International Institute of Research & Studies, Faridabad, Haryana, India

8.1 Introduction

Agriculture has evolved for centuries; this process continues as humanity looks for ways of producing more food for the increasing population while tapping into such challenges as climatic change, scarcity of resources, and food insecurity. Similarly, within the past few years, AI has become an innovative and revolutionary technology in agriculture, offering an excellent opportunity to upturn the whole agricultural farming system in terms of productivity, profitability, innovation, and sustainability in the agricultural production system [1]. Computer and Internet technology have made it easy to learn about new careers, and this has also affected agriculture. Food production is under pressure to provide enough supply for today's increasing global population, which calls for effective strategies that would enhance productivity and ecological farming. Analysis using AI, which has the features of analyzing large volumes of data and making rational decisions, has great potential in this industry. After practicing agriculture for centuries, along with relying on traditional techniques and manual labor, a lot of change is observed. New ICT being developed in agriculture is being applied to address issues such as climate risks, natural resource dangers, and unwanted crops. These technologies have multiple opportunities to apply in intelligent production and various sectors such as farming and prediction, precise manufacturing, robots, and others [2]. Also, drone techniques under AI and self-driving tractors are coming into farming techniques as well. These machines are capable

Emerging Smart Agricultural Practices Using Artificial Intelligence, First Edition.
Edited by Ashish Kumar, Jai Prakash Verma, and Rachna Jain.

Published 2025 by John Wiley & Sons, Inc.

of sowing seeds, applying insecticides, and monitoring the crops' well-being for a given period with minimal human assistance and little or no likelihood of wrongdoing. Not only that, electronic systems, supported by AI applications, can also identify soil quality for the appropriate application of fertilizers for better nutrient distribution and improved soil structure and fertility.

AI implementation in the agricultural sector also has the potential to improve farming output and profitability, in addition to having a positive impact on the sustainability of farming. Advancements in technology and the application of artificial intelligence (AI) in the farming sector are depicted in the Figure 8.1 below. In order to explore the different areas of AI in agriculture and learn about their real-world existential issues, problems, and the future prospects of their evolution, this chapter goes deeper into covering the following subtopics:

AI technology adoption in agriculture is deemed necessary due to the need to minimize losses and maximize profits, as well as the requirement for effective resource management, crop monitoring, and predictive analytics application [3]. Modern farming practices often rely on manpower and simple human instincts, which may not suffice when the world is going through drastic changes in the environment and markets. Through the application of robotic process automation, data science, and AI to improve agricultural methods, AI has been demonstrated

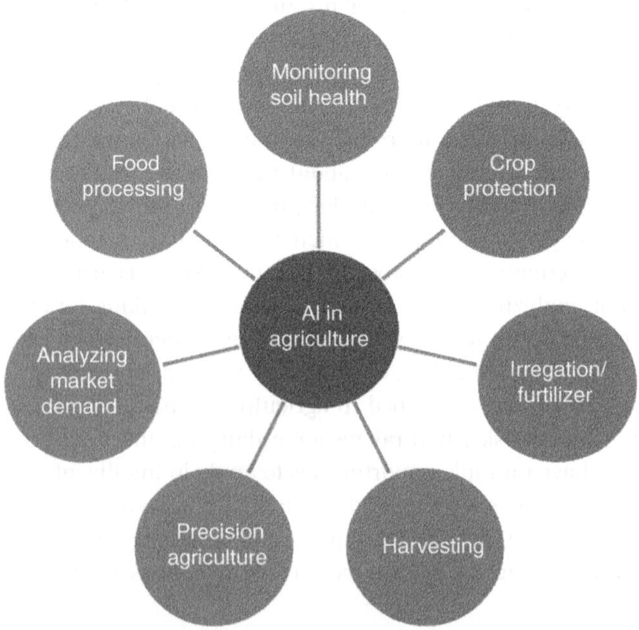

Figure 8.1 Applications of AI in agriculture.

to offer a likely solution to the issue. As a result, the goal of this chapter is to provide a comprehensive overview of the application of AI in agriculture, from the farm to the farmer's plate. It seeks to ascertain how advancements in supply chains, pest management, agricultural and animal production, and predictive modeling are made possible by AI technologies such as smart sensors, robots, and remote sensing. The chapter also compares how AI applications affect the bottom line of farms, the environment, and the social situation in developed and developing agrifood systems. The chapter is organized to provide a systematic analysis of various AI applications in agriculture, as follows:

- **Precision Farming**: Examines how the concepts of AI contribute to modern precision agriculture, including accurate planting, water management, and the examination of soil PH levels [4].
- **Crop Monitoring and Management**: Exploring how AI is applied in crop surveillance, disease identification, yield estimation, and input recommendations, which are fertilizers and pesticides.
- **Livestock Management**: Discusses current and futuristic uses of AI, which include animal health monitoring, behavior analysis, automated feeding systems, and so on.
- **Supply-Chain Optimization**: Discusses how AI functionality can solve numerous supply-chain issues, including agricultural production, storage, transport, and disbursement, thus reducing many losses that are encountered along the supply chain.
- **Socioeconomic Implications and Future Directions**: Considers the economic impact and consequences of the use of AI in agriculture, including job opportunities, rural population, and technology accessibility. It also discussed the trends and future developments related to agricultural evolution enabled by AI [5].

8.2 Precision Agriculture: Transforming Farming Practices

Smart farming, or precision farming, is a revolutionary approach to agricultural management in which technology integrates data-driven solutions to improve crop management at the individual field level. Precision agriculture is an excellent shift from the traditional approaches of farming, where technology is applied to increase efficiency while improving profitability. This involves the application of more approaches to data analysis to ensure valid control of variation in fields and to ensure that crops are fertilized with the necessary requirements at the right time and in the right proportions [6]. They also align product development and production with specific clients, thereby minimizing waste, improving efficiency, and ultimately increasing the profitability of farms while decreasing the harm to the environment. Precision agriculture is the art of managing farms where various techniques like data analytics, satellite images, IoT devices, and GPS are used to ensure automation, efficiency, and productivity in the farming sector.

8.2.1 Key Techniques in Precision Agriculture

Various techniques of precision farming have been discussed below and are illustrated with the help of Figure 8.2.

- Precision agriculture uses GPS information to provide accurate field locations, allowing for precise planting, feeding, and harvesting. Farmers who are aware of the geographical position of various parts of their fields can apply production inputs in the right proportions, thereby lowering costs while at the same time enhancing productivity [6].
- Remote sensing and satellite imagery are used to help with data about the status of the soil, crops, and even the weather. The use of satellites and drones possessing multispectral and hyper spectral cameras helps them identify issues such as pest attack, lack of nutrients, and water scarcity at an early stage, which could be remedied.
- **IoT and Sensors**: Based on the data existing in fields with IoT devices and sensors, several factors indicating environments such as soil moisture, temperature, and moisture are available. This data collection enables farmers to make proper decisions on the amount of water to be supplied, fertilizer to be used, or pesticides to be applied depending on the individual crop requirements.
- **Variable Rate Technology (VRT)**: VRT offers the means for accurate targeting of inputs that include fertilizer, herbicides, and water. This is because, through proper utilization of the information from a field, farmers may enhance efficiency and reduce application costs in a way that minimizes the environmental impact.

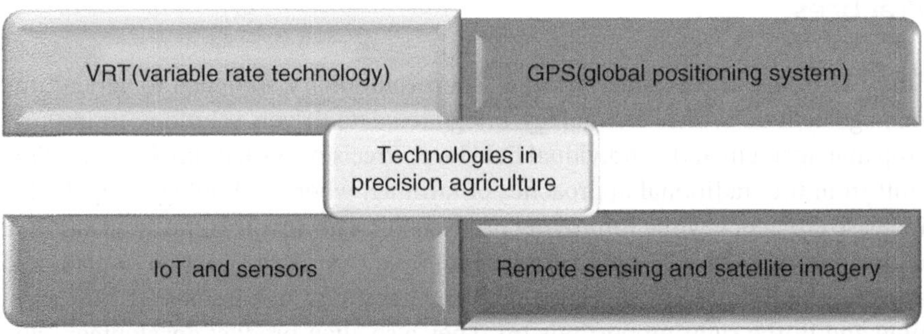

Figure 8.2 Techniques in precision farming.

8.2.2 Advantages of Precision Agriculture

- **Increased Efficiency and Productivity**: Precision agriculture does not use inputs where they are not wanted since it only applies the inputs necessary depending on what the crops need to have optimal growing conditions. This serves to boost yields and produce better-quality food [7].
- **Cost Savings**: Precision agricultural technologies help to lower overall farming costs by lowering the consumption of seeds, fertilizers, pesticides, and water. Efficient resource usage leads to significant cost savings over time.
- **Environmental Sustainability**: Less input utilization means fewer chemicals are released into the environment, which reduces pollution and promotes biodiversity. Precision irrigation systems can conserve water, which is a valuable resource in many agricultural areas.
- Precision agriculture's data-driven approach gives farmers broad insight into their farming operations, allowing for better decision-making. This enables more effective planning of strategies, risk management, and adaptive procedures in the face of changing environmental conditions.
- **Labor Efficiency**: Automation and precision machinery lessen the demand for human labor, permitting farmers to manage larger regions with fewer employees. This is especially advantageous in areas with labor shortages [8].

8.2.3 Role of AI in Precision Agriculture

AI is critical to advancing precision agriculture. AI systems can detect patterns, trends, and insights by analyzing massive volumes of data collected from multiple sources such as satellites, drones, sensors, and machines. This allows farmers to make informed decisions. Specifically, machine learning algorithms have the ability to adjust and acquire knowledge from data, continuously improving their models to provide more precise suggestions for maximizing farming operations.

8.2.4 Soil Analysis and Management: Application of AI in Precision Farming

Soil analysis and management is one of the most important applications of AI in precision agriculture. AI-powered systems can process soil data and calculate its composition, fertility, and moisture levels with unparalleled precision. This can be used by farmers to adjust their fertilization and irrigation to meet the exact needs of various parts of a field, leading to increased crop yields and decreased use of resources. In addition, the AI can predict erosion, compaction levels, and nutrient depletion in soils, thereby preventing soil degradation and reduced yields [9]. Integration of AI in soil analysis and management has the potential to revolutionize agricultural practices. Farmers and crop growers can enhance their soil management processes, use them to improve soil yield and quality worldwide, and also provide an ecological approach to the problem by applying AI technologies. An implication of soil analysis is further demonstrated in Figure 8.3 below.

8.2.4.1 Implications in Soil Analysis

- **Soil Health Monitoring**: Machine learning algorithms can be used to analyze large amounts of data from soil sensors in order to determine the soil status in real time. These include measuring relative humidity, pH level, temperature, and nutrient concentration, among other things. AI systems require such characteristics to analyze patterns and provide farmers with fast information on the state of their soils [10].
- **Soil Mapping**: Traditional soil mapping requires significant labor and time. AI enables the production of high-resolution soil maps from satellite photos, drones, and remote sensing data. Machine learning models analyze this data to identify various soil types and their features across huge areas. This facilitates precise land management and crop planning.
- **Nutrient Management**: AI can anticipate soil nutrient requirements by studying past data and current soil conditions. Machine learning models can recommend the best type and amount of fertilizer to use, avoiding excess or underuse. This not only increases agricultural productivity but also reduces the environmental impact of excessive fertilizer use [11].
- **Precision Agriculture**: AI-driven precision agriculture manages soil and crop needs at the micro level. Drones outfitted with AI-powered cameras may survey fields and give precise images to assist in detecting regions with specific soil difficulties. This permits tailored treatments, such as localized irrigation or fertilization, which improves resource efficiency and crop yield.

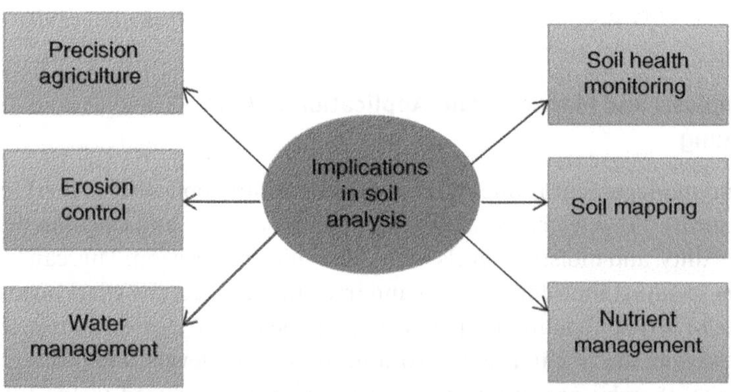

Figure 8.3 Implications in soil analysis.

- **Erosion Control**: Soil erosion is a major issue that impacts both agricultural output and environmental health. AI algorithms can forecast erosion patterns by assessing topography, rainfall data, and soil composition. These forecasts aid in the development of successful soil conservation methods, such as contour cultivation and the installation of buffer strips, to minimize erosion.
- **Water Management**: Agriculture relies heavily on efficient water management. AI systems can optimize irrigation schedules by assessing soil moisture, weather forecasts, and crop requirements. AI-powered automated irrigation systems can regulate water distribution in real time, ensuring that crops receive the appropriate amount of water while saving this valuable resource [12].

8.2.4.2 Benefits of AI in Soil Analysis and Management

- AI automates data collection and analysis methods, resulting in improved soil analysis and management. This saves time and effort, freeing farmers to concentrate on more critical tasks.
- AI algorithms improve accuracy for soil analysis and management decisions. This leads to higher crop yields, lower input costs, and less environmental impact.
- AI-driven soil management approaches enhance sustainable agriculture by optimizing resource utilization, including water and fertilizer. This promotes soil health and prevents deterioration.
- AI technologies give farmers with real-time insights, allowing for rapid decision-making. This can dramatically improve crop management and prevent possible difficulties from escalating [13].

8.2.5 Case Studies: Successful Implementations of AI in Precision Agriculture

Several case studies show the real benefits of incorporating AI into precision agriculture.

8.2.5.1 John Deere's See and Spray Technology

John Deere, a global leader in agricultural machinery, has developed an AI-powered system known as See & Spray. This technology leverages computer vision and machine learning to differentiate between crops and weeds in real time. To illustrate, See and Spray directs herbicides to weeds exclusively, thereby cutting back the volume of chemicals to a great extent, reducing the outlay and the harm the chemicals cause to the ecosystem. A recent study on the field use of the system has shown that, in the worst-case scenario, herbicide use was slashed by up to 90% [14], portraying the effectiveness of the system and the possibility it holds for improving sustainable agriculture.

8.2.5.2 IBM Watson and the Wine Industry

As it has been clearly evidenced, IBM's Watson, for example, an innovative AI-oriented platform, has been implemented effectively in the sphere of viticulture. For instance, with E. and J. Gallo Winery using Watson's AI algorithms, data from weather stations, soil sensors, and satellite imagery is fed to improve irrigation and anticipate vineyard conditions. It assists in controlling water resources in an efficient manner and, therefore, impacts the quality of grapes produced. The adoption of these smart solutions results in increased crop productivity and a 25% decrease in water consumption, proving that AI can advance productivity while also ensuring sustainability in this particular sector.

8.2.5.3 The Climate Corporation's FieldView Platform

Some of the products that are under the Bayer company are: The Climate Corporation, which has FieldView, an AI-based, acclaimed farmer's sidekick platform. This one combines satellite imagery with weather data and soil data to create precise recommendations on when to plant or fertilize, as well as marking the harvesting time. Computer-assisted farming through FieldView has also been said to help increase farmers' yield rates and make better use of inputs. For example, in the Corn Belt of the United States, farmers who practice genetically modified crops have reported an increase in yields of about 10% and reduced expenses by way of fertilizer and pesticides, among other things [15].

8.2.5.4 Blue River Technology's LettuceBot

Blue River Technology, acquired by John Deere, developed the LettuceBot, an AI-driven machine designed for lettuce farming. The LettuceBot uses computer vision to identify and thin out lettuce plants with high precision. This process, traditionally done manually, is time-consuming and labor intensive. LettuceBot automates the thinning process, ensuring optimal plant spacing and improving crop quality and yield. The implementation of LettuceBot has resulted in labor cost reductions and more consistent crop production, highlighting the transformative impact of AI on labor-intensive agricultural tasks.

8.2.5.5 Prospera Technologies' Digital Farming Solutions

Prospera Technologies offers AI-driven solutions that monitor and analyze plant development, health, and stress conditions. Their systems use high-resolution imagery and sensor data to provide actionable insights into crop conditions. In a notable case, a greenhouse operation utilizing Prospera's technology achieved a 20% increase in crop yield and a significant reduction in resource use. The ability to detect diseases early and manage crops with precision has proven invaluable for farmers aiming to maximize efficiency and sustainability [16].

8.2.6 Challenges and Future Prospects of Precision Agriculture

While precision agriculture has numerous advantages, some barriers must be overcome before it can be widely adopted. Small- and medium-sized farms may face challenges due to high initial investment expenses. Furthermore, the complexity of integrating numerous technologies, as well as the necessity for technical competence, can dissuade farmers from adopting precision farming practices.

Prospects for the further evolution of precision agriculture: The future of precision agriculture based on machine learning and AI, mathematical tools, and blockchain technologies is promising. In particular, AI and ML are considered to enhance forecasting and decision-making capabilities, while blockchain is considered to enhance flow transparency as well as the possibility of tracing the origins of the corresponding products [17].

Overall, precision agriculture has a measureable improvement in farming practices as compared to traditional farming. In this way, technology is not only improving the efficiency and productivity of the agricultural industry but it is also promoting the use of environmentally friendly practices that will guarantee the sustainability of the agricultural industry as well as the environment. With the advancement of technology, this paper is going to conclude that precision agriculture will remain a crucial element in shaping the future of agriculture and food production around the globe.

8.3 Crop Monitoring and Management Techniques

The potential for AI to monitor and manage crops necessitates a new level of agriculture, particularly because it provides farmers with a unique picture of their fields. Crops can be tracked from development to maturity using AI, telling farmers about their crops' state, including their health and any external factors that may be influencing them. Using information from remote-controlled sensors, meteorological data, and databases, an AI system can diagnose diseases, pests, or nutrient deficits in their early stages. This makes it easier for farmers, through the use of pattern analysis, to adopt early interventional measures like targeted pesticide spraying or adjustments to irrigation regimes to stop or limit production losses [18].

With the help of such diverse approaches as satellite imaging, aerial drones, and ground sensors, AI environments can collect and process actual information about the state of crops, the availability of moisture and nutrients, and the development of plant life cycles. The collected data is analyzed using machine learning algorithms to identify a problem based on pest infestations, diseases, and nutrient deficiencies, among others. Crop monitoring and management techniques mean that the farmer's input can only be provided where it is needed and only in the necessary amounts as needed through such practices as targeted pest control or water irrigation. The outcome is better utilization of the facilities; costs are lowered; and productivity is seen in the crops produced. Also, the use of AI in predictive analyses allows for the identification of future crop yields and possible difficulties in this process, helping farmers fine-tune their production consistently. These enhanced monitoring and management features are mutually beneficial for improved productivity while adhering to sustainable and environmentally sound agriculture [19]. It is quite clear that farmers can also benefit from the application of AI in that they can be able to enhance their farming procedures and production rates so they can feed the nation and make sure that farming is done in a sustainable manner. Some of the main AI-enabled strategies that are transforming crop management are listed in Figure 8.4.

8.3.1 Strategies for Monitoring and Managing Crops

- **Remote Sensing and Data Analysis**: AI-powered remote sensing systems use drones, satellites, and ground-based sensors to collect detailed data on crop health, soil conditions, and environmental variables. Such data is processed with the help of sophisticated analytical tools ranging from machine learning algorithms to present actual-time results [1]. For instance, plant stress, pest infestation, and nutritional deficiency observed through multispectral and

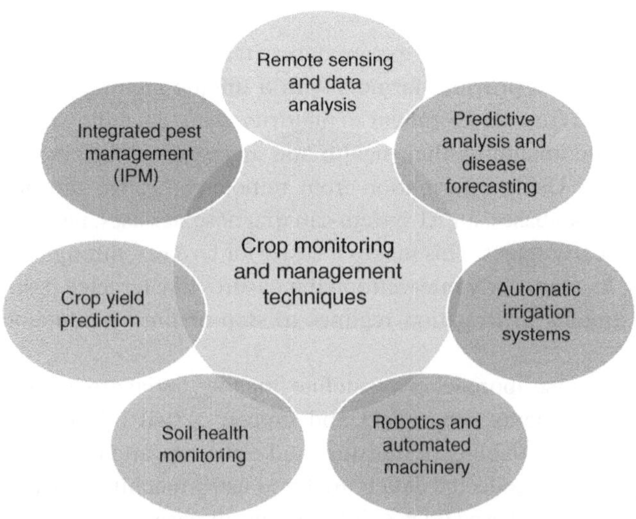

Figure 8.4 Strategies for monitoring and managing crops.

hyperspectral images captured by the drone are not easily visible to the human naked eye. These insights help in conducting necessary early interventions that can reduce cases of crop failure and enhance production.

- **Predictive Analysis and Disease Forecasting**: AI-based predictive models of weather, insect infestation, and disease prevalence involve analyzing present and past data to come up with forecasts on possibilities on the shared planet. This way, farmers can prevent these disasters from happening as they work towards preserving their crops. For instance, milling technology in crop farming can predict when a fungal outbreak is probable given the climatic conditions of humidity and temperature, and farmers can apply fungicides as needed rather than a fixed time schedule. The fertilization approach to cropping systems also helps in reducing the usage of chemicals and enhancing crop yield [2].
- **Automatic Irrigation Systems**: Sophisticated technology in the contemporary world is used in automated systems that schedule irrigation using the available data from the moisture content in the soil, the forecasts of the climatic conditions, and the rates of water consumption by crops. Such systems enable the proper application of water for irrigation, thus avoiding wastage of water and cases where plants receive excess water as well [3]. Automated decision-making plans are dynamic and vary on a daily basis depending on the stage of crop growth and maturity. Additionally, advanced details concerning weather aid in water usage and enhance productivity.
- **Robotics and Automated Machinery**: It is also necessary to note that in the framework of current canning, agricultural robotics is based on the use of AI technologies, including tractors and harvesters and plant-planning devices [4]. These machines are equipped with vision and AI; they are used for plowing through fields, sowing seeds at the right depths, and harvesting crops while keeping them intact. Through such methods, meanings that require human labor can be executed through machines that incorporate AI; this eradicates inconsistencies, forges speed, and negates the issue of scarcity of labor in the agricultural sector.
- **Soil Health Monitoring**: Thus, there is a need to ensure that the soil has the right nutrients it requires for proper fertility. AI technologies deploy sensor networks and collect soil samples to analyze the type of soil, its pH, and levels of nutrients, among other factors [5]. Using the collected data, these models calculate precise fertilization schedules to meet the crop's nutrient requirements. Regarding the SOP chosen above, this precision agricultural strategy enhances the quality of soil, enhances the food productivity rate, and has less harm to the external environment than other production procedures.
- **Crop Yield Prediction**: An accurate yield forecast is critical because it determines the priorities for business and resource utilization in the market. Weather IoT data is collected using historical data, climate parameters, and actual crop

growth performance figures to make a precise yield forecast [20]. These projections help the farmers make the right decisions on storage, marketing, and distribution, thereby yielding better results.

- **Integrated pest management (IPM)**: That's why the option of applying AI ensures more efficient integrated pest management with the help of sensor data and the visual image of pests. Automatic systems recognize pest species and the extent of concern, providing strategy options such as using natural enemies or applying precise pesticide treatments [6]. This reduces dependence on wide-spectrum chemicals and supports the use of organic agricultural practices.

We can state that modern and progressive approaches to managing crop production through AI systems hold great potential. Applying AI to farming practices can help raise productivity, make farming more eco-friendly, and ensure that farmers make good profits from it. These technologies not only try to make the best use of the available resources and crops produced but also aim to achieve the necessary goal of feeding the increasing world population through sustainable means. Researchers expect that with AI developments, the application in agriculture will increase as it continues to change the field and offer new solutions to the challenges faced by farmers globally.

8.4 Revolutionizing Livestock Management Through AI

AI is a part of technology that is evolving and changing many aspects of our society, including cattle management. AI improves livestock care through the application of methods that raise the profitability, quality, and efficiency of the sector. Wearable sensor technology, IoT devices, machine learning algorithms, and cattle health data can all be tracked in real time by AI systems [21]. These systems can normally identify a tipping point of ailment, stress, or unusual behavior and prevent further escalation of the problem. Moreover, the effectiveness of fertilization and reproduction, in general, can be improved by the predictive capabilities of an insemination schedule based on the breeding cycles of the animals. Farmers deploy AI in precision livestock farming techniques to monitor the health and other details of the animals, which is advantageous from an economic standpoint as well as encouraging humane animal participation in livestock farming [22]. This innovation minimizes environmental damage while reducing resource losses, which helps to move animal production toward more sustainable techniques. Here's how AI is transforming livestock management.

- **Breeding and Genetic Improvement**: AI is also altering the breeding process. Advanced algorithms evaluate genetic data to identify the finest mating couples, with the goal of improving desirable qualities like illness resistance, growth rates, and milk supply. This data-driven approach to breeding accelerates genetic advancements and aids in the development of more resilient and productive cattle herds [9]. Furthermore, AI can help track and manage reproductive cycles. Machine learning models anticipate an optimal breeding time, which improves the success rates of artificial insemination and natural mating.

- **Behavior Analysis and Welfare**: Understanding animal behavior is critical to guaranteeing their welfare and production. AI systems with video surveillance and behavioral analysis software can continuously monitor livestock [23]. These systems detect abnormal behavior, like as changes in movement patterns or social interactions, which can signal stress, disease, or discomfort. By giving insights into animal behavior, AI assists farmers in enhancing living conditions, lowering stress, and improving overall animal welfare. This, in turn, can result in increased production and higher-quality products.
- **Operational Efficiency and Decision Support**: AI-powered management solutions integrate data from multiple sources, giving farmers' detailed insights into their operations. These technologies aid decision-making by accessing data on feed utilization, health records, production levels, and environmental circumstances. Farmers may use AI to make more educated decisions about resource use, productivity, and operating expenses [24]. Additionally, AI can automate mundane operations such as feeding, milking, and cleaning, allowing farmers to focus on strategic areas of farm management. Automation boosts productivity while also ensuring uniformity and precision in daily activities.

The incorporation of AI into livestock management is altering how farmers manage their herds. Farmers may use AI technologies to improve animal health, optimize nutrition, improve breeding methods, and assure improved animal comfort. As AI advances, its applications in livestock management are likely to expand, driving more advances and efficiencies in the agricultural sector [25]. This technological revolution promises the assurance of more environmentally friendly and productive livestock production, which will meet the expanding worldwide need for animal products while also protecting the animals and the environment.

8.4.1 Challenges in Livestock Management

Animal illnesses and inadequate feed intake are two of the major issues they deal with when managing livestock. Even if technology is advancing and agricultural techniques are changing much more quickly in the modern world, there are still a number of obstacles to overcome in the process of managing livestock that could have an impact on future sustainability, animal welfare, and production [26].

8.4.1.1 Disease Control and Biosecurity

Livestock diseases are significant challenges in the livestock field, as they can quickly affect the animals and lead to large losses. Diseases, on the other hand, have to be contained through biosecurity measures, which can be less easily managed and more challenging for big operations. As bacteria develop antibiotic resistance, diseases like those caused by bacteria are harder to treat.

8.4.1.2 Animal Welfare

More specifically, it is never an easy effort to make sure that the animals – which are commonly called livestock – are properly cared for. It's critical to house animals in buildings that can provide the right atmosphere for them to develop healthy bodies and behaviors. The level of health in an organization, overcrowding, inadequate rationing, and improper handling of birds can cause stress, injuries, and sicknesses that, in turn, influence the productivity of the birds. However, high levels of welfare mean that standards need to be checked regularly and, where appropriate, remedial action taken, and this can be expensive [13].

8.4.1.3 Resource Management

These inputs include feed, water, and land; their use should therefore be optimally utilized in order to make sustainable use of these natural resources. Feed has often been considered one of the biggest production expenses, and obtaining high-quality feed that is also readily accessible can be a very complicated matter [27]. Another challenge that needs to be solved in order to achieve the efficiency of the method mentioned above is the shortage of fresh water and the influence of cattle farming on the environment, namely, the depletion of the land and the production of GHGs.

8.4.1.4 Labor Shortages

One of the major challenges competition brings to the livestock industry is that the industry struggles, especially in sourcing skilled labor. There are several factors that contribute to worker absenteeism; these include the nature of work, which may entail long hours of work and physically challenging duties in carrying out the operations of livestock farming [28]. Some of these challenges can be reduced through automation and AI, although their implementation poses a major demand on faculty investment, technological capabilities, and employee training [14].

8.4.1.5 Climate Change

If not addressed adequately, climate change will have a great impact on feed and water, among other issues concerning the livestock farming system. Heat stress has effects that include reduced fertility, especially when heat stress affects animals during the breeding season, and reduced milk and beef production for the duration of heat stress incidences such as droughts and heat waves [29]. To overcome these challenges, farmers have to adopt some flexible farming practices and structures, which may not be easy to put in place, especially due to the high costs involved.

8.4.1.6 Market Volatility

Poultry producers and other livestock producers are particularly exposed to market risks that include variations in feed prices and demand due to changing consumer tastes and production capacity among producers within global market systems. Milton & Walker also argued that managing financial stability in such an environment necessitates the implementation of risk management policies and programs, as well as the flexibility to make necessary changes quickly [15].

8.4.1.7 Regulatory Compliance

Adherence to the best standards in developing animal health, welfare, and environmental standards is crucial; however, it is not always easy. These regulations are now tighter as farmers are forced to keep abreast of these measures and adapt to them depending on the prevailing circumstances, which can sometimes be extremely time-consuming and expensive [16].

These challenges within contemporary business management can be addressed by the application of various techniques based on technological advancement, sustainable management, and ICT. With regards to improvements in AI and other tools, it enables livestock farmers to be more effective in the performance of their tasks and thereby formulate more resilient livestock farming systems.

8.4.2 Case Studies: Implementations of AI in Livestock Management

Some examples based on real-life use cases show the positive impact of AI in cattle rearing. For instance, a dairy farm in California used the sensors to monitor cattle movement patterns for disease diagnosis, including mastitis, before it occurred. Likewise, a European-based chicken firm was able to improve feed formulas using AI algorithms, resulting in a notable improvement in feed conversion ratios and the overall performance of the chicken farm. A few prominent case studies are:

- **Connecterra's Intelligent Dairy Farmer's Assistant (IDA)**: Connecterra is an innovative firm specializing in the dairy sector, whose flagship product is the Intelligent Dairy Farmer's Assistant, or IDA, an AI-based tool for efficient farming. With the help of wearable sensors on the cows, IDA gathers information on several factors like locomotion, eating, and rumination durations. In analyzing this data, the AI identified the early signs of health disorders, the cycles of estrus, and other aspects of the health of a specific group of cows or the entire herd. For example, when using IDA, Duos, a Dutch dairy farm, eliminates 50% of antibiotic use and 20–30% of cow mortality decline.

- **Cainthus' Visual Recognition System**: Cainthus is an agri-tech company that deploys AI computer vision to observe dairy herds. Surveillance cameras placed both within and outside the farm provide images that are processed by an AI system that helps identify individual livestock and their activity patterns. It is also capable of recognizing individuals based on their physical characteristics, such as cows, and offering information on feeding patterns, levels of socialization, and health records. A relevant case of a dairy farm in the United States exposed that by implementing the AI system, the production of milk increased by 20%, along with a cut in labor expenses because the system helped in the accurate tracking of the bulls.
- **FarrPro's Haven Technology**: FarrPro has therefore ventured into the development of Haven, which is an AI-driven system aimed at boosting the survival rates of piglets. For the farrowing pens, the system for measuring the temperature and movement of newborn piglets incorporates infrared optics technology. Management that involves maintaining a warm environment protects piglets from the threat of cold that leads to hypothermia, resulting in high mortality rates in piglets. This chapter is an actual example of using Haven technology in a large pigs' farm, in which, in simple arithmetic, piglet death was cut by a third, and the dosage was given a marked boost by the application of an AI plan in order to improve levels of welfare and production standards of pigs.
- **Allflex Livestock Intelligence's SenseHub**: Allflex Livestock Intelligence, a livestock sector giant, has introduced SenseHub, which uses ear tags to track animals. Such sensors include but are not limited to, temperature, activity level, and other health-related parameters. This information is processed by the AI system for the purpose of ensuring early signs of illness, the estrus cycle, and even the general health of the cattle in the herd are all noticed early enough. In a trial conducted on a beef cattle operation in Australia, SenseHub helped farmers achieve a 25% reduction in veterinary costs and a 15% increase in calving rates, demonstrating the system's effectiveness in improving herd management and reproductive performance.

These case studies highlight the transformative impact of AI in livestock management. The successful implementations demonstrate how AI can address key challenges in livestock management, ultimately leading to more efficient and resilient farming operations.

8.5 Innovations in Food Supply Chains with AI

AI is driving remarkable innovations in food supply chains, enhancing efficiency, transparency, and sustainability from farm to fork. AI-powered systems optimize logistics by predicting demand, managing inventory, and automating route planning, which minimizes food waste and reduces transportation costs. Machine learning algorithms analyze vast datasets to forecast market trends and consumer preferences, enabling producers and retailers to align their operations with real-time demand [17]. Blockchain technology, integrated with AI, ensures traceability and food safety by recording every transaction in the supply chain, from production to delivery, thereby preventing fraud and ensuring compliance with regulations. AI also facilitates precision agriculture, ensuring optimal harvest timing and quality, which translates into fresher products reaching consumers faster [18]. Additionally, predictive maintenance of equipment and machinery using AI reduces downtime and enhances productivity. These innovations collectively contribute to a more resilient, efficient, and sustainable food supply chain, capable of adapting to market fluctuations and ensuring food security.

Food supply chains are intricate networks with numerous players, including producers, farmers, sellers, distributors, and consumers. Understanding the complexities of these supply networks is critical for increasing efficiency, reducing waste, and maintaining food quality and safety.

8.5.1 Role of AI in Enhancing Food Traceability

AI technologies, such as machine learning and blockchain, are transforming food traceability by allowing for real-time monitoring and tracking of food products from farm to table. By employing AI, stakeholders can immediately identify the source of contamination or spoiling, improving food safety and lowering the risk of foodborne illness. AI significantly enhances food traceability by providing advanced solutions that ensure safety, quality, and transparency across the food supply chain. IoT devices and sensors integrated into AI systems gather data on weather conditions, temperature, and humidity during product storage and transportation to minimize adequate handling. The merging of AI with blockchain technology ensures that there is an end-to-end declaration of each stage, starting from the farm or field and concluding at the table. AI-driven predictive analytics identify potential risks and vulnerabilities, allowing for proactive measures to prevent contamination and logistical issues. AI has the ability to immediately show flagged batches and their whereabouts and origins in the event of contamination or when a batch is being recalled. Third, AI enhances the consumer's ability to find reliable information regarding the sources and treatment of foods that they consume through technological devices and platforms, making it easier for consumers to make better decisions when procuring foods in the market. In conclusion, the role of AI in improving the efficiency of food traceability is revolutionary, with higher degrees of food safety, traceability, and supply chain sustainability being the outcomes. Few important topics to consider are discussed in the following sections.

8.5.1.1 Real-Time Monitoring and Data Collection

AI-enabled sensors and IoT devices are used to collect real-time data on various parameters such as temperature, humidity, and handling conditions during transportation and storage [30, 31]. This data is continuously monitored and analyzed by AI systems to ensure that food products are stored and transported under optimal conditions, reducing the risk of spoilage and contamination. For example, the shipments of sensitive commodities such as dairy or meat may be traced to guarantee the specified temperature is achieved in the distribution channel.

8.5.1.2 Blockchain Integration for Immutable Records

By applying AI to the blockchain, the model for traceability improves by making impenetrable records of each transaction and movement of food products. From manufacturing to processing, sale, and redistribution, all the details are recorded on a blockchain, which ensures the history's transparency and immutability [33, 34]. This integration also enables the various stakeholders in the supply chain to prove the authenticity of the food product that they are sourcing, processing, and transporting since it comes with a paper trail.

8.5.1.3 Predictive Analytics for Risk Management

Another area is the analysis of the potential problems in the supply chains. The application of AI will allow one to consider historical data and analyze the tendencies that could make problems evident in further supply chain planning. For instance, in supply chain management, AI can forecast the probability of contamination instances or even the occurrence of a few logistical constraints so that relevant players can apply preventive measures to these events. It ultimately enables provisions to be made regarding the purity and safety of foods and against spoilage.

8.5.1.4 Rapid Response to Contamination and Recalls

Another advantage shown by the use of AI technology is that in case of a contamination breakout or a product recall, the AI system is able to identify the affected batches and their location in the supply chain. With a reasonable and fast understanding of the sources and the magnitude of the contamination, AI is helpful in managing the problem within a short period of time and limiting its outcomes. It is essential to sustainability and ensures the safety of the public and the maintenance of confidence in products.

8.5.1.5 Enhanced Transparency for Consumers

AI also makes shopping easier because it provides consumers with comprehensive information about the food products on the market. For instance, through quick response codes or any other related advancements in the same field, the customers are in a position to acquire statistical information that includes the source, manner of preparation, and how the food products have been handled. Furthermore, this transparency helps to establish much-needed consumer confidence while also promoting better and more responsible consumerism.

8.5.2 Case Studies: AI Applications in Food Supply Chains

In light of the following real-world examples, many industries can attest to the efficacy of AI in improving food supply chain logistics. For example, a global food retailer used AI demand forecasts to improve stockouts by 30% and excess inventory by 25%. Example: A food manufacturer discovered that it could use AI-based quality control systems to improve product homogeneity and reduce the rejection rate. Some of the prominent case studies are:

8.5.2.1 IBM Food Trust and Walmart

IBM Food Trust is a blockchain-based food supply tracking system that adds AI to increase food transparency and the proper supply chain. It is ironic that Walmart, which sources fresh produce from Mexico, had to turn to IBM Food Trust this year to help it test its tracing systems using leafy greens. The efficiency of applying AI and incorporating blockchain technology can be evidenced by the example of Walmart, which used to take several days to trace where specific batches of produce were grown; introducing the two technologies allowed the company to achieve this goal in a matter of seconds. This capability greatly improves food safety because it allows grocery sellers to quickly identify contaminated food products and recall them from the market, thereby reducing the likelihood of foodborne illnesses.

8.5.2.2 Nestlé and AI-Driven Quality Control

Nestlé has implemented AI to improve quality control across its supply chain. The company uses machine learning algorithms to analyze data from various stages of production and distribution, identifying patterns that indicate potential quality issues [32]. For instance, AI systems analyze data from factory sensors and production line cameras to detect anomalies in real time, such as packaging defects or contamination. This proactive approach has led to improved product quality and reduced recall incidents, ensuring that only safe and high-quality products reach consumers.

8.5.2.3 Zest Labs and Freshness Management

Zest Labs, a provider of AI-driven freshness management solutions, works with major retailers and growers to extend the shelf life of perishable goods. Their Zest Fresh platform uses IoT sensors to monitor environmental conditions and product handling throughout the supply chain. AI analyzes this data to provide real-time insights into the freshness and remaining shelf life of produce. A notable implementation with Walmart resulted in a significant reduction in waste and spoilage, as well as improved inventory management, by ensuring that the freshest products are prioritized for sale.

8.5.2.4 Alibaba and Blockchain for Food Safety

Alibaba has leveraged AI and blockchain to enhance food traceability and safety in China. The company's Food Trust Framework combines AI with blockchain to track the movement of food products through the supply chain, from farms to consumers. For dairy products, the AI scanned data from the various sensors in the farms and depots, which provided information such as temperature and humidity to meet the ideal storage conditions. The blockchain layer offered the structural framework for recording and verifying every transaction, thus increasing security.

8.5.2.5 GrainChain and AI for Commodity Traceability

GrainChain, an agricultural technology company, uses AI and blockchain to improve traceability and efficiency in the commodity supply chain. Their platform tracks the journey of grains from farms to end-users, providing real-time data on quality and handling. AI algorithms analyze data from sensors and smart contracts to optimize logistics, reduce fraud, and ensure compliance with regulatory standards. In a case study involving coffee farmers in Honduras, GrainChain's system improved transparency and reduced transaction times, leading to fairer prices for farmers and more reliable sourcing for buyers.

AI technologies, particularly implemented in the food sector, might help stakeholders discover new ways to improve the performance, product quality, and sustainability of supply chain management systems. These technologies are also revolutionizing methods of cultivating and supplying food while at the same time improving the global food security of most food available for consumption.

8.6 Addressing Ethical and Regulatory Considerations

In this case, we need to carefully study ethical and regulatory consideration domains regarding the consequences of farming methods with the help of AI and apply AI appropriately as researchers. This chapter focuses on four main areas: legal and regulatory concerns of AI in agriculture, steps toward ensuring that AI culturally embodies equality, ways of looking at various success stories and case studies of AI use in agriculture that have been recorded in order to learn from them and emulate the best practices recorded, as well as a look at the key trends consistent across the use of AI in agriculture that need strategic attention.

8.6.1 Ethical Implications of AI in Agriculture

In addition, there are some other issues to discuss, including data rights and privacy issues, since AI deals tremendously with large amounts of data collected from farms. In light of these issues, it is necessary for researchers to pay attention to the concerns of permission, visibility, and responsibility so that the possible harm that may result from the deployment of AI in agriculture can be abridged while guaranteeing that all the stakeholders in the agriculture value chain stand to benefit.

8.6.2 Regulatory Frameworks for AI Adoption

Governmental regulations are an important aspect in ensuring that authorities control and guide the appropriate application of AI in farming. The farming techniques that involve the use of AI must be separately and particularly given strict parameters by the governments and regulatory bodies in the testing and implementation of the same. These laws should contain the provisions of safety, responsibility, and balance, but they should also keep the incentives for innovations and competition in the sphere of agriculture. Some of the solutions that could be given to the problem include the following: By coming up with a good regulatory system, researchers could be in a position to ethically maneuver through the complex world of using AI.

8.6.3 Ensuring Equity and Inclusivity in AI-driven Agriculture

Therefore, it is necessary to remain cautious to avoid negative impacts on the farming community and, at the same time, embrace the use of AI in agriculture in a way that benefits all the different groups within this sector. Tackling the research questions and developing solutions that would benefit smallholder farmers and other vulnerable groups will require researchers to be more involved with the communities and explore how best to make research accessible to such people. To ensure that farmers who are currently unable to access technology can imbibe the benefits of the new advanced technology, researchers should focus on developing suitable AI systems that are readily available and affordable to farmers, taking into consideration existing barriers in different parts of the world.

8.6.4 Case Studies: Lessons Learned and Best Practices

Drawing real-world case studies is crucial to understanding their implications for end-users, academics, and practitioners. It is significant to analyze prior AI implementations in farming to determine the factors that foster AI effectiveness and longevity. Additionally, the evaluation of failures and obstacles is crucial since it reveals best practices, opportunities, and ways to avoid such circumstances. The analysis of cases by researchers incorporating theoretical frameworks can help create critical parameters, recommendations, and benchmarks for subsequent AI interventions in agriculture for the purpose of impact and innovation.

Addressing ethical and regulatory concerns is critical for reaching AI's full potential in agriculture. By prioritizing ethical principles, following regulatory guidelines, promoting equity and inclusivity, and learning from real-world experiences, researchers can navigate the complexities of AI adoption with integrity and responsibility, ultimately advancing sustainable and inclusive agricultural development.

8.7 Conclusion

In this chapter, we summarized our findings and examined the consequences of using AI in agriculture. The chapter illustrated several major takeaways:

a) AI technologies have the potential to improve several parts of agriculture, including crop management, disease detection, supply chain optimization, and market forecasting.
b) Integrating AI techniques into traditional agricultural operations can boost production, efficiency, and sustainability, resulting in better food security and economic growth.
c) Successful AI application in agriculture necessitates resolving issues such as data protection, access to technology in rural areas, and guaranteeing inclusivity across socioeconomic groups.
d) Collaboration among stakeholders, such as farmers, researchers, policymakers, and technology developers, is critical for the successful acceptance and implementation of AI solutions in agriculture.
e) Continued improvements in machine learning algorithms and computational methodologies will allow for more precise and individualized agricultural decision-making.
f) The integration of AI with other revolutionary technologies, such as the Internet of Things (IoT), blockchain, and remote sensing, will provide synergies and open up new opportunities for precision agriculture.

g) The development of AI-powered autonomous farming systems and robotics has the potential to transform labor-intensive jobs, lowering reliance on manual labor and alleviating labor shortages in agriculture.

h) The growing emphasis on sustainability and climate resilience is projected to stimulate the development of AI-powered solutions for climate-smart agriculture, such as drought prediction, soil health monitoring, and adaptive crop management.

8.8 Future Directions

While significant progress has been made in the application of AI in agriculture, several areas warrant further investigation and research:

a) Long-term impact assessments are required to assess the socioeconomic, environmental, and ethical consequences of AI adoption in agriculture, ensuring that the benefits are dispersed equitably and sustainable in the long run.
b) Interdisciplinary research endeavors that span the gap between agriculture, computer science, and social sciences can allow for more holistic approaches to addressing difficult challenges in agricultural development and rural livelihoods.
c) Research into building AI solutions that are suited to the specific demands and limits of smallholder farmers in low-resource settings is critical for assuring inclusivity and fostering equal access to technology.
d) The search for innovative uses of AI, including smart farming practices, agroecology, biophysical variety conservation, and inclusive processes for defining agriculture's future, would clarify the potential role of AI across the industry.
e) There are several implications that arise from the integration of AI in agriculture, and these call for interdisciplinary research on the compatible utilization of AI in agriculture, ethical considerations, and approaches to making the best use of AI in enhancing agriculture for sustainable development. With this kind of approach, it would be possible to develop cohesion and a learning organization and culture in order to maintain and build up a new type of agriculturally shaped future for the coming generations.

A holistic approach suggesting improvements in the education and training of farmers, the development of inexpensive AI applications, and data protection policies and guidelines will need to be employed to overcome the challenges. Since AI can be greatly impacted by the variability of the environment and conditions in agriculture, future studies should try to enhance the flexibility of AI technologies in the agricultural field. Moreover, further research should explore the ways AI can be combined with other innovative technologies, such as blockchain, to enhance the efficiency and trustworthiness of the food supply chain.

References

1 Ahmad, M., Abbas, S., Fatima, A. et al. (2023). AI-Driven livestock identification and insurance management system. *Egyptian Informatics Journal* 24 (3): 100390.

2 Dawn, N., Ghosh, T., Ghosh, S. et al. (2023). Implementation of artificial intelligence, machine learning, and internet of things (IoT) in revolutionizing agriculture: a review on recent trends and challenges. *International Journal of Experimental Research and Review* 30: 190–218.

3 De Abreu, C.L. and van Deventer, J.P. (2022). The application of artificial intelligence (AI) and internet of things (IoT) in agriculture: a systematic literature review. *Southern African Conference for Artificial Intelligence Research*, pp. 32–46. Springer: Cham, 5–9 December 2022.

4 Eli-Chukwu, N.C. (2019). Applications of artificial intelligence in agriculture: a review. *Engineering, Technology & Applied Science Research* 9 (4): 4377.

5 Esau, T.J., Hennessy, P.J., MacEachern, C.B. et al. (2023). Artificial intelligence and deep learning applications for agriculture. In: *Precision Agriculture* (ed. Q. Zaman), 141–167. Academic Press.

6 Javaid, M., Haleem, A., Khan, I.H., and Suman, R. (2023). Understanding the potential applications of artificial intelligence in agriculture sector. *Advanced Agrochem* 2 (1): 15–30.

7 Kovalenko, O., Yashchenko, L., and Verbytskyi, S. (2019). Smart technologies of food production in the system of global technological waves. *The Scientific Journal of Cahul State University "Bogdan Petriceicu Hasdeu" Economic and Engineering Studies* 5 (1): 56–63.

8 Kutyauripo, I., Rushambwa, M., and Chiwazi, L. (2023). Artificial intelligence applications in the agrifood sectors. *Journal of Agriculture and Food Research* 11: 100502.

9 Linaza, M.T., Posada, J., Bund, J. et al. (2021). Data-driven artificial intelligence applications for sustainable precision agriculture. *Agronomy* 11 (6): 1227.

10 Marwaha, S., Deb, C.K., Haque, M.A. et al. (2023). Application of artificial intelligence and machine learning in agriculture. In: *Translating Physiological Tools to Augment Crop Breeding* (ed. M.H. Masthigowda, K. Gopalareddy, R. Khobra, et al.), 441–457. Singapore: Springer Nature Singapore.

11 Megeto, G.A.S., Silva, A.G.D., Bulgarelli, R.F. et al. (2020). Artificial intelligence applications in the agriculture 4.0. *Revista Ciência Agronômica* 51 (spe): e20207701.

12 Mohan, S.S., Venkat, R., Rahaman, S. et al. (2023). Role of AI in agriculture: applications, limitations and challenges: a review. *Agricultural Reviews* 44 (2): 231–237.

13 Raj, E.F.I., Appadurai, M., and Athiappan, K. (2022). Precision farming in modern agriculture. In: *Smart Agriculture Automation Using Advanced Technologies: Data Analytics and Machine Learning, Cloud Architecture, Automation and IoT* (ed. A. Choudhury, A. Biswas, T.P. Singh, and S.K. Ghosh), 61–87. Singapore: Springer Singapore.

14 Suma, K.V., Koppad, D.B., Awasthi, K., and Vikas, R. (December 2022). Application of AI models in agriculture. *2022 4th International Conference on Circuits, Control, Communication and Computing (I4C)*, pp. 387–390. IEEE, 21–23 December 2022, Bangalore, India.

15 Tripodi, P., Nicastro, N., Pane, C., and Cammarano, D. (2022). Digital applications and artificial intelligence in agriculture toward next-generation plant phenotyping. *Crop and Pasture Science* 74: 597–614.

16 Wakchaure, M., Patle, B.K., and Mahindrakar, A.K. (2023). Application of AI techniques and robotics in agriculture: a review. *Artificial Intelligence in the Life Sciences* 3: 100057.

17 Zhao, B. (June 2020). The application of artificial intelligence in agriculture. *Journal of Physics: Conference Series*, Vol. 1574, No. 1, p. 012139. IOP Publishing, First International Conference on Computer Applied Science and Information Technology (ICCASIT2020), 15–17 May 2020, Dalian, China.

18 Singh, N. and Gupta, N. (2017). Bayesian network for decision-support on pest management of tomato fruit borer, H. armigera. *International Journal of Engineering and Technology* 6 (4): 168–170.

19 Singh, N. and Gupta, N. (2020). Bayesian network for development of expert system in pest management. *Internet of Things and Analytics for Agriculture* 2: 45–65.

20 Finger, R., Swinton, S.M., El Benni, N., and Walter, A. (2019). Precision farming at the nexus of agricultural production and the environment. *Annual Review of Resource Economics* 11: 313–335.

21 Singh, N. and Gupta, N. (2017) Bayesian network based decision-support for leaf minor pest management in tomato.

22 Singh, N. and Gupta, N. (2017). Effect of weather on fruit borer, Helicoverpa armigera (Hub) activity in tomato. *Journal of Agriculture and Veterinary Sciences* 4 (10): 414–417.

23 Singh, N. and Gupta, N. (2016). ICT based decision support systems for integrated pest management (IPM) in India: a review. *Agricultural Reviews* 37 (4): 309–316.

24 Li, X. (2024). Prospects of artificial intelligence applications in future agriculture. *Advances in Resources Research* 4 (2): 171–180.

25 Mandal, S., Yadav, A., Panme, F.A. et al. (2024). Adaption of smart applications in agriculture to enhance production. *Smart Agricultural Technology* 7: 100431.

26 Jeffrey, L. and Bommu, R. (2024). Innovative AI solutions for agriculture: enhancing cropmanagement and yield. *International Journal of Advanced Engineering Technologies and Innovations* 1 (3): 203–221.

27 Mahibha, G. and Balasubramanian, P. (2023). Impact of artificial intelligence in agriculture with special reference to agriculture information research. *Current Agriculture and Research Journal* 11 (1): 287–296.

28 Singh, D. and Kumar, A. (2024). A deep recurrent neural network for plant disease classification. *SN Computer Science* 5 (8): 1053.

29 Krishna, I. (2023). Examining the applications of artificial intelligence in agriculture. *Journal of Informatics Education and Research* 3 (2): 2026–2037.

30 Kumar, A. and Jha, R. (2022). IoT in smart farming: a boon to farmers. In: *Internet of Everything: Smart Sensing Technologies* (ed. T. Kavitha, V.A. Devi, S.N. Pari, and S. Ramanathan), 112–143. Nova Publisher.

31 Kumar, A., Mamgai, R., and Jain, R. (2023). Application of IoT-Enabled CNN for natural language processing. In: *IoT-enabled Convolutional Neural Networks: Techniques and Applications* (ed. M. Naved, V.A. Devi, L. Gaur, and A.A. Elngar), 149–177. River Publishers.

32 Malhotra, M., Jaiswar, A., Shukla, A. et al. (2023). Application of AI/ML approaches for livestock improvement and management. In: *Biotechnological Interventions Augmenting Livestock Health and Production* (ed. C.S. Mukhopadhyay, R.K. Choudhary, H. Panwar, and Y.S. Malik), 377–394. Singapore: Springer Nature Singapore.

33 Purohit, R.M., Verma, J.P., Jain, R., and Kumar, A. (2025). FedBlocks: federated learning and blockchainbased privacy-preserved pioneering framework for IoT healthcare using IPFS in web 3.0 era. *Cluster Computing* 28 (2): 1–15.

34 Kumar, A., Sachan, R.K., Singh, D., and Jain, R. (2024). Analysis of block chain applications in asset backed securitization system. *Procedia Computer Science* 233: 147–153.

9

Smart Agriculture: Predictive Modeling of Fertilizer Requirements Using Neural Networks

Heet Dave and Jai Prakash Verma

Department of Computer Science and Engineering, Institute of Technology, Nirma University, Ahmadabad, Gujarat, India

9.1 Introduction

Technological development in agriculture yields sustainable food produced well and fast using present methods of farming better has increasingly merged its goals. Precision agriculture – its main feature being the use of the latest technology – has pathways for improved resource conservation and economic returns through enhanced resource use efficiency. The use of fertilizers stands out as one critical aspect of precision agriculture because it determines both crop quality and productivity. Many times, the right amounts of crop nutrients are not used when fertilizers are usually applied, which wastes material and pollutes the environment.

This research work addresses the contemporary problem of "Smart Agriculture," which implies a changing of mind about the fertilizer recommendation systems through the application of neural networks and predictive modeling. The main aim of this project is to build a forecast model for determining various types of crops' fertilizer requirements based on machine learning techniques, including neural networks.

Despite the age-old application of AI in agriculture, the presence of extensive agricultural datasets and computational power offer great benefits for enhancing precision and the predictive models' accuracy. This work examines how neural networks might foresee fertilizer needs by examining various factors that affect them, such as soil composition and climate conditions of plant types, in order to participate in the sustainable agriculture debate.

Emerging Smart Agricultural Practices Using Artificial Intelligence, First Edition.
Edited by Ashish Kumar, Jai Prakash Verma, and Rachna Jain.

Published 2025 by John Wiley & Sons, Inc.

Inefficient agriculture practices and environmental degradation result from the imprecisions of conventional fertilizer application techniques: the task consists of maximizing the fertilizers used for any given plant to satisfy the particular nutrient needs of that crop.

9.1.1 Objective

The purpose of this research is to invent and appraise an anticipating format that applies neurons to optimizing fertilizer quantities employed in farming cultivation within a paracontact. The present project applies algorithms from machine learning so as to build a scalable and personalized system aimed at executing precision agriculture: projecting the optimal fertilizer amount for various crops based on their types, soil quality and structure, and weather patterns. The project aims to enhance sustainable agricultural practices by giving farmers a dependable tool applicable for fertilizer with sense through combining state-of-the-art computer methods with vast agricultural databases.

Machine learning methods such as neural networks are used to analyze vast agricultural datasets. Crop types, soil composition, and weather patterns are considered in the model, which makes optimized predictions to allow for several types and inclusive fertilizing prescriptions.

The aim of this chapter is to understand the efficiency level of neural network predicting models when they are optimizing for fertilizer needs. Its discretion of nutrient prerequisites among various crops ensures its predictability, making it a highly potential precision agricultural qualm.

The chapter's findings related to agriculture are important for farmers to have a tool that helps them in deciding on fertilizer quantity application. By using smart farming techniques, an anticipatory model has been created, which creates an opportunity to raise yields and lessen environmental harm while advocating for sustainable farming practices.

9.2 Related Work

Other researchers use temporal analysis to understand dynamic nutrient demands over time, bring together different data sources (such as soil sensors and climate data) to enhance prediction accuracy, or incorporate remote sensing technologies (like drones or satellites) to monitor crop health. While some researchers are exploring how open data initiatives could be utilized to deploy IoT devices to get real-time field data, others are concentrating on ensemble learning techniques that combine the advantages of several models. This teamwork will improve sustainable farming practices through offering trustworthy, data-driven tools to farmers that will facilitate precise fertilizer application rates so as also boost crop yields without harming the environment in the long run.

Comparison of work done by different researchers by different parameters

Reference	Objective	Approach	Methodology	Findings	Limitation
[1]	To increase crop productivity and profitability in agriculture, the study tries to forecast the optimum type and amount of fertilizer for diverse environmental, soil, and plant conditions	The Random Forest (RF) algorithm, a machine learning technique, is utilized in this study to anticipate fertilizer requirements using historical data and farmer experience	The study employs machine learning techniques to create models based on historical data on crop productivity, soil composition, and other relevant parameters. These computers can examine large data sets and identify trends that humans would overlook, resulting in increased agricultural productivity and more accurate projections	The study found that the RF method forecasts fertilizer more correctly than other strategies, such as linear regression and K-Nearest Neighbors	The performance of the fuzzy logic model is not compared to that of other machine learning approaches in this study. The performance of the SVR model can only be evaluated under certain soil and climate conditions
[2]	The chapter's purpose is to assist farmers in selecting the best crop and fertilizer for their specific needs while taking into account local weather, soil, sowing season, and other factors. This is intended to reduce soil erosion while enhancing agricultural income and productivity	Machine learning approaches like RF, Decision Tree, Naive Bayes, and KNN, are the proposed method that recommends crops and fertilizers that are based on environmental factors and soil composition	The authors obtained crop and fertilizer datasets from Kaggle.com, which included soil pH, temperature, humidity, rainfall, NPK levels, crop type, and fertilizer names	The crop suggestion classification techniques have an accuracy rate of 98.63% for Decision Trees, 98.33% for KNN, and 99.54% for Gaussian Naive Bayes. Fertilizer suggestions techniques have accuracy ratings of 100% (RF), 95% (KNN), and 97.56% (Decision Trees)	Seasonal fluctuations lead crop nutrient requirements to change over time, and current models may struggle to account for these temporal dynamics

(Continued)

(Continued)

Reference	Objective	Approach	Methodology	Findings	Limitation
[3]	Machine learning algorithms that are trained on these digital footprints to identify patterns and predict outcomes, the paper aims to investigate how using digital footprints might improve agricultural practices and productivity	The study looks at different machine learning techniques, namely RF optimization, to improve farming operations and increase productivity. It forecasts agricultural production depending on variables such as weather patterns, soil quality, and fertilizer use using RF optimization	The study predicts agricultural yield with a 99.97% accuracy rate using RF optimization, a powerful machine learning technique. It also proposes the need for further research and investment to develop machine learning algorithms designed to the unique needs of the agriculture industry	The results show the accuracy with which the RF optimization technique predicts crop yield, giving farmers valuable knowledge to improve overall productivity through prudent resource management	Certain models rely significantly on historical data, which might become less accurate as agricultural methods or environmental factors change
[4]	The goal of the study is to increase soil fertility and reduce the risk of leaching and soil runoff by creating a model that can predict the quantity of nutrients required for different crops by evaluating rainfall patterns and crop fertility	Based on time-series data, the study suggests an improved version of the RF method. To choose the most prominent and vital aspects for assessment, feature selection is used. To build an accurate prediction model, the model uses k-fold cross-validation and RF regression. It also includes extracting pertinent meteorological data, such as temperature, humidity, and rainfall, with the help of weather API	The procedure consists of dividing the dataset into training and test datasets, training the labels with the training dataset, and then applying RF regression to each nutrient value. This approach needs inputs such as the crop and location. It then applies the recommended strategy until a warning message arrives, suggesting that severe rains are imminent	The study demonstrates that, when meteorological conditions are considered, the proposed technique can provide information about the amount of nutrients needed by crops for better development and output. It also provides weather warnings and communications, which include alarms in the event of adverse weather	Inadequate insights may result from certain current systems' ineffective integration of disparate data sources, including satellite imaging, soil sensors, and climate data

[5]	The research aims to develop a Multimodal Machine Learning Based Crop Recommendation and Yield Prediction (MMML–CRYP) model that can accurately estimate crop yield and provide effective crop recommendations	The MMML–CRYP model that has been suggested utilizes the RF approach for predicting crop production and the equilibrium optimizer (EO) plus kernel extreme learning machine (KELM) technique for recommending crops	The MMML–CRYP model makes use of the RF technique for crop yield prediction and the EO with the KELM technique for crop recommendation. Many benchmark datasets and simulations are used to assess the model	The MMML–CRYP method performs noteworthy in crop recommendation and yield prediction, with a maximum accuracy of 97.91% Considering a score of 98.54, the MMML–CRYP technique exceeds other current algorithms in this regard

9.3 Proposed Research Work

This chapter aims to change farming systems by creating practical models for guiding fertilizer application rates in smart agriculture, with attention given to neural networks and modern machine learning techniques in addressing any inadequacies present in previous manure application strategies. For producing accurate and personalized fertilizer recommendations, the study will combine a variety of data sources like soil sensors, climate data, or satellite images to come up with an all-inclusive model. Greater attention should be paid to the temporal dynamics aimed at making sure that the model changes correspond to those distinct growth periods in fertilizer demand for crops.

The main goal of this project is to provide farmers with a tool that can help them make fertilizer decisions that are sustainable. The anticipated result shows a brand new but better-performing strategy in using neural network prediction models within the context of smart agriculture to improve agricultural output, efficient use of resources as well as environmental sustainability.

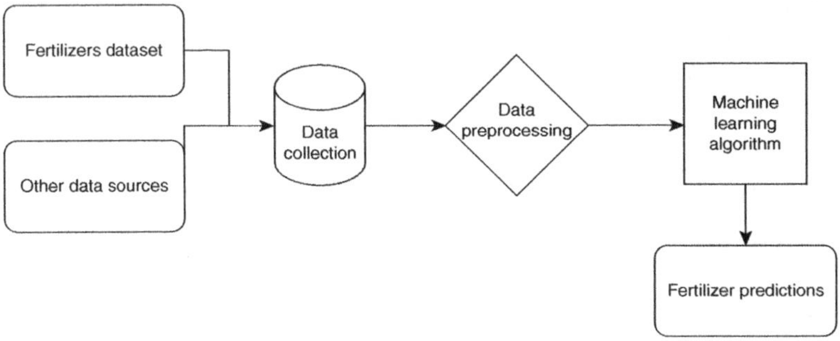

System Architecture

Let us understand the components of system architecture

- Fertilizers Dataset
  The basic building block is a vast dataset that contains information about different types of fertilizers, previous application data, and their compositions. It serves as the knowledge base necessary for training and validating the machine learning model.

- Other Data Sources
  There are other data sources like meteorological data, soil sensor data, and satellite imagery that are also used as supplementary information regarding the dataset. These are called such because they, e.g. enhance precision and flexibility through several features that affect crop health and nutrient requirement levels.
- Data Collection
  This approach involves the careful collection of data from state-of-the-art sensing devices together with conventional data sources in farming. This way, a large and complex data set is obtained, which depicts instabilities in various soil, crops, and climatic conditions.
- Data Preprocessing
  Techniques such as feature engineering, cleaning, and normalization are used to preprocess data prior to training the model. They help fix mistakes in the dataset, thus making it suitable for different machine learning algorithms.
- Machine Learning Algorithm
  Applying a neural-network-based machine learning approach to determine complex patterns in the dataset. The predictive model for fertilizer recommendations is constructed by training the algorithm on historical data, where different factors such as soil type, weather conditions, and crop type are included.
- Fertilizer Predictions
  The algorithm that is trained and learned is now applied to input data given by farmers to predict the amounts of fertilizer that they should be using. This suggestion is customized to suit every farmer depending on their type of crop and its stage of growth to improve the quality of agriculture.

9.4 Methodology and Concepts

A new concept of combining smart agriculture with up-to-date ideas and technology. The predictive modeling system uses neural networks and incorporates long short-term memory (LSTM) nets. This technology is important as it allows us to process time series data while detecting temporal trends in developing crops and the need for fertilizers. The aim of this is to combine numerous data sources like soil sensor measurements and satellite images, this helps us to obtain an encompassing view of the farming environment. This combination of concepts and technology guarantees a holistic approach toward the quality of fertilizer suggestions.

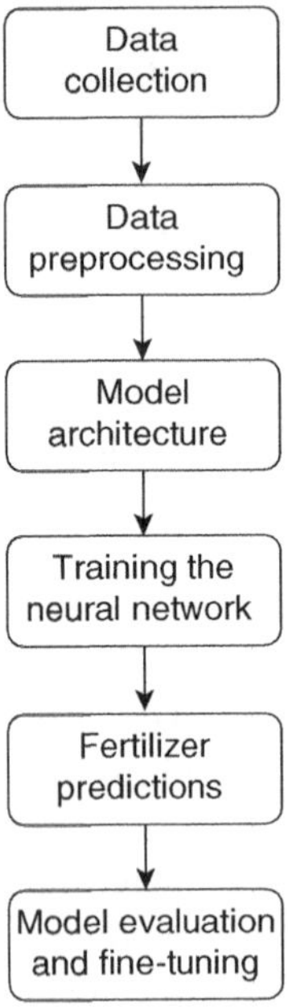

Execution Flow

For training and validation of the neural network, the collection of satellite imagery files and information from ground-based sensors like soil moisture levels, weather forecasts and historical records indicating locations of previous applications must be done. This is the training set.

You should clean and standardize collected in such a way data consistent without noise before performing any feature engineering, which will improve its value making it relevant when training with neural networks.

The structure of a neural network should blend regular layers with LSTM layers, which will enable it to perceive the dynamism in nutritional demands during different times.

Assign randomly produced initial values to the weights and biases of the neural network. Integrate the preprocessed data into the network, thereafter, apply the gradient descent method and backpropagation to train it. To that end, the network adapts itself so that the disparity between projected and actual fertilizer needs might reduce.

Once it has been trained and integrated, use the neural network to predict fertilizer. Sent in the network are historical data together with contemporary environmental conditions, which are used to compute the optimal type and quantity of fertilizer needed for a specific crop and soil type.

Validate the model with test data that is separate from the training datasets. Modify the artificial intelligence (AI) so as to improve accuracy and generalize well, ensuring that it can adapt efficiently to incorporate new data and give consistent fertilizer recommendations.

9.5 Implementation and Execution flow

9.5.1 Data Selection

We opted for Kaggle's fertilizers dataset, which has such attributes as temperature, humidity, moisture, soil, crop, nitrogen, potassium, phosphorus, and name of fertilizer. It has 205 rows and 9 columns.

Dataset source: https://www.kaggle.com/datasets/coder0311/crops/data?select=fertilizer.csv

9.5.2 Data Analyzing

By using EDA methods, we can understand the distribution of data, correlations as well as statistical characteristics. The identification of patterns and trends in fertilizer use, crop health, and environmental characteristics requires the use of sophisticated statistical techniques and visualization tools. This profound research forms the basis for feature engineering and selection, which control the subsequent phases of model building. We must extract the appropriate information from the data to create a neural network that can quickly train and forecast the optimal fertilizer recommendations for a variety of farming situations.

Dataset Sample

	Temparature	Humidity	Moisture	Soil type	Crop type	Nitrogen	Potassium	Phosphorous	Fertilizer name
0	26	52	38	Sandy	Maize	37	0	0	Urea
1	29	52	45	Loamy	Sugarcane	12	0	36	DAP
2	34	65	62	Black	Cotton	7	9	30	14-35-14
3	32	62	34	Red	Tobacco	22	0	20	28-28
4	28	54	46	Clayey	Paddy	35	0	0	Urea

Let us understand some statistical information about the dataset feature

	Count	Mean	Std	Min	25%	50%	75%	Max
Temparature	205.0	30.204878	3.592882	25.0	27.0	30.0	33.0	38.0
Humidity	205.0	58.990244	5.973791	50.0	54.0	60.0	64.0	72.0
Moisture	205.0	43.590244	11.236665	25.0	34.0	42.0	51.0	65.0
Nitrogen	205.0	18.814634	11.273115	4.0	10.0	14.0	24.0	42.0
Potassium	205.0	3.219512	5.571778	0.0	0.0	0.0	7.0	19.0
Phosphorous	205.0	18.634146	13.494654	0.0	9.0	18.0	30.0	42.0

9.5.3 Data Preprocessing

Data preprocessing is important for importing the raw dataset and making it easier for the neural network to train. There are three steps involved in this process, which include identifying unusual/unique items from a group and eliminating them; identifying ways of dealing with missing items such as filling them or removing them altogether; and finally, making sure that numerical data has similar weight when undertaking statistical analysis by standardizing it. Feature engineering is applied to generate new variables describing complex relations among many factors influencing the need for fertilizer, thus enhancing the demographic appropriateness of the dataset. Checking for the Null values in the dataset

Checking outliers for several features

Temparature	0
Humidity	0
Moisture	0
Soil type	0
Crop type	0
Nitrogen	0
Potassium	0
Phosphorous	0
Fertilizer name	0
dtype: int64	

<class 'pandas.core.frame.DataFrame'>

RangeIndex: 205 entries, 0 to 204

Data columns (total nine columns):

#	Column	Non-Null Count	Dtype
0	Temparature	205 non-null	int64
1	Humidity	205 non-null	int64
2	Moisture	205 non-null	int64
3	Soil type	205 non-null	object
4	Crop type	205 non-null	object
5	Nitrogen	205 non-null	int64
6	Potassium	205 non-null	int64
7	Phosphorous	205 non-null	int64
8	Fertilizer name	205 non-null	object

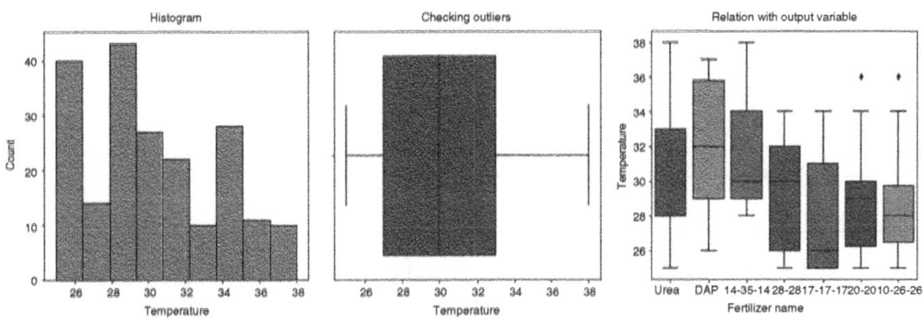

Temperature Feature

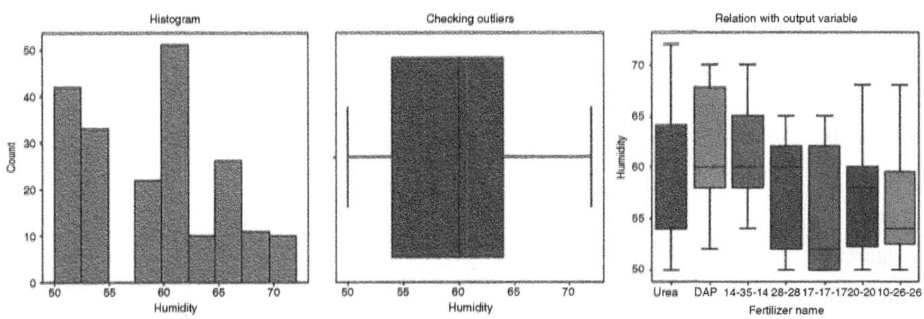

Humidity Feature

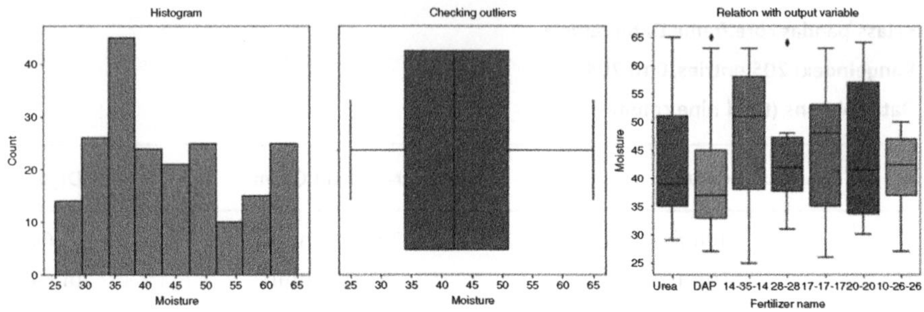

Moisture Feature

Now Label Encoding Soil Type and Crop Type

Count Plot and Distribution of Fertilizer Name (Target Feature)

	Encoded
Original	
Black	0
Clayey	1
Loamy	2
Red	3
Sandy	4

	Encoded
Original	
Barley	0
Cotton	1
Ground Nuts	2
Maize	3
Millets	4
Oil seeds	5
Paddy	6
Pulses	7
Sugarcane	8
Tobacco	9
Wheat	10

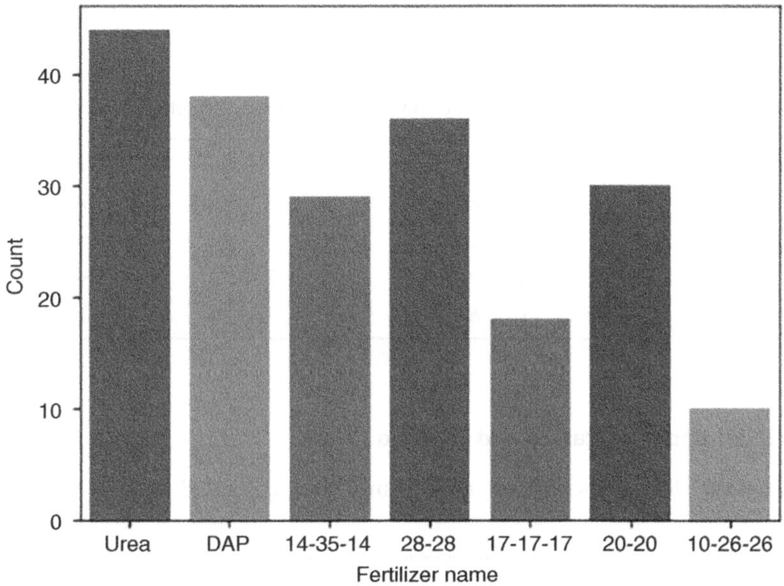

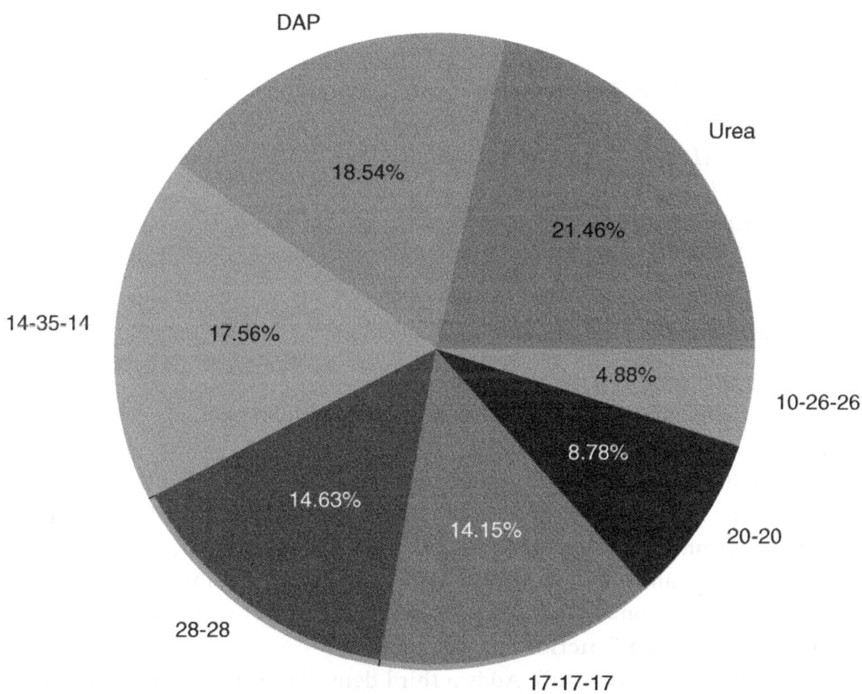

We are also **applying the synthetic minority over-sampling technique (SMOTE)** for good accuracy because of an imbalanced dataset.

Sample Dataset After Preprocessing

	Temparature	Humidity	Moisture	Soil type	Crop type	Nitrogen	Potassium	Phosphorous	Fertilizer name
0	26	52	38	4	3	37	0	0	Urea
1	29	52	45	2	8	12	0	36	DAP
2	14	65	62	0	1	7	9	30	14-35-14
3	32	62	34	3	9	22	0	20	28-28
4	28	54	46	1	6	35	0	0	Urea

9.5.4 Model Implementation and Training

Neural network. Now let us understand our implemented model

Algorithm 9.1 Neural Network Model [6]

Require: $num\_classes = 7$
1: $model \leftarrow Sequential()$
2: $model.add(Dense(128, activation = \text{'}relu\text{'}, input\_shape = (x\_train.shape[1],)))$
3: $model.add(Dense(64, activation = \text{'}relu\text{'}))$
4: $model.add(Dense(32, activation = \text{'}relu\text{'}))$
5: $model.add(Dense(num\_classes, activation = \text{'}softmax\text{'}))$
6: $model.summary()$

Let us understand the above model in detail

1) Defining the Neural Network Architecture
   model = Sequential() initializes an empty sequential model.
2) Adding Layers to the Model
   Dense(128, activation='relu', input_shape=(x_train.shape[1],)): Adds the first dense layer with **128 neurons**, using the rectified linear unit **(ReLU) activation function.** The shape of the input data is determined by the number of features in x_train, and it is defined by the input_shape parameter.
   Dense(64, activation='relu'): Adds a second dense layer with **64 neurons** and a **ReLU activation function.**
   Dense(32, activation='relu'): Adds a third dense layer with 32 neurons and a **ReLU activation function.**

3) Output Layer
   Dense(num_classes, activation='softmax'): Adds the output layer with a number of neurons equal to the number of classes (num_classes). The **softmax activation function** is used for multi-class classification problems, providing probability distributions over the different classes.
4) Model Summary
   model.summary() prints a summary of the model, including the type of each layer, the number of parameters, and the output shape at each step. This helps to visualize the architecture and understand the parameters involved in the neural network.

Model: "sequential"

Layer (type)	Output shape	Param #
dense (Dense)	(None, 128)	1152
dense_1 (Dense)	(None, 64)	8256
dense_2 (Dense)	(None, 32)	2080
dense_3 (Dense)	(None, 7)	231
Total params: 11,719		
Trainable params: 11,719		
Non-trainable params: 0		

9.5.4.1 Long Short-Term Memory (LSTM)

Keras designs neural networks through LSTM. It employs a sequential model composed of three LSTM layers. All these three LSTM layers have 64, 64, and 32 neurons with ReLU activation functions. To reduce overfitting, dropout layers are also applied. After the last LSTM layer, a flattened layer then transforms the 3D output to a 2D format. Adding two layers with 50 units each, followed by an additional one that has more units but less than "num_classes" neurons will do fine; all we need at last is the "Softmax" activation function, which shows how many classes were involved because sometimes it becomes unclear when nothing is written before on the number of categories; this helps one know what is happening most of the time, thus making it easier for those using them, especially when doing multiple class classification. By constructing your model, you are using the Adam optimizer if there is no specification about it, then by default, we can use sparse categorical cross-entropy as our loss function. The description of the architecture, such as layer types, output shapes, and total parameters, is available for this model using the summary() function.

Model: "sequential_1"

Layer (type)	Output shape	Param #
lstm (LSTM)	(None, 1, 64)	18688
dropout (Dropout)	(None, 1, 64)	0
lstm_1 (LSTM)	(None, 1, 64)	33024
dropout_1 (Dropout)	(None, 1, 64)	0
lstm_2 (LSTM)	(None, 32)	12416
flatten (Flatten)	(None, 32)	0
dense_4 (Dense)	(None, 50)	1650
dense_5 (Dense)	(None, 7)	357
Total params: 66,135		
Trainable params: 66,135		
Non-trainable params: 0		

9.6 Results

For neural network

```
5/5 [==============================] - 0s 2ms/step -
loss: 3.5734e-04 - accuracy: 1.0000
```

```
Test loss: 0.00035733584081754O9
```

```
Test accuracy: 1.0
```

Accuracy: 100%

Classification reports

Precision	**100%**
Recall	100%
F1-score	100%
Accuracy	100%

For LSTM

```
5/5 [==============================] - 0s 3ms/step -
loss: 0.0076 - accuracy: 1.0000

Test loss: 0.007603965699672699

Test accuracy: 1.0
```

Accuracy: 100%

Classification report

Precision	**100%**
Recall	100%
F1-score	100%
Accuracy	100%

Test failure

Neural network failures	**0**
LSTM failures	0

Comparisons with other machine learning algorithms

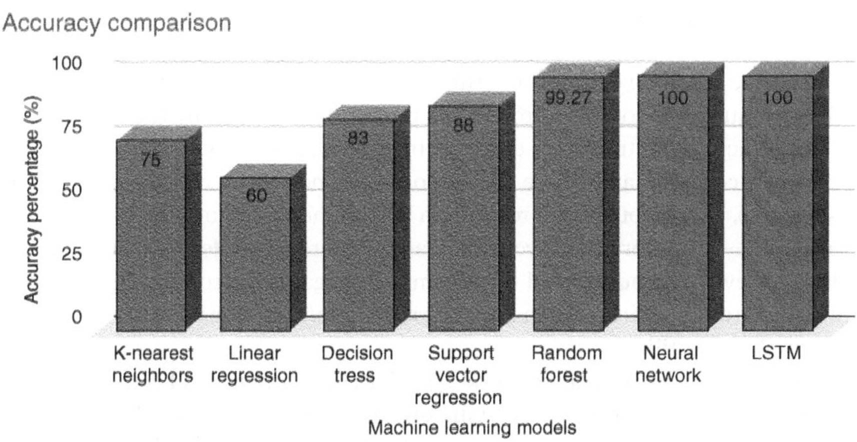

9.7 Discussion

The performance of the proposed model in various agricultural settings determines its accuracy as well as generality levels. It is crucial to mention that this model's strength lies in its capacity to handle alternate soil compositions and weather patterns changes such as correcting an uneven supply of fertilizers. The incorporation of time-varying needs through the growing period within a crop has thus required modification on some of the outputs made by the model.

The conversation explores the impact of data preparation on model performance through domain analysis, pointing out the parts played by normalization, feature engineering, and temporal structure in enhancing the resilience of neural networks. It also covers difficulties presented by skewed agricultural datasets, such as SMOTE oversampling for handling class imbalances.

The openly discussed disadvantages of the methodology relate to the variance in model accuracy caused by data points that can alter results greatly and how acquiring up-to-date data is quite problematic. Avenues for future research are considered, including the combination of transparent AI systems and the addition of more data for enhanced interpretation by models [26, 27].

By bringing the context of the proposed neural network model's contributions into the picture, comparisons have been made with existing machine learning approaches and traditional fertilizer recommendation methods. Precision farming and sustainable agriculture, in general, are other broader research implications that are also scrutinized with an emphasis on increasing resource efficiency and cutting down on negative environmental effects [24–27].

9.8 Conclusion

Sophisticated machine learning algorithms are useful in predicting the amount of fertilizer required for optimal crop development. This is seen by the fact that its 100% accuracy, in contrast with different methodologies, shows how precise and stable such concepts are. There are important conclusions we can make from these results, which could see a revolution within the nature of precision agriculture, all in favor of more adaptive and data-concentrated strategies. This research adds to an ever-expanding pool of information regarding eco-farming methods available; through it farmers can acquire credible information they need to know when applying fertilizer.

The reason why the neural network–LSTM architecture is great is its capability to find intricate patterns in time-varying diverse datasets such as soil characteristics, weather conditions, and past crop yield. Moreover, to tackle class imbalance issues, over-sampling techniques, including SMOTE have been used, resulting in improved capability of the model to make precise suggestions for many kinds of fertilizers. The research discovered that advanced machine learning techniques can revolutionize smart agriculture and help increase agricultural productivity, resource efficiency, and environmental sustainability.

References

1 Devi, O.R., Lakshmi, P.N., Babu, S.N., and Bai, K.V.S. (2023, April). Fertilizer forecasting using machine learning. In: *2023 International Conference on Inventive Computation Technologies (ICICT)*, 24–27. IEEE.

2 Sanush, B.R., Vijay, S., Soujan, W.D. et al. (2023). Intelligent farming techniques using machine learning. *International Research Journal on Advanced Science Hub* 5 (05): 248–253.

3 Pabitha, C., Benila, S. and Suresh, A., 2023. A digital footprint in enhancing agricultural practices with improved production using machine learning.

4 Durai, S.K.S. and Shamili, M.D. (2022). Smart farming using machine learning and deep learning techniques. *Decision Analytics Journal 3*: 100041.

5 Gopi, P. and Karthikeyan, M. (2023). Multimodal machine learning based crop recommendation and yield prediction model. *Intelligent Automation & Soft Computing* 36: 313–326.

6 https://scikit-learn.org/stable/modules/neural_networks_supervised.html.

7 https://www.tensorflow.org/api_docs/python/tf/keras/layers/LSTM.

8 https://www.kaggle.com/.

9 Elbasi, E. et al. (2023). Crop prediction model using machine learning algorithms. *Applied Sciences* 13 (16): 9288.

10 Sumathi, P. et al. (2023). Improved soil quality prediction model using deep learning for smart agriculture systems. *Computer Systems Science and Engineering* 45 (2).

11 Gupta, A. and Nahar, P. (2023). Classification and yield prediction in smart agriculture system using IoT. *Journal of Ambient Intelligence and Humanized Computing* 14 (8): 10235–10244.

12 Kwaghtyo, D. and Eke, C. (2023). Smart farming prediction models for precision agriculture: a comprehensive survey. *Artificial Intelligence Review* 56 (6): 5729–5772.

13 Alex, N., Sobin, C., and Ali, J. (2023). A comprehensive study on smart agriculture applications in India. *Wireless Personal Communications* 129 (4): 2345–2385.

14 Baburao, P. et al (2023). Review of machine learning model applications in precision agriculture. *International Conference on Applications of Machine Intelligence and Data Analytics (ICAMIDA 2022)*, pp. 916–930.

15 Saranya, T., Deisy, C., Sridevi, S. and Anbananthen, K.S.M., 2023. A comparative study of deep learning and Internet of Things for precision agriculture. *Engineering Applications of Artificial Intelligence, 122,* p.106034.

16 Ahmad, A., Saraswat, D., and El Gamal, A. (2023). A survey on using deep learning techniques for plant disease diagnosis and recommendations for development of appropriate tools. *Smart Agricultural Technology* 3: 100083.

17 Punithavathi, R. et al. (2023). Computer vision and deep learning-enabled weed detection model for precision agriculture. *Computer Systems Science and Engineering* 44 (3): 2759–2774.

18 Bouguettaya, A. et al. (2023). A survey on deep learning-based identification of plant and crop diseases from UAV-based aerial images. *Cluster Computing* 26 (2): 1297–1317.

19 Mummigatti, K., Chandramouli, S., and Ramachandra, D. (2023). Deep neural network system using ontology to recommend organic fertilizers for a sustainable agriculture. *Ingenierie des Systemes d'Information* 28 (2).

20 Thorat, T., Patle, B., and Kashyap, S. (2023). Intelligent insecticide and fertilizer recommendation system based on TPF-CNN for smart farming. *Smart Agricultural Technology* 3: 100114.

21 Thendral, R. and Vinothini, M. (2023). Crop and fertilizer recommendation to improve crop yield using deep learning. *Second International Conference on Electronics and Renewable Systems (ICEARS)* 2023: 1123–1130.

22 Aruna, P. (2023). Crop yield prediction and fertilizer recommendation using artificial intelligence.

23 Kolagani, R., Venkatram, J.R., Kemmasaram, V. et al. (2023). Easy agriculture: crops' disease detection, pesticide, fertilizer and crop recommendations. In: *E3S Web of Conferences*, vol. 391, 01055. EDP Sciences.

24 Kumar, A. and Jha, R. (2022). IoT in smart farming: a boon to farmers. In: *Internet of Everything: Smart Sensing Technologies* (ed. T. Kavitha, V.A. Devi, S.N. Pari, and S. Ramanathan), 185–206. Nova Publisher.

25 Singh, D. and Kumar, A. (2024). A Deep Recurrent Neural Network for Plant Disease Classification. *SN Computer Science* 5 (8): 1053.

26 Vohra, R. and Kumar, A. (eds.) (2024). *Advanced Geospatial Practices in Natural Environment Resource Management.* IGI Global.

27 Vohra, R., Kumar, A., and Rongali, G. (2024). Multi-scale extraction and spatial analysis of growth pattern changes in urban water bodies using sentinel-2 MSI imagery: a study in the central part of India. *Stochastic Environmental Research and Risk Assessment* 38 (6): 2397–2423.

10

Reviewing Advances in Image-Based Plant Disease Detection

Gautmi Tomar, Yuvraj Ahuja, Yogita Arora, and Neera Agarwal

Department of Electronics and Communication, Bharati Vidyapeeth's College of Engineering (BVCOE), GGSIPU, Delhi, New Delhi, India

10.1 Introduction

Even after giving the fields useful nutrients, the presence of illnesses in the planted plants drastically lowers the yield [1]. This highlights the importance of effective methods for identifying plant diseases. Farmers are worried about several plant diseases. Due to their potential to significantly disrupt the global food supply, plant diseases pose a special threat to small-scale farmers [2]. To create effective detection and preventive measures, it is imperative to identify the type of plant disease that is present as soon as possible. Recently, a great deal of work has gone into identifying plant diseases that are affecting a wide range of crops globally [2].

Nowadays, crop production makes up a sizable portion of agricultural output. The yield of crops has also drastically changed. A little modernization of agricultural agriculture has occurred with the development of knowledge technology. To increase productivity, modern agronomy makes use of the most sophisticated technological tools and procedures. The application of cutting-edge techniques to swiftly ascertain the optimal conditions for greater agricultural production has grown [3]. To encourage growth, a lot of fertilizer and insecticides are employed. Larger-scale genetically engineered seed trials are also being carried out in an attempt to increase global yield overall [4].

Emerging Smart Agricultural Practices Using Artificial Intelligence, First Edition.
Edited by Ashish Kumar, Jai Prakash Verma, and Rachna Jain.

Published 2025 by John Wiley & Sons, Inc.

Animals and humans both need agricultural products to suit their nutritional needs [5]. Everybody's life has been impacted by agriculture, whether directly or indirectly. Agriculture has advanced significantly as a result of the growth of technology in all facets of life. In addition to the range of crops cultivated, other pursuits like farming and livestock rearing were started. Plant disease detection is a critical component of integrated disease management in agriculture [6]. Farmers and agricultural stakeholders can take prompt action to limit the effects of illnesses when they are discovered in their early stages.

Not only can early illness diagnosis lessen dependence on chemical treatments, but it also adds to sustainable agriculture. Furthermore, early identification improves crop production efficiency by reducing resource waste and optimizing yield potential. The identification of factors that cause disease is making significant progress. Fungi infections induce diseases in certain cases, viral infections cause diseases in others. This becomes a serious problem when the original element cannot be found until it has spread to important production sections [7]. Table 10.1 tabulates the various segmentation techniques for plant disease detection.

The study identifies several key research gaps in the field of image-based plant disease detection, including challenges related to scalability, environmental variability, model generalization, and the limited availability of diverse datasets.

Table 10.1 A collection on plant disease segmentation techniques.

Segmentation techniques	Merits	Demerits
Clustering method	• Quicker when it comes to computation • *K*-means performs better the smaller the value of *K*	• It demands clusters of comparable sizes
Regional method	• The shift from the inner point to the outer region creates clearer entity boundaries • It outperforms other approaches in terms of reliability	• User seed selection that is noisy triggers improper segmentation • The splitting technique used in the region causes the splitting segments to seem square
Edge detection method	• It works best with images that have a high contrast between regions	• Worked poorly on a photo with a lot of edges
Threshold method	• Any previous understanding of the image is not required • It's simple to use and appropriate for real-world situations	• Extra sensitive to noise

Future research objectives are also explored, including the integration of multimodal imaging techniques, the development of strong frameworks for transfer learning, and the implementation of portable, reasonably priced imaging devices for field use. The study of the chapters helps to fill in these gap. Figure 10.1 illustrates the various plant disease detection and crop administration. The introduction emphasizes the promise of image-based techniques and stresses the significance of plant disease identification for the security of the world's food supply [8]. The literature study offers a thorough summary of the approaches, technologies, and constraints used today. The methodology chapter outlines the methods and strategies utilized to fill in the gaps that have been found by

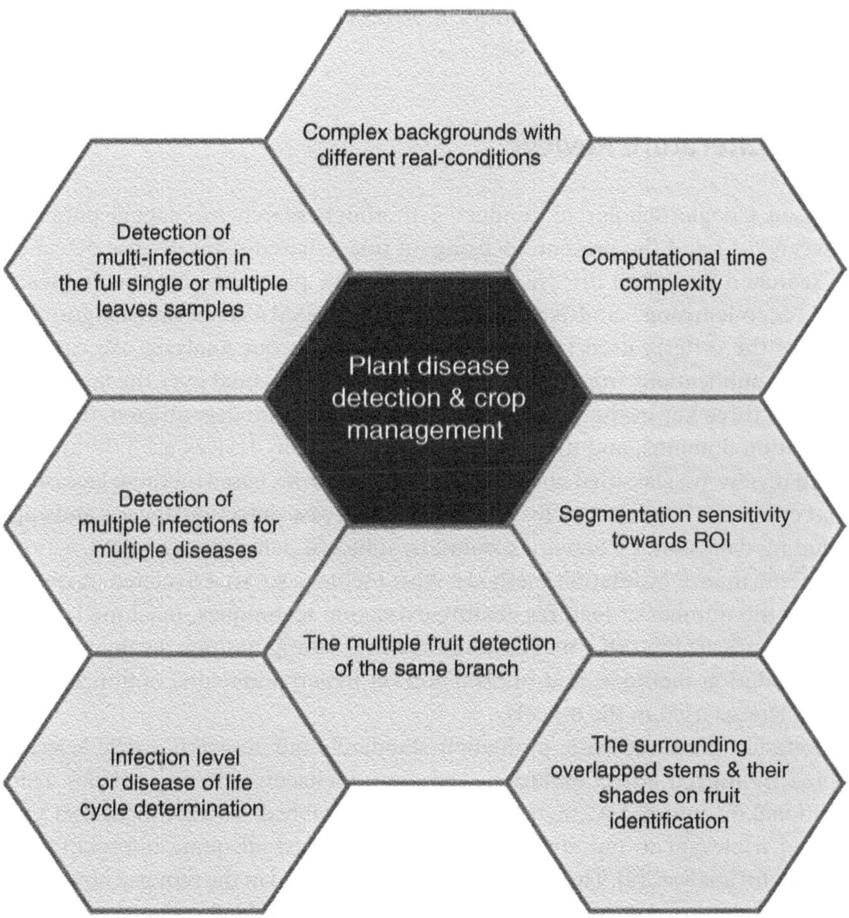

Figure 10.1 Present concerns in plant illness diagnosis and crop administration. *Source:* Hasan et al. [10]/MDPI/CC BY 4.0.

describing the experimental setup and the creation of new models and methods. To improve model robustness, the chapter on scalability and environmental variability looks at the difficulties in handling environmental variations and scaling detection systems. The chapter on model generalization investigates ways to make models more flexible in a variety of scenarios and datasets. The chapter on dataset availability discusses efforts to create and organize big datasets and stresses the value of comprehensive datasets for model training. The chapter on future directions delves into the possibilities of multimodal imaging and transfer learning. It suggests creating and implementing low cost, lightweight imaging equipment for real-world agricultural applications. The study concludes by summarizing its contributions and conclusions and highlighting the possible effects of image-based plant disease identification on crop health management and the security of food supply worldwide.

10.2 Literature Review

We used Google Scholar to conduct a thorough search for papers published between 2012 and the present, focusing on image-based plant disease detection. We searched for "plant disease detection," "image processing," "machine learning," "deep learning," and "sensor-based." The PRISMA diagram in Figure 10.2 displays the criteria used to include publications in our analysis. We carefully chose 49 publications from a pool of 1636 articles discovered over the last decade based on three key evaluation criteria: the types of AI models utilized, the target application domains, and the variety of data collection devices used. In our suggested review, we classified qualities into two categories based on these key assessment criteria. It covers seven features for plant disease detection devices and apps, including data sources, activities, datasets, subjects, detection scenarios, actions, and performance evaluation methods. Also includes seven AI-related attributes, such as the number of features, feature extraction techniques, machine learning (ML)/deep learning (DL) models, architectural configurations, evaluation metrics, validation methods, and information on hyperparameters, optimizers, and loss functions used in the models.

Alongside the three basic evaluation standards, our research article selection approach utilized three exclusion and four inclusion criteria. Articles using traditional or outdated AI methodologies were scrubbed, as were those that were deemed irrelevant to our study objectives and lacked adequate information for meaningful analysis [9]. These exclusion criteria resulted in the removal of 915, 213, and 44 articles (marked as eX1, eX2, and eX3 in the PRISMA flowchart), totaling 49 articles. Nonredundant articles were included, as well as those subjected to detailed abstract and conclusion screening, articles meeting eligibility criteria based on

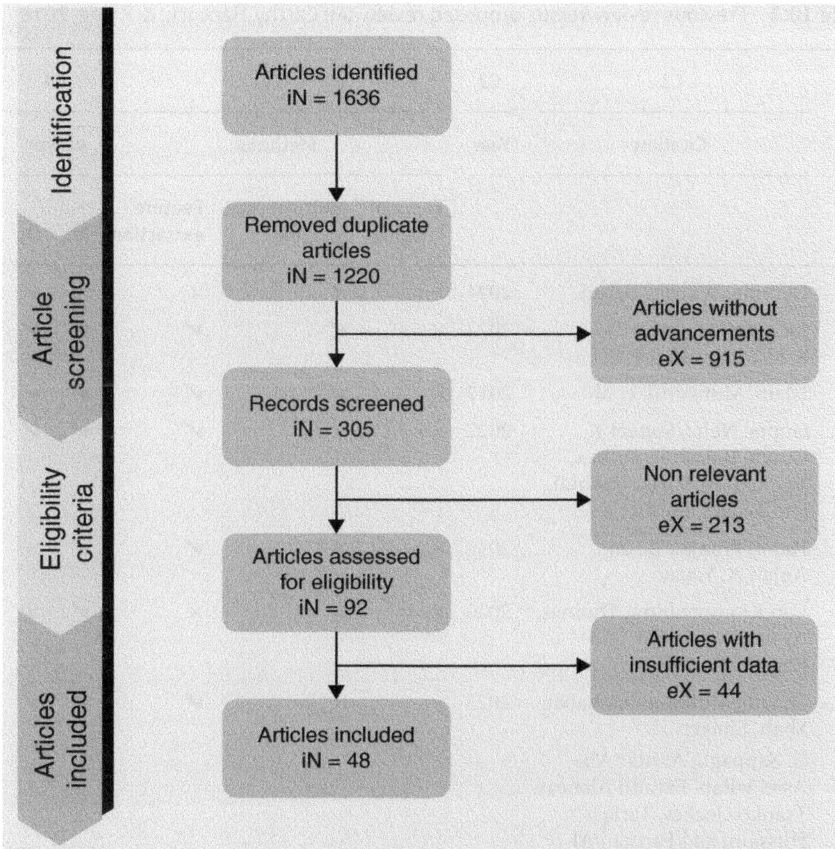

Figure 10.2 PRISMA model for study selection. (Jackulin & Murugavalli, 2022).

advanced AI techniques, target domain relevance, and device type, and articles undergoing qualitative synthesis, taking into account factors such as journal impact factor and authors' contributions to the plant disease domain. The inclusion criteria are labeled iN1, iN2, iN3, and iN4 in the PRISMA flow diagram [10].

In our suggested review, we conducted a thorough examination of plant disease detection, focusing on the interplay of AI techniques, device types, and application domains. One of our study's key findings is the presence of a symbiotic relationship between plant disease and AI techniques. We discovered that the decision type has a significant impact on the selection and effectiveness of AI approaches utilized in the system, whereas advances in AI techniques have allowed for the development of more sophisticated solutions that appeal to a wide range of application domains [44, 49, 51, 52]. This mutual link underscores the importance of taking a complete

Table 10.2 Previous review versus proposed review (McCarthy, Hancock, & Raine, 2010).

	C1	C2	C3			C4	
	Citation	Year	Methods			AI type	
			Thermal imaging	Multispectral imaging	Feature extraction	ML	DL
R1.	Demilie, Wubetu Barud	2024	×	×	×	×	✔
R2.	Jackulin, C., and S. Murugavalli	2022	×	✔	✔	✔	✔
R3.	Islam, Monzurul, et al	2017	×	×	✔	✔	✔
R4.	Gupta, Neha, Suneet K. Gupta, Rajesh K. Pathak, Vanita Jain, Parisa Rashidi, and Jasjit S. Suri	2022	×	×	✔	✔	✔
R5.	Padol, Pranjali B., and Anjali A. Yadav	2016	×	×	✔	✔	×
R6.	Van Klompenburg, Thomas, Ayalew Kassahun, and Cagatay Catal	2020	×	×	×	✔	×
R7.	Shoaib, Muhammad, Babar Shah, Shaker Ei-Sappagh, Akhtar Ali, Asad Ullah, Fayadh Alenezi, Tsanko Gechev, Tariq Hussain, and Farman Al	2023	×	×	✔	✔	✔
R8.	Proposed	2024	✔	✔	✔	✔	✔

approach to building and implementing plane disease detection systems, taking into account the intricate interdependence of technology components and real-world events [11]. Table 10.2 compares the various review articles in the domain.

10.3 Imaging Techniques of Plant Disease

10.3.1 Magnetic Resonance Imaging

This technique first creates a pulse and then transmits it to the body part being studied. After that, they absorb the energy that is given to them and spin in a different direction. This is referred to as the magnetic resonance imaging (MRI)'s resonance process. It is most commonly referred to as an MRI device, although it is also known as an NMR or nuclear magnetic resonance scanner. These magnets produce a detectable signal that is spatially recorded with distinct body representations by efficiently polarizing and activating the concentrated proton that is individually implanted in the water molecules in the tissue.

10.3.2 Photoacoustic Imaging

The technique known as photoacoustic imaging was created from hybrid biomedical imaging and is based on the photo acoustic effect. It is combined with many advantages, such as optical absorption contrast and ultrasonic spatial resolution enabling deep imaging of diffusive and other regimes. According to the research, photoacoustic imaging has a wide range of uses, including disease diagnosis, brain activity imaging, blood oxygen level mapping, and tumor analysis.

10.3.3 Thermography

Contact thermography, dynamic angio-thermography, and tele-thermography are the three most widely used methods. The notion of metabolic activity is employed by imaging thermographic digital technology. Vascular circulation around a spreading plant disease is also studied to determine its increased relevance. The use of noninvasive sensors for plant disease detection is covered in the study [12]. We are analyzing and contrasting hyperspehumanctral sensors, thermography, and chlorophyll fluorescence. Large volumes of data are recorded by hyperspectral sensors, which means that different strategies must be used to get the desired outcomes. Temperature is the most significant thermographic parameter. There is a challenge because these technologies haven't reached their full potential yet.

The article [13] discusses the many plant disease classification systems. Flow cytometry, polymerase chain reaction, enzyme-linked immunosorbent assay, and immunofluorescence are examples of direct approaches. Although the direct approach is easily obtainable, it necessitates the expertise of personnel, is difficult to operate, and takes time to analyze data. Furthermore, they are unsuitable for a number of examinations. Although indirect approaches are useful for on-site illness detection, their specificity for many diseases is limited. Since the development of nanotechnology, sensitive biosensors have advanced dramatically; the detecting component of these sensors can be enhanced further by employing DNA or antibiotics.

10.3.4 Spectroscopic and Imaging Technologies

The study [14] compares spectroscopy-based imaging technologies with profiling-based methods for assessing leaves for both normal health and disease. Using these technologies has the benefit of properly identifying plant diseases. The challenges associated with these approaches are figuring out which treatment is most appropriate for a given plant disease and automating methods for ongoing plant disease surveillance.

This work [15] provides a modern method of plant disease detection based on protein and nucleic acid analysis. This study describes a variety of lateral flow devices and mobility spectrometers that can directly identify early infections in fluid. It also describes how spectroscopy-based techniques and remote sensing technologies work together to produce excellent spatialization, which helps in the early diagnosis of plant diseases. The study describes all these technologies and their potential applications in the search for different plant diseases, including serological and nucleic acid-based methods.

10.3.5 Hyperspectral Imaging

Hyperspectral imaging makes use of spectroscopy in addition to numerous traditional imaging techniques to simultaneously merge diverse spectrum data. Finding the spectrum for the affected pixel in the image under review is the aim of this technique. A hyperspectral imaging device is employed for wavelength dispersion in addition to the standard computer vision system, and a transportation stage is provided [7, 16].

Stressors to plants can be classified as biotic or non-biotic, with biotic stressors being generated by organisms like fungi, viruses, and others. The study [16] uses thermal and hyperspectral imaging to detect fungal-induced plant stress in oilseed rape. Initially, temperature distributions were examined in oilseed rape thermal images. Additionally, hyperspectral images were taken. Kernel density curves were also used to compare temperature distributions. We performed another spectral analysis on the leaves with different reflectance values.

The article [17] suggests creating a weed detection system that uses hyperspectral imaging in order to control weed detection. Depending on their genus, plants are categorized as either crops or weeds. First, hyperspectral imaging is used to separate plants from the soil surface. Next, species and other criteria are used to classify the cropped image pixels. Different classes of variables are created and used for feature extraction and data compression. The fact that this classification system is practical is one of its biggest benefits.

The electromagnetic spectrum has wavelengths between 400 and 700 nm that are visible to the human eye (Figure 10.3) [18]. Hyperspectral imaging frequently

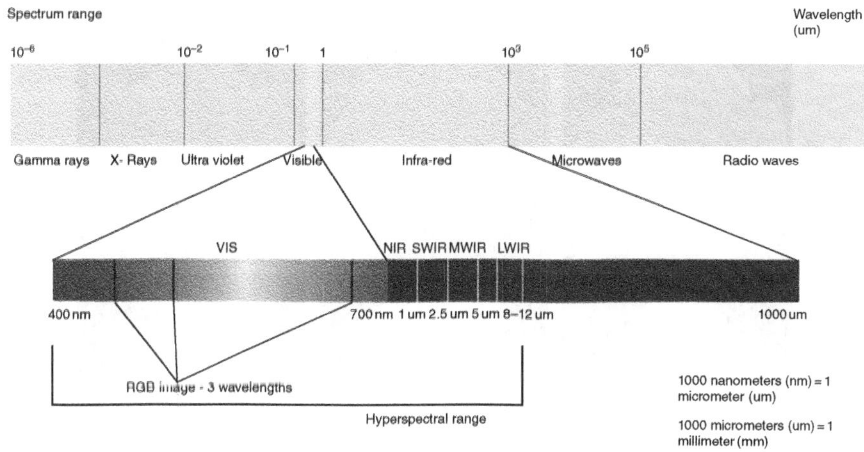

Figure 10.3 The electromagnetic spectrum's hyperspectral range. *Source:* Amy Lowe et al. [22]/with permission of Springer Nature.

includes several limited wavelength bands over a spectral range. This offers a large spatial resolution (thousands of data pixels per leaf) and a color dataset with meaningful information. The article discusses techniques for identifying different types of plant stressors, emphasizing hyperspectral imaging techniques for early disease detection and plant health prediction. Everyday, a number of indicators rise in importance and are essential for identifying particular features of the vegetation. The vegetation index, for instance, can be used to assess a plant's average health; however, the drawback lies in the fact that we are unable to apply the index set for one plant to data from another plant.

10.3.6 Fluorescence Techniques

Plant photosynthetic activity has often been studied using fluorescence methods (Figure 10.4). This technique is crucial for crop monitoring because it helps us minimize stress early on, which dramatically lowers yield losses.

This research [19] compares and contrasts fungal and viral infection in plant leaves using thermography and chlorophyll fluorescence imaging. This study compares pathogen–plant interactions under various growth conditions using both imaging techniques. First, the imaging of tobacco leaves is analyzed to look for lesions caused by TVM. Color reflectance is also used to check the use of salicylic acid treatment on tobacco leaves. The Cercospora Beticola spore solution is then applied to the beetroot once again. The use of several imaging techniques for plant-pathogen pairings to identify early indicators is covered in this study.

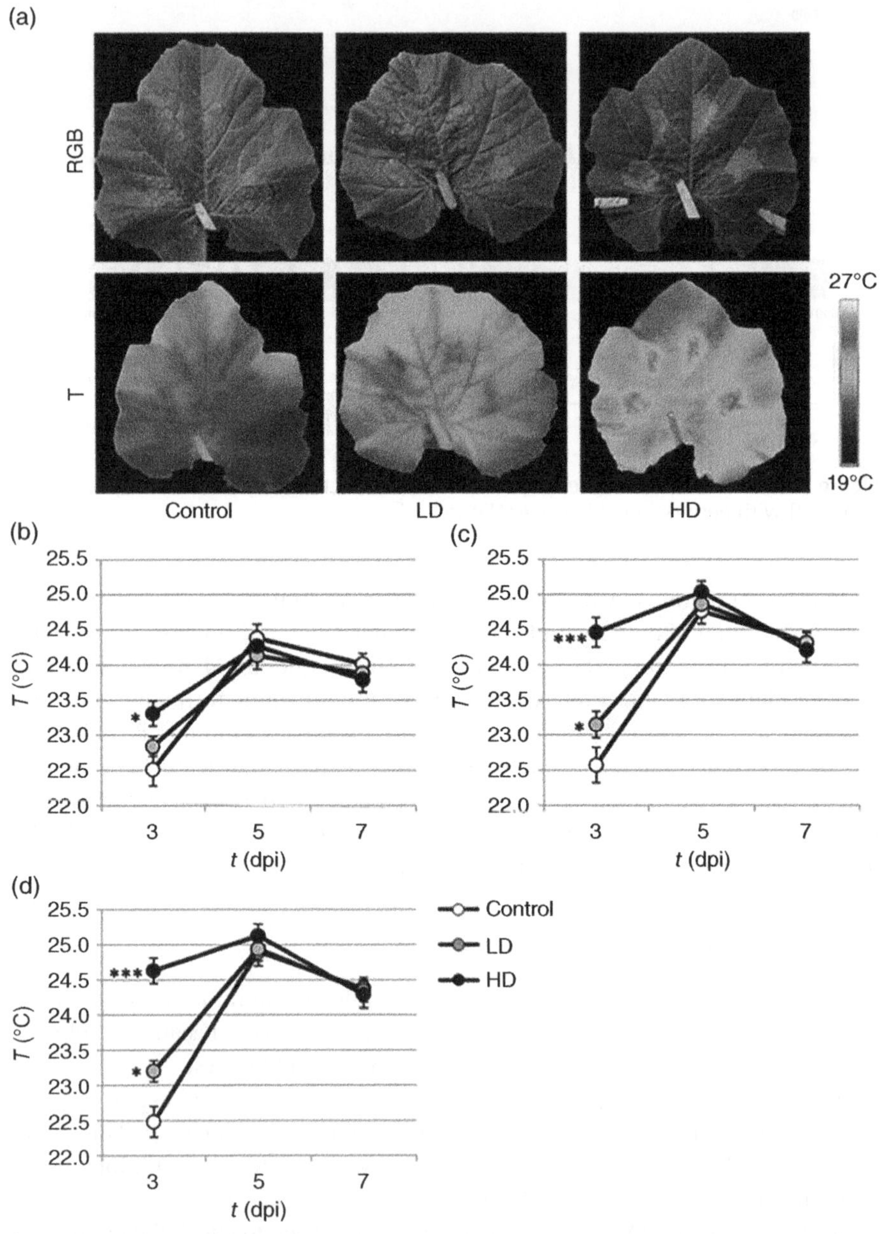

Figure 10.4 (a) Standard thermal images of mock-control, LD, and HD-inoculated zucchini leaves at 3 dpi, and their corresponding RGB images. (b) Evolution of temperature in I (b), N (c), and D (d) areas of the leaves along the period of study. Symbols * and *** stand for $p < 0.1$ and 0.001, respectively. *Source:* Pérez-Bueno et al. (2016)/ Frontiers Media S.A/CC BY 4.0.

In the work [20], it is discussed how fusarium head blight in wheat may be identified in the field and in the lab using hyperspectral imaging and chlorophyll fluorescence. Furthermore, a comparison is made between hyperspectral imaging and chlorophyll fluorescence imaging (CFI) analysis as two techniques for fusarium identification. While hyperspectral imaging helps with pertinent illness monitoring by using many wavelengths in image processing, the CFI methodology is helpful during the initial stage of inspection. Hyperspectral imaging is helpful in addressing the visibility of disease symptoms. Two passages will be required for any imaging technique on field recording of head blight, depending on whether the imaging equipment is used alone or in conjunction with other tools.

10.3.7 Thermal Imaging

It was first applied in defense-related fields, it is now widely used in many other fields, including agricultural engineering techniques. There are numerous tools and techniques available for thermal imaging of plants. The use of thermal imaging technology is especially advantageous for a wide range of agricultural processes both before and after harvest. In this site-specific context of crop management and precision farming, where the link between plants, soil, and water has been thoroughly studied by a number of researchers, thermal imaging is an important phenomenon [21]. Here, several fluorescence effects are described. This study addresses the application of several imaging techniques for plant–pathogen pairings. Figure 10.5 represents the various thermal techniques.

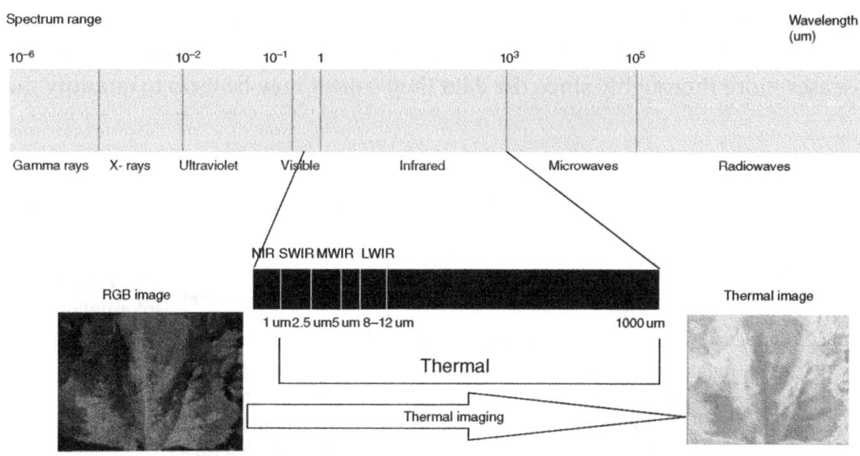

Figure 10.5 Thermal image technique. *Source:* Singh et al. 2020/with permission of Elsevier.

Since temperature measurement is directly correlated with a decrease in water quantity through stomatal aperture openings, digital infrared thermography can be used to detect plant diseases. This study [22] investigates if *Pseudoperonospora cubensis*-infected plants exhibit any abnormalities. First, the leaves were examined visually for signs of disease, and several flaws were noted to increase awareness of the possibility of disease. Later, spectral images were used to look into the presence of disease based on a number of variables, including stomata apertures and transpiration rate. Finally, an analysis is conducted of the known correlation between electrolyte loss and transpiration rate for the presence of disease.

10.3.8 3D Imaging

In two-dimensional (2D) imaging, information is gathered from two different dimensions that divide certain plant characteristics, such as growth, total height, and yield estimation. Therefore, it appears that three-dimensional (3D) imaging is essential for automatically identifying plant diseases. A variety of methods exist for obtaining 3D photographs [23–25], as well as 2D and 3D machinery for agricultural purposes [26, 27]. A 3D imaging system's block diagram is displayed in Figure 10.6.

Surface representations and volume representations are the two categories of 3D representations. The first includes the surface element, the depth data, and the numerous spots denoted by their dimension coordinates. Along with the volume, the coordinates of the model's volumetric and frequency components are also defined. There are now several kinds of low-cost sensors that are accessible.

Additionally, they demonstrate technological breakthroughs in a number of areas, including biomass estimation, crop presence, growth level, nutritional content, and height and health. These sensors are being used to study plant leaf diseases more thoroughly since the data they collect may be used to quantify the

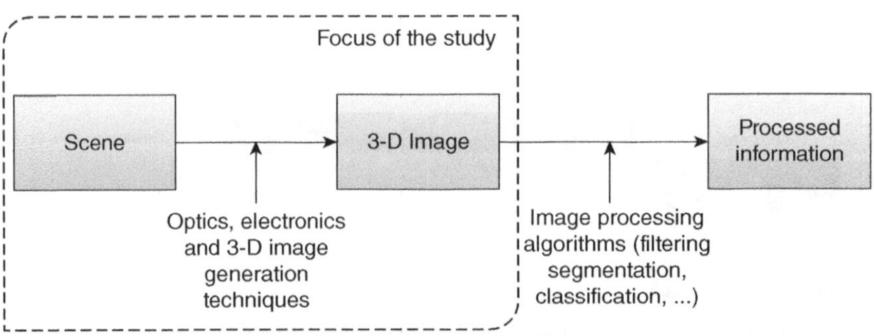

Figure 10.6 Block diagram of 3D imaging. *Source:* Vázquez-Arellano et al. [31]/CC BY.

variety of production characteristics that have previously been identified. Light velocity is used by sensors – like different kinds of LiDARs and Time-of-flight (TOF) cameras – to measure depth [28, 29].

Using this technique, one of the most prevalent diseases detected in cotton leaves, Fusarium wilt, may be easily identified. Other diseases, such as root rot, gray mildew, leaf blight, bacterial blight, bolt rot, and leaf curl, have also been identified.

Arivazhagan talked about a four-step early disease detection method in [31]. It also shows how software is used for automated identification. The various procedures discussed here are aimed at producing a color alteration for the specified picture. These are then followed by the deletion of particular threshold values as well as segmentation. Additionally investigated are the features extraction and texture information computation. It is at the moment with a 94% accuracy rate, and was tested on 500 plant samples, including lemon, tomato, beans, and bananas.

Pomegranate leaves can be used to identify disease, as shown in the article [31]. The method exhibited also helps in determining the precise deteriorating stage of the illness. This study describes the application of fuzzy logic, Bayesian networks, and decision tree learning.

Sladojevic [32] describes in their research a deep neural network method for identifying plant diseases from leaf photos. The developed methodology makes plant disease diagnostics easier. Furthermore, a way to distinguish them has been demonstrated. Using Caffe, deep convolution neural network (CNN) training is provided. The trained model's final accuracy was 96.3%. The issue that is brought up here is the requirement for image acquisition in order to improve model accuracy and to carry out the necessary augmentation.

Since disease spots in plants have varied hues, this research [32] advises color-transforming RGB photos to better disease spot segmentation. This allows for the identification of disease spots in affected areas. The CIELAB model is distinguished from other color models, such as YCbCr and CIELAB, by its "A" component. The OTSU threshold is used to segment the spots in the color-transformed image after a median filter has been applied. This method's benefit is that background, vein, and camera flash noise are eliminated thanks to the CIELAB model. The sole remaining task is to distinguish veins from areas of similar color.

Figure 10.7 illustrates the procedural steps for plant disease detection and classification. Agriculture and plant disease management are two fields in which artificial intelligence (AI) is crucial. AI-powered solutions can be used by farmers and agricultural specialists to identify, treat, and manage plant diseases, improving crop health and production. To identify illness signs early on, AI systems employ a variety of data sources, including images, sensor data, weather patterns, and historical records. This makes it possible to take prompt action to stop the spread of illness and lower agricultural losses, such as crop rotation or targeted pesticide

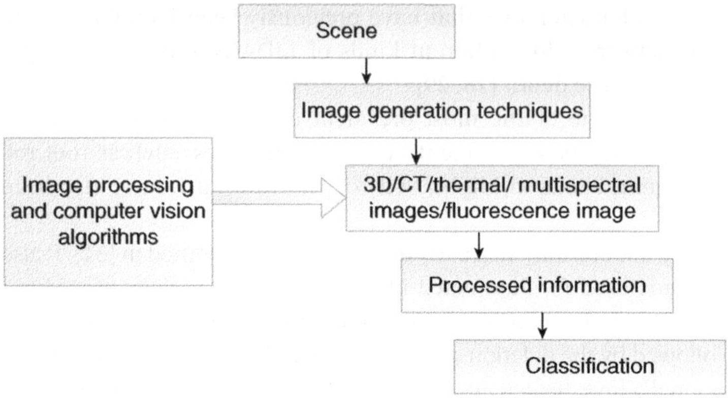

Figure 10.7 Plant disease detection and classification. *Source:* Sannakki et al. [37]/ CC BY 4.0.

use. Proactive management tactics can also be enabled by AI-powered predictive models, which can predict illness outbreaks and offer preventive actions. In general, AI gives farmers valuable information and instruments to help them make decisions, which supports resilient and sustainable farming.

A study on DL methodology-based tomato plant leaf disease identification was proposed by the authors "SurampalliAshok'et al." for 2020 [33]. This suggested method is offered as a way to diagnose crop leaf infection using picture preparation techniques, with an emphasis on tomato crops. It uses bunching, picture segmentation, and open-source techniques to deliver a dependable, safe, and precise arrangement of leaf disease.

The fundamental purpose of a CNN is to recognize images from a specific perspective. It deviates dramatically from earlier neural network techniques [34]. In general, CNNs employ minor preprocessing contrasts when computing various images.

This indicates that the classification learns the channel, which is typically hand-built. The core square of a CNN is represented by its convolutional layer structure. Figure 10.8 illustrates the basic architecture of convolutional neural network. The parameters layer consists of a collection of learnable channels (or pieces), however having a small open field results in full information volume complexity.

ML and DL are two independent strategies for solving data-driven challenges. Feature engineering is important in ML because it trains algorithms like decision trees, support vector machines, and random forests with custom features derived from domain expertise. These models perform best in settings with well-defined problems and small datasets, producing interpretable results while using fewer computer resources [36].

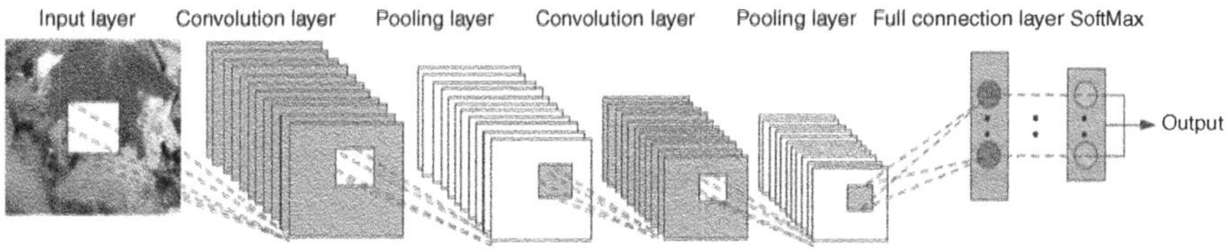

Figure 10.8 The basic structure of CNN. *Source:* Liu and Wang [42]/Springer Nature/ CC BY 4.0.

10.4 Critical Discussion

10.4.1 Strength

First of all, it emphasizes the significance of early identification by acknowledging the complex nature of plant diseases and their effects on productivity. The book also recognizes how modern technology, including IoT, AI, and UAVs, has revolutionized disease detection methods. Additionally, it highlights how early diagnosis can optimize resource utilization and lessen reliance on pharmaceutical treatments – two factors that support sustainability [30, 50]. The paragraph also acknowledges the challenges associated with controlling diseases, such as determining their etiology and addressing their global spread, which is exacerbated by the effects of climate change. Last but not least, the use of DL models emphasizes the text's grasp of cutting-edge technologies and their potential to improve detection accuracy in agricultural situations.

10.4.2 Weakness

It first calls attention to the dearth of specific examples or case studies that elucidate real-world uses of disease detection technologies. The lack of concrete examples makes it more difficult for the reader to understand how the concept can be used in real-world circumstances. Second, the use of pesticides and genetically modified seeds are two significant ethical problems that are contentious in the agricultural community and are ignored in the text. These problems have to do with farming methods. Third, while the text notes that disease dynamics are impacted by climate change, it does not go into great detail on how these changes impact the transmission of disease or advocate for the adoption of adaptive management strategies.

Furthermore, there isn't a thorough examination of the security and privacy risks related to the use of IoT and AI technologies in illness diagnosis. At last, even though the article mentions the difficulties DL models encounter, it fails to critically analyze these constraints and the negative effects of using DL techniques in agricultural contexts.

10.4.3 Benchmarking

10.4.4 Challenges and Future Trends

This publication has already disclosed significant research conducted in the past few years examining a variety of methodologies. An article shows how to use a method to detect disease in pomegranate plants [37]. There are numerous training methods that have been found by using the back propagation method. Fruit rot and bacterial blight were two of the main illnesses discovered. It also described how to generate image clusters using *k*-mean clustering. This illustrates the importance of early disease identification. First, the background was hidden in order to extract the region of interest. Additionally, using color and texture characteristics, a multicast support vector classifier is being used to distinguish between the presence and absence of disease [4]. Afterward, particular attributes like homogeneity, correlation contrast, and energy are extracted using the Gray Level Co-occurrence technique.

The challenge with this method is that it will eventually require sensors for real-world use. The SVM classification algorithm in the publication [38] is used to detect and further categorize grape leaf disease. This article discusses two fungus infections that can harm grape leaves: powdery mildew and downy mildew. Additionally, classification is done all throughout [39].

10.4.5 Limitations

In the subject of agricultural research, it is imperative to recognize the various constraints that are inherent in this essential area of study. Despite being essential to provide for both human and animal nutritional needs, agriculture faces a number of difficulties. With agriculture frequently being connected to problems like soil erosion, water pollution, deforestation, and greenhouse gas emissions, environmental sustainability is an urgent concern. These effects contribute to the larger problem of climate change and have far-reaching effects on ecosystems. In addition, the resource-intensive character of agriculture – specifically, the high demand for land, energy, and water – becomes more troublesome as the world's population grows, placing further pressure on scarce resources [35, 43]. Given the predilection for monocultures in contemporary farming techniques, which can disturb ecosystems and harm beneficial species like pollinators, biodiversity loss is a related concern. In addition, there are still obstacles to overcome in order to guarantee food security, including uneven resource distribution, crop failures brought on by climate change, and barriers to market access. Many farmers, especially those who operate on a smaller scale, face financial risks due to economic obstacles such as variable crop prices and high input expenses. Consumers and those involved in agriculture alike are concerned about the health effects of synthetic fertilizers and pesticides, which may also have an influence on the quality of food. Furthermore, the susceptibility of agriculture to the consequences of climate change, limited water resources, disputes over land use, and moral and societal concerns like labor rights and work conditions highlight the intricacies present in this area of study. In order to establish resilient and sustainable agricultural techniques that can fulfill the increasing demands of a changing and dynamic world, it is imperative that these limits be addressed.

10.5 Conclusion

In this chapter, numerous approaches for diagnosing ailments are outlined. It also provides a brief overview of the various imaging modalities that are helpful in early plant disease detection. We draw attention to the most recent advancements and challenges in the detection of plant diseases through the application of modern imaging and computer vision techniques. Thermal, hyperspectral, fluorescence, multispectral, and 3D imaging are some of these techniques. We also discussed several approaches to early detection and classification of plant diseases. KNN, SVM, *k*-means clustering, and DL are some of the important techniques. This analysis concludes that, given the costs involved, an effective approach is necessary. Enhancements in agriculture will also benefit from the deployment of a robust and effective sensor to verify that the plant health standards are met [40].

Subsequently, plant diseases constitute a significant threat to global agricultural output, disproportionately affecting small-scale farmers and compromising food security. Efforts to improve plant disease detection are crucial for lowering production losses and promoting sustainable agricultural methods. The convergence of modern technologies, such as IoT, AI, and UAVs, provides promise for improved sickness detection and management [41]. Early detection not only decreases the need for chemical interventions but also encourages sustainable agriculture practices by optimizing resource allocation and reducing waste. Furthermore, developments in DL models hold significant promise for increasing sickness detection accuracy and efficiency [42]. However, challenges persist, particularly when dealing with the complexities of illness dynamics influenced by factors such as climate change and global growth. Continued research and innovation are required to create efficient strategies for eliminating plant diseases and ensuring agricultural productivity for future generations.

References

1 Singh, V., Sharma, N., and Singh, S. (2020). A review of imaging techniques for plant disease detection. *Artificial Intelligence in Agriculture* 4: 229–242.

2 Bradford, G.E. (1999). Contributions of animal agriculture to meeting global human food demand. *Livestock Production Science* 59 (2–3): 95–112.

3 Buja, I., Sabella, E., Monteduro, A.G. et al. (2021). Advances in plant disease detection and monitoring: from traditional assays to in-field diagnostics. *Sensors* 21 (6): 2129.

4 Sladojevic, S., Arsenovic, M., Anderla, A., Culibrk, D., and Stefanovic, D. (2016). Deep neural networks based recognition of plant diseases by leaf image classification. *Computational Intelligence and Neuroscience* 2016: 3289801.

5 McCarthy, C.L., Hancock, N.H., and Raine, S.R. (2010). Applied machine vision of plants: a review with implications for field deployment in automated farming operations. *Intelligent Service Robotics* 3: 209–217.

6 Grift, T. et al. (2008). A review of automation and robotics for the bio-industry. *Journal of Biomechatronics Engineering* 1 (1): 37–54.

7 Fang, Y. and Ramasamy, R.P. (2015). Current and prospective methods for plant disease detection. *Biosensors* 5 (3): 537–561.

8 Rubio, L., Galipienso, L., and Ferriol, I. (2020). Detection of plant viruses and disease management: relevance of genetic diversity and evolution. *Frontiers in Plant Science* 11: 1092.

9 Jackulin, C. and Murugavalli, S. (2022). A comprehensive review on detection of plant disease using machine learning and deep learning approaches. *Measurement: Sensors* 24: 100441.

10 Hasan, R.I., Yusuf, S.M., and Alzubaidi, L. (2020). Review of the state of the art of deep learning for plant diseases: a broad analysis and discussion. *Plants* 9 (10): 1302.

11 Esteva, A. et al. (2019). A guide to deep learning in healthcare. *Nature Medicine* 25 (1): 24–29.

12 Gupta, N. et al. (2022). Human activity recognition in artificial intelligence framework: a narrative review. *Artificial Intelligence Review* 55 (6): 4755–4808.

13 Still, C., Powell, R., Aubrecht, D. et al. (2019). Thermal imaging in plant and ecosystem ecology: applications and challenges. *Ecosphere* 10: e02768.

14 Ishimwe, R., Abutaleb, K., and Ahmed, F. (2014). Applications of thermal imaging in agriculture – a review. *Advances in Remote Sensing* 3: 13.

15 Mahlein, A.-K. et al. (2012). Recent advances in sensing plant diseases for precision crop protection. *European Journal of Plant Pathology* 133: 197–209.

16 Sankaran, S. et al. (2010). A review of advanced techniques for detecting plant diseases. *Computers and Electronics in Agriculture* 72 (1): 1–13.

17 Martinelli, F. et al. (2015). Advanced methods of plant disease detection. A review. *Agronomy for Sustainable Development* 35: 1–25.

18 Rumpf, T. et al. (2010). Early detection and classification of plant diseases with support vector machines based on hyperspectral reflectance. *Computers and Electronics in Agriculture* 74 (1): 91–99.

19 Li, X. et al. (2018). Hyperspectral imaging and their applications in the nondestructive quality assessment of fruits and vegetables. In: *Hyperspectral Imaging in Agriculture, Food and Environment* (ed. D.P. Agrawal, S.K. Gupta, and R. Akerkar), 27–63.

20 Baranowski, P. et al. (2015). Hyperspectral and thermal imaging of oilseed rape (Brassica napus) response to fungal species of the genus Alternaria. *PLoS One* 10 (3): e0122913.

21 Okamoto, H. et al. (2007). Plant classification for weed detection using hyperspectral imaging with wavelet analysis. *Weed Biology and Management* 7 (1): 31–37.

22 Lowe, A., Harrison, N., and French, A.P. (2017). Hyperspectral image analysis techniques for the detection and classification of the early onset of plant disease and stress. *Plant Methods* 13 (1): 80.

23 Chaerle, L. et al. (2004). Thermal and chlorophyll-fluorescence imaging distinguish plant-pathogen interactions at an early stage. *Plant and Cell Physiology* 45 (7): 887–896.

24 Bauriegel, E. and Herppich, W.B. (2014). Hyperspectral and chlorophyll fluorescence imaging for early detection of plant diseases, with special reference to Fusarium spec. infections on wheat. *Agriculture* 4 (1): 32–57.

25 Vadivambal, R. and Jayas, D.S. (2011). Applications of thermal imaging in agriculture and food industry – a review. *Food and Bioprocess Technology* 4: 186–199.

26 Oerke, E.-C. et al. (2006). Thermal imaging of cucumber leaves affected by downy mildew and environmental conditions. *Journal of Experimental Botany* 57 (9): 2121–2132.

27 Bellmann, A., Hellwich, O., Rodehorst, V., and Yilmaz, U. (2007). IEEE conference on computer vision and pattern recognition. *A Benchmarking Dataset for Performance Evaluation of Automatic Surface Reconstruction Algorithms*, pp. 1–8. IEEE: Minneapolis, MN, USA, 17–22 June, 2007.

28 Zhang, N. et al. (2020). A review of advanced technologies and development for hyperspectral-based plant disease detection in the past three decades. *Remote Sensing* 12 (19): 3188.

29 Sankaran, S. et al. (2013). Huanglongbing (citrus greening) detection using visible, near infrared and thermal imaging techniques. *Sensors* 13 (2): 2117–2130.

30 Singh, D. and Kumar, A. (2024). A deep recurrent neural network for plant disease classification. *SN Computer Science* 5 (8): 1053.

31 Vázquez-Arellano, M. et al. (2016). 3-D imaging systems for agricultural applications – a review. *Sensors* 16 (5): 618.

32 Underwood, J.P. et al. (2013). A robot amongst the herd: Remote detection and tracking of cows. *New Zealand Spatially Enabled Livestock Management Symposium*.

33 Lachat, E. et al. (2015). First experiences with Kinect v2 sensor for close range 3D modelling. *The International Archives of the Photogrammetry, Remote Sensing and Spatial Information Sciences* 40: 93–100.

34 Kumar, A., Jain, R., Vairamani, A.D., and Nayyar, A. (2023). *Object Tracking Technology: Trends, Challenges and Applications*. River Publishers.

35 Kumar, A., Singh, D., and Vohra, R. (2023). Improving learning abilities using AI-based education systems. In: *AI-Assisted Special Education for Students With Exceptional Needs* (eds. A. Kumar, A. Nayyar, R. Sachan, and R. Jain), pp. 137–155. IGI Global.

36 Arivazhagan, S. et al. (2013). Detection of unhealthy region of plant leaves and classification of plant leaf diseases using texture features. *Agricultural Engineering International: CIGR Journal* 15 (1): 211–217.

37 Sannakki, S.S. et al. (2011). A hybrid intelligent system for automated pomegranate disease detection and grading. *International Journal of Machine Intelligence* 3 (2): 36–44.

38 Kruft, D. (2001). Impacts of genetically-modified crops and seeds on farmers. *The Agricultural Law Resource and Reference Center* 1–11.

39 Kumar, A., Mamgai, R., and Jain, R. (2023). Application of IoT-Enabled CNN for natural language processing. In: *IoT-enabled Convolutional Neural Networks: Techniques and Applications* (ed. R. Akerkar, A. Kumar, and M. Sharma), 149–177. River Publishers.

40 David, H.E. et al. (2021). Literature review of disease detection in tomato leaf using deep learning techniques. *2021 7th International Conference on Advanced Computing and Communication Systems (ICACCS)*, Vol. 1. IEEE, Virtual Conference (due to the COVID-19 pandemic), 7–9 January, 2021.

41 Bhagat, M. et al. (2020). Plant leaf disease classification using grid search based SVM. *2nd International Conference on Data, Engineering and Applications (IDEA)*. IEEE, 21–22 February 2020, Pune, India.

42 Liu, J. and Wang, X. (2021). Plant diseases and pests detection based on deep learning: a review. *Plant Methods* 17: 1–18.

43 Vohra, R. and Kumar, A. (eds.) (2024). *Advanced Geospatial Practices in Natural Environment Resource Management*. IGI Global.

44 Kumar, A., Walia, G.S., and Sharma, K. (2020). Robust object tracking based on adaptive multicue feature fusion. *Journal of Electronic Imaging* 29 (6): 063001.

45 Demilie, W.B. (2024). Plant disease detection and classification techniques: a comparative study of the performances. *Journal of Big Data* 11 (1): 5.

46 Sharma, P., P. Hans, and S. C. Gupta. 2020. "Classification of Plant Leaf Diseases Using Machine Learning and Image Preprocessing Techniques." In *Proceedings of the 2020 10th International Conference on Cloud Computing, Data Science & Engineering (Confluence)*, Noida, India, December 17–19, 2020. IEEE.

47 Sharma, P., Hans, P., and Gupta, S.C. (2020). Classification of plant leaf diseases using machine learning and image preprocessing techniques. *2020 10th International Conference on Cloud Computing, Data Science & Engineering (Confluence)*. IEEE, Noida, India, 17–19 December, 2020.

48 Klompenburg, V., Thomas, A.K., and Catal, C. (2020). Crop yield prediction using machine learning: a systematic literature review. *Computers and Electronics in Agriculture* 177: 105709.

49 Kumar, A., Walia, G.S., and Sharma, K. (2020). Recent trends in multicue based visual tracking: a review. *Expert Systems with Applications* 162: 113711.

50 Kumar, A. and Jha, R. (2022). IoT in smart farming: A boon to farmers. In: *Internet of Everything: Smart Sensing Technologies* (eds. T. Kavitha, V. Ajantha Devi, S. Neelavathy Pari, and S. Ramanathan), (pp. 185–206). Nova Publisher.

51 Kumar, A. (2023). *Visual Object Tracking Using Deep Learning*. CRC Press.

52 Kumar, A., Vohra, R., Jain, R. et al. (2024). Correlation filter based single object tracking: a review. *Information Fusion* 112: 102562.

53 Pérez-Bueno, M.L., Pineda, M., Cabeza, F.M., and Barón, M. (2016). Multicolor fluorescence imaging as a candidate for disease detection in plant phenotyping. *Frontiers in plant science* 7: 1790.

11

Leveraging ResNeXt50 and LSTM for Enhanced Plant Disease Detection: A Hybrid Model Proposal

Jaspreet Singh[1] *and Shashi Tanwar*[2]

[1] *Professor, Computer Science Department, Christ University, Ghaziabad, Uttar Pradesh, India*
[2] *Professor, Computer Science & Engineering, Anangpuria School of Management & Technology, Faridabad, Haryana, India*

11.1 Introduction

11.1.1 Overview

Plant infection is a serious issue that has an impact on plant patterns and how humans visually perceive them when they are afflicted. Traditionally, farmers rely on agricultural professionals who study plants with their naked eyes and determine the illnesses using precise testing methods. This approach faces the limitation of requiring a large number of agricultural professionals to cover vast farming areas. Consequently, specialists often study plants already suffering, allowing the sickness to spread to other plants. Agriculture, a rigorous and challenging profession, is heavily reliant on any country's economy. Plant diseases decrease the quality and quantity of agricultural produce. Current strategies for eradicating plant diseases include the physical removal of diseased plants, mechanized cultivation, and the use of various pesticides.

The agronomists, who must recognize and diagnose plant diseases by visually observing diseased plant leaves, would benefit immensely from creating an automated algorithmic disease detection and diagnosis approach. Manually removing afflicted plants is not only a lengthy and costly but also a labor-intensive task. Applying pesticides is another approach, but excessive use can promote plant growth while degrading plant quality. Advanced computing technologies such as machine learning (ML), computer vision, and big data analytics can help classify

Emerging Smart Agricultural Practices Using Artificial Intelligence, First Edition.
Edited by Ashish Kumar, Jai Prakash Verma, and Rachna Jain.

Published 2025 by John Wiley & Sons, Inc.

and diagnose plant diseases, apply pesticides in necessary quantities, and avoid excessive use. Farmers may not always be able to afford to monitor their plants continuously, particularly when it comes to crops or trees in isolated rural locations. An inexpensive, very accurate substitute is provided by remote monitoring using detectors, drones, and other tools.

Farmers need efficient systems to detect all forms of plant illnesses quickly, reducing labor requirements and pesticide use. Scientists have developed various methods to measure agricultural yields, tested them in laboratories, and created systems for effective plant leaf disease analysis. Crop damage leads to significant output declines, adversely affecting the economy. Plant leaves, being the most sensitive parts, are the first to show disease signs. Monitoring crops for illness throughout their life cycle is essential. Formerly, crop fields were inspected with the naked eye, but this method not only takes patience but requires experts to practically examine the fields. Many automatic and semiautomated systems for the detection of plant diseases have been developed in recent years. Leaves, stems, and fruits are the principal sites for disease diagnosis, making leaves the primary focus for disease identification.

Figure 11.1 Some diseased plant leaves [1]. *Source:* Kaggle, Inc.

11.1.2 Issues and Challenges

Several significant concerns, difficulties, and techniques related to plant disease and technology are identified and discussed below:

- **High-Quality Photographs of Plant Leaves**: Capturing clear, high-resolution images is crucial for accurate disease identification.
- **Large Number of Data Points**: Considering numerous data points is essential for comprehensive analysis.
- **Background Data and Noise**: Background information and noise can affect the quality of acquired images.
- **Precise Location Identification**: Determining the exact location on a leaf that corresponds to a significant illness is challenging.
- **Training and Testing Samples Preparation**: Preparing samples from input images for training and testing is necessary.

11.1.3 Plant Diseases Analysis and Its Symptoms

Agricultural science extensively uses RGB picture feature pixel counting techniques. Image analysis is employed for various applications, including the following:

- **Identifying Diseases**: Detecting illnesses of leaves, stems, and fruits.
- **Quantifying Disease-Affected Areas**: Measuring the affected area.
- **Controlling Impacted Area Borders**: Managing the borders of the affected area.
- **Ascertaining Hue**: Determining the color of the afflicted area.
- **Establishing Size**: Assessing the size of the affected area.

11.1.4 Current Approaches and Challenges

Traditionally, plant diseases are managed through methods such as physical removal of diseased plants, mechanized cultivation, and pesticide application. However, these methods have limitations:

1) **Manual Inspection**: Time-consuming and requires skilled professionals, leading to delays in disease detection.
2) **Pesticides**: Overuse can degrade plant quality, harm the environment, and pose health risks due to a lack of precise dosage information.

11.1.5 The Need for Automation

An automated system for detecting and diagnosing plant diseases can significantly aid agronomists. Such automated systems have a high probability of using use big data analytics, ML, and artificial intelligence technologies to advance technologies like computer vision, ML, and big data analytics to monitor large fields efficiently and cost-effectively. This automation can help in:

1) **Early Detection**: Identifying diseases at a premature stage to prevent spread and reduce crop loss.
2) **Precision Agriculture**: Applying pesticides accurately and in the required quantities, avoiding excess use.
3) **Remote Monitoring**: Utilizing sensors and drones for continuous surveillance, especially in remote areas.

11.2 Literature Review

New developments in the identification of plant diseases have seen the integration of deep learning models to enhance accuracy and efficiency. Vamshi et al. [2] suggested a unique strategy and proposed a novel approach using a hybrid deep learning strategy for identifying diseases in plant leaves. This method integrates multiple models for deep learning to leverage their strengths, resulting in enhanced accuracy and robustness. The hybrid model processes and classifies plant leaf images more effectively compared to traditional single-model methods. This study focuses on the potential of hybrid deep learning strategies in evolving agricultural disease diagnostics, thereby providing a better reliable and productive Solution for order to identify diseases early and management.

Tanwar and Singh [3] suggested a convolutional neural network (CNN) long short-term memory (LSTM) model for plant disease classification based on ResNext50.Published in "Multimedia Tools and Applications," their research demonstrates the effectiveness of combining deep learning models to enhance disease detection accuracy. The model of ResNext50 extracts spatial features from plant images, while the LSTM model captures temporal dependencies, resulting in a robust system for identifying plant diseases. Their approach achieved significant accuracy improvements, highlighting the potential of advanced neural network techniques in agricultural applications.

Kalaivani et al. [4] proposed an innovative indices-based histogram intensity segmentation strategy for the division of agricultural leaf blight disease. Published in Multimedia Tools and Applications, their study focused on segmenting diseased areas in leaf images by using histogram intensity values. This method proved effective in isolating diseased regions, facilitating accurate disease detection, and contributing to improved disease management practices.

Trivedi et al. [5] investigated the role of ML in the identification of plant leaf diseases, presented at the International Conference on Electronics, Communication, and Networking's Emerging Technology Trends. Their research focuses on evolving ML models to diagnose disease detection accurately from leaf images.

Elaraby et al. [6] optimized a deep learning model that uses a particle swarm optimizer to detect plant diseases. Published in "Computers, Supplies, and Continua," their work focuses on improving deep learning models' performance through optimization techniques.

Atila et al. [7] employed ML in the classification of plant leaf diseases using the Efficient Net deep learning model as presented in "Ecological Informatics." Their study highlights the efficiency with which the Efficient Net Deep Learning Model accurately categorizes diseases of plants.

Additionally, Wagle [8] presented a method grounded in deep learning for identifying and verifying the tomato leaf disease in "Traitement du Signal." These studies collectively demonstrate the usefulness of various ML models in deep learning in plant disease detection and classification. Recent advancements in plant disease detection have increasingly utilized deep learning and artificial intelligence techniques, demonstrating significant improvements in accuracy and as well as efficiency. Kumar [9] investigated the utilization of CNNs to detect plant diseases, showcasing The accuracy with which the model can accurately identify diseases originating from leaf images due to its powerful feature extraction and classification abilities. Nigam and Jain [10] provided diseases, emphasizing the benefits of deep learning models over conventional methods regarding precision and processing speed.

Chohan et al. [11] further explored this domain, highlighting the efficiency of deep learning techniques in identifying diseases in plants. Their research demonstrated that models for deep learning could obtain a high degree of precision in identifying various plant diseases, making them valuable tools for early detection and prevention.

Similarly, in Sujatha et al. [12], in comparison to deep learning models for conventional ML techniques, it was concluded models of deep learning with high precision and reliability in ML algorithms for plant leaf disease detection. Altogether, all of this research indicates the possibility of deep learning to revolutionize large-scale agriculture management, providing versatile and accurate applications for maintaining plant health. These last several years have witnessed radical developments in the field of plant disease classification and detection.

Dey et al. [13] conducted research on the practice of image processing procedures to identify the leaf rot disease in betel vines (Piper Betle L.). Their research, published in Procedia Computer Science, includes techniques for feature extraction, classification, and image segmentation to identify diseased regions in leaf images.

Further advancements in the field are evident in the work of Hridoy et al. [14], who explored ML approaches for the purposes of recognizing betel plant diseases. CNNs were employed for their study, which was published in the book, "Evolutionary Computing and Mobile Sustainable Networks," to classify different diseases from leaf photos.

This deep learning model accomplished high precision in the recognition of disease, outperforming traditional image processing techniques. Complementing these studies, Shah et al. [15] provided comprehensive research in their survey on rice plant disease detection and classification. Delivered at the IEEE International Conference on Current Trends in Advanced Computing (ICCTAC), 2016, IEEE reviewed numerous techniques comprising computer learning and image processing methods. The authors emphasized the effectiveness of combining these methods to increase illness detection's dependability and accuracy. In [16] Singh et al. develop two probabilistic prediction models, which use two types of text representation: Bag of Words and Term Frequency-Inverse Document Frequency to extract relevant features. In [17] Rastogi et al. propose an innovative greedy algorithm using majority voting, which uses three searching criteria: forward search, backward search, and recovery search to select the optimal classifiers. The proposed method yields significantly better results than the conventional search-based selection of classifiers.

11.3 Research Methodology

11.3.1 Overview

Plant disease study is focused on plant patterns and how people interpret them visually when they are diseased. Farmers often use a technique in which agricultural specialists check the plants with their eyes and use precise testing to identify plant illnesses.

Automatic identification of plant diseases is the development of a programmed computational system for the identification and diagnosis of plant illnesses, which would be of great benefit to the agronomist who is tasked to carry out such diagnoses via visual examination of affected plant leaves [18, 19].

Automatic identification of plant diseases is a significant task because it is useful for farmers to monitor large fields of plants and detect diseases utilizing advanced computing tools such as computer vision, big data analytics, and ML in various agricultural applications. This chapter reviews the research methodology, which uses several techniques in methodologies such as CNN and LSTM.

11.3.2 Research Design

This research has four stages of plant disease detection, such as plant village dataset (input data), ResNeXt50 CNN (future extraction), LSTM classifiers, and classification of plant disease. The ResNeXt50 + LSTM model is used to improve the dataset's efficiency in the process of classification.

Table 11.1 Literature review table.

Authors and year	Approach	Advantages	Disadvantages	Focus
Vamshi et al. [2]	Hybrid deep learning strategy	High accuracy, effective in complex scenarios	Computationally intensive	Plant leaf diseases
Tanwar and Singh [3]	ResNext50 + LSTM	Good at capturing temporal features	Requires significant computational resources	Plant disease classification
Kalaivani et al. [4]	Histogram intensity segmentation	Simple implementation, effective segmentation	Limited to specific types of blight diseases	Leaf blight disease
Trivedi et al. [5]	Machine learning	Faster training time, interpretable models	Lower accuracy compared to deep learning	Plant leaf diseases
Elaraby et al. [6]	Deep learning + particle swarm optimizer	Enhanced model performance, reduced training time	Complexity of implementing optimizer	Plant disease detection
Atila et al. [7]	EfficientNet	High efficiency, state-of-the-art performance	Requires substantial computing power	Leaf disease classification
Wagle [8]	Deep learning approach for tomato diseases	High accuracy, robust model	Dataset-specific, may not generalize well	Tomato leaf disease
Kumar [9]	CNN for plant diseases	Good accuracy, well-established method	May require large datasets for training	Plant disease detection
Nigam and Jain [10]	Deep learning review	Comprehensive overview, identification of trends	Limited to theoretical insights, no new model has been proposed	Plant disease identification
Chohan et al. [11]	Deep learning	High accuracy, generalizable models	High computational cost requires large datasets	Plant disease detection
Sujatha et al. [12]	Deep learning versus machine learning	Comparative analysis, useful insights	Limited by the scope of the study	Leaf disease detection
Dey et al. [13]	Image processing	Specific to betel vine, good accuracy	Limited generalizability to other plants	Leaf rot disease
Hridoy et al. [14]	DNN for betel plant diseases	High accuracy, effective for specific diseases	Dataset-specific, may not apply to other plants	Betel plant diseases
Shah et al. [15]	Survey on rice plant diseases	Comprehensive survey identifies gaps	Does not propose new methods	Rice plant diseases

CNN is a kind of neural network that is used for more accurate visual pattern detection and categorization [20, 21]. This technique helps in enhancement in classification of plant diseases.

11.3.3 Research Framework

A Hybrid Model combines the power of ResNext50 and LSTM layers. ResNext50, a robust CNN, is employed for image feature extraction, while LSTM layers enable the model to capture sequential dependencies, which is particularly useful for understanding temporal aspects in plant health progression.

11.3.4 Simulation Setup

The Proposed Hybrid model (ResNeXt50 + LSTM) is applied for plant disease classification to improve efficiency. The dataset description, system configuration, data collection, data conversion, parameter settings, and performance metrics are given in this section. To increase efficiency, the Proposed Hybrid (ResNeXt50 + LSTM) model is used for plant disease categorization. This part included information on the dataset description, data collection, data conversion, system configuration, parameter settings, and performance metrics.

11.3.4.1 Dataset

The dataset of plant disease for performance analysis was acquired from Kaggle (https://www.kaggle.com/vipoooool/new-plantdiseases-dataset) for this research. A duplicate of Plant Village leaf disease is present in this dataset. In total, 87,867 healthy and unhealthy augmented leaf photos are included in this dataset, which are separated into 38 categories of diseases and species. The whole dataset contains validation and training sets in the ratio of 80/20. Consequently, there are 70,295 total training photos and 17,572 total validation images. Furthermore, 33 test photos are included in this dataset to evaluate the proposed model.

11.3.4.2 Data Collection

The process of automatic disease detection in plants is possible with the assistance of images or data. Thus, it is required to collect appropriate data that can be used at every step in the process. So, various datasets have been reviewed, and the dataset used in this development is a replica of PlantVillage leaf Disease. A few categories are Apple-_Cedar_Rust, Apple_Scab, Apple_Healthy, Potato_Early_Blight, Potato_Healthy, Tomato_Early_Blight, Tomato_Healthy, and Tomato_Yellow_Curl_Virus. It presents some images of healthy and also unhealthy leaves from the used dataset with diseases specified in the bracket. This dataset contains 33 test images to test the designed model. Figure 11.2 shows the images of some of the different varieties of plants from the dataset as given below:

Figure 11.2 Images of different varieties of plants from the dataset [1]. *Source:* Kaggle, Inc.

11.3.4.3 Data Conversion

The conversion of data or preprocessing of collected data is completed to raise the number of images being utilized to increase the precision of the disease detection process in plants. The data is big enough, but still using simple tools like flip/rotation, brightness, zoom, etc., to create new images, which further aid in improving the precision of the model. Such tools or handling are termed data augmentation and normalization. The dataset used here has already been restructured utilizing offline augmentation from the real dataset.

11.4 A Proposed Hybrid Model Using ResNext50 & LSTM for Plant Disease Detection

This phase covers the first objective of the research: To develop a hybrid model for the classification and prediction of diseased plants with high accuracy.

11.4.1 Proposed Methodology: Hybrid Model (ResNext50-LSTM)

The proposed approach marries the supremacy of ResNeXt50 with the latter's ability to handle sequence data (LSTM) in order to sort out plant leaf disease classification and segmentation. ResNeXt50 serves as a useful tool for this process, given its efficiency in extracting more detailed features from the images of plant leaves, which play a key role in identifying diseases. At the same time, LSTM structures are capable of capturing time evolution because they can be used to depict sequences within the dataset. This combined model will not only take advantage of hierarchical representation capabilities offered by ResNeXt50 but also exploit LSTM memory retention and sequential learning abilities thereby, increasing accuracy and contextual awareness about the plant leaf diseases. In other words, both of these architectures are combined together by this model, which aspires to unfold spatial features and temporal patterns interplay that shall ultimately enhance performance in disease classification and segmentation tasks over plant leaves as a whole. CNN models, on the other hand, are generally used for image classification and pattern recognition.

11.4.2 Layer Information of Proposed Hybrid Model

Convolution layers, pooling layers, and fully connecting layers make up the hybrid model, as per the description given in Figure 11.3.

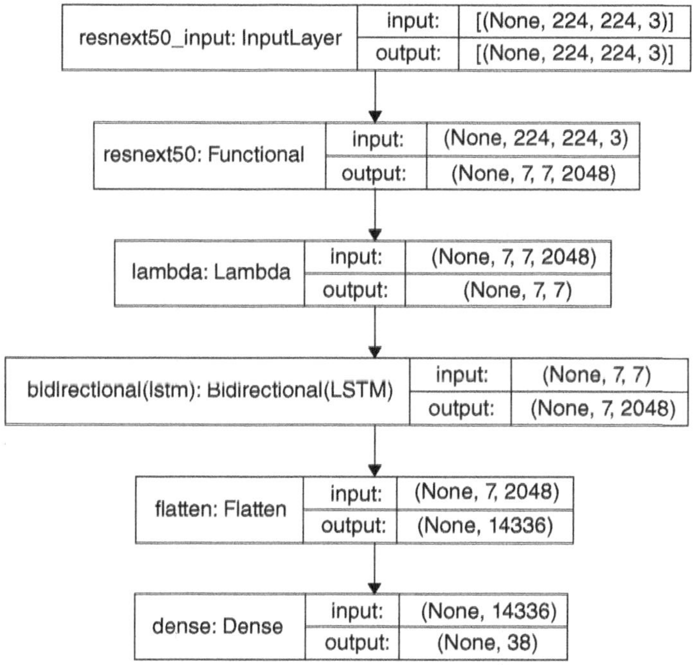

Figure 11.3 Layer information of hybrid model(ResNeXt50+LSTM).

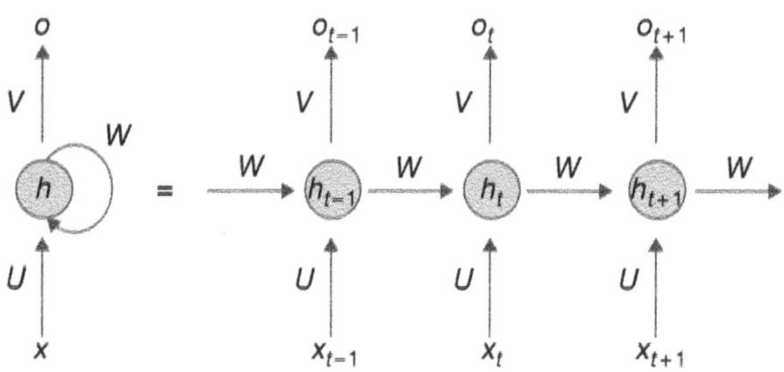

Figure 11.4 Architecture of CNN.

11.5 Results and Implementation

Different performance assessment methods and testing datasets were utilized to analyze to the ResNeXt50+LSTM model's results. Furthermore, these results are considered as the performance of the ResNeXt50+LSTM model. These results are compared to the performance of other techniques, which are ResNeXt101, Inception V3, AlexNet, Sqeeze Net, VGG19, VGG16, ResNet50, Xception, Inception ResNetV2,

ResNeXt50, NasNet Mobile, and ResNeXt50 + LSTM. The plant disease dataset was used to train these models on a DL server system, and then those models were evaluated with the testing dataset. Therefore, the models will provide their finest results when applied to images that are positioned with their faces facing upward.

11.5.1 Result of Hybrid Model(ResNeXt50 + LSTM)

In these results, training and validation of the parameters are required, with the training signified by the blue line in the figure and the validation represented by the orange line. The proposed model has had a maximum training accuracy of 99.74% and a validation accuracy of 95.44%. The minimum accuracy required for training is 76.4%, while the accuracy required for validation is 80.12% in 0 epochs. For the proposed model, more than 14 epochs are required to achieve a higher performance in validation data. Figure 11.5 shows the ResNeXt50 + LSTM model accuracy value for various epochs as given below:

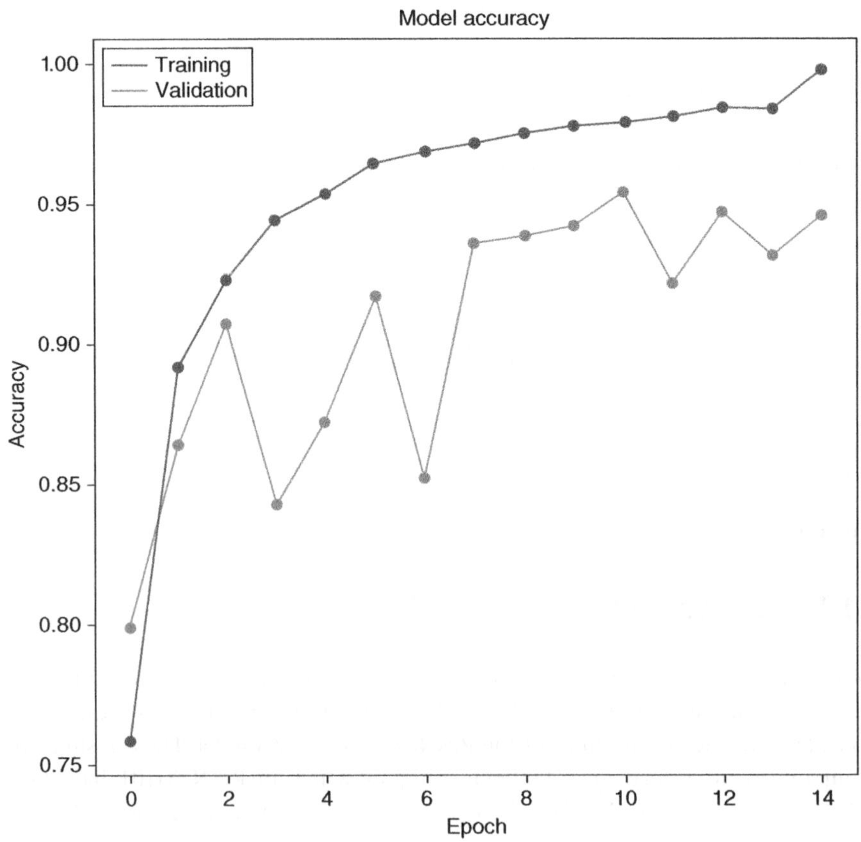

Figure 11.5 Hybrid model(ResNeXt50 + LSTM) accuracy value for various epochs.

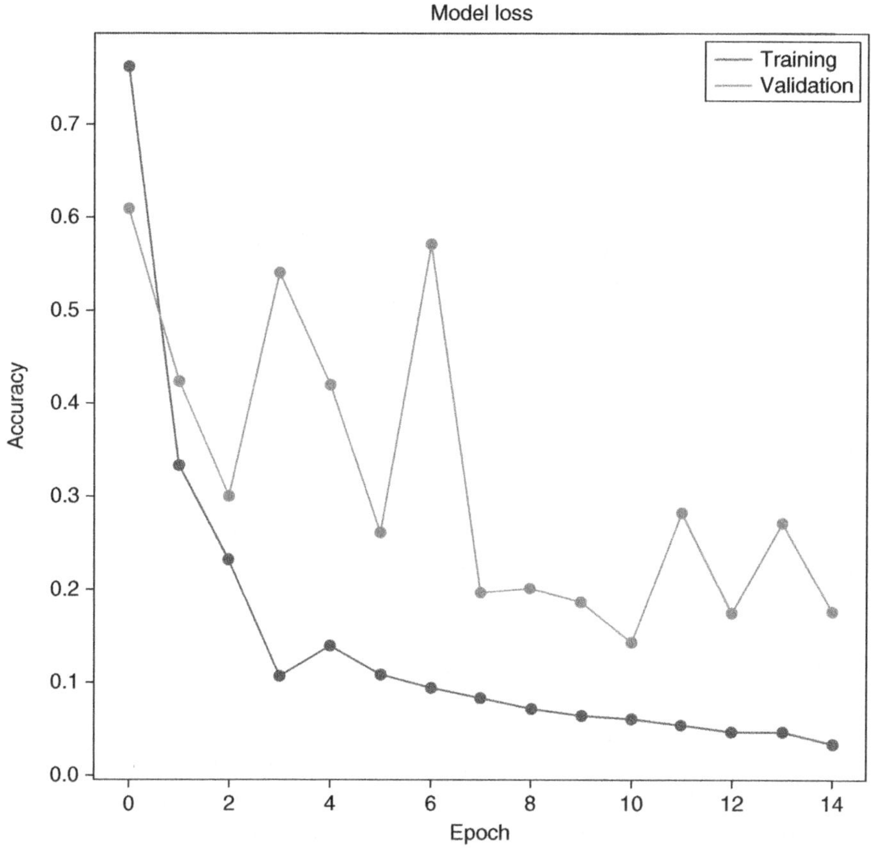

Figure 11.6 Hybrid model(ResNeXt50 + LSTM) loss value for various epochs.

Figure 11.6 shows the proposed Hybrid Model (ResNeXt50 + LSTM) model loss value for varying epochs given above. The accuracy of the loss model is shown in several different epochs. The maximum training accuracy of the loss model was found to be 0.7940, whereas the validation accuracy was calculated to be 0.6044. The minimum accuracy required for training is 0.1063, while the accuracy required for validation is 0.1431 in 14 epochs. The suggested model has an accuracy total of 96%, a recall score of 94%, and an F_1 score of 94%.

11.6 Conclusion and Future Work

11.6.1 Overview

The most important activities in agriculture aimed at improving the quality of plant yield for economic advancement include the identification and classification of plant lesions. Early identification of plant illnesses is the first and probably the most crucial step in agriculture's quest to improve the quality of plant output for future economic expansion. Plant diseases are caused by a diversity of pathogens, including bacteria, fungi, and viruses. Plant diseases are among the major challenges that directly undermine the quality of agricultural products produced. The main aim of improving the quality of plants for economic development is the diagnosis and categorization of plants' lesions.

Most verification is still done physically, which makes it difficult to accurately diagnose the problem and decide. Automation has had a revolutionary impact on every industry, and it is now making its way into the agriculture sector in a variety of scenarios. An automated system capable of recognizing plant diseases with a substantially better degree of precision can be constructed with the use of DL.

11.6.2 Conclusion

The DL models AlexNet, GoogLeNet, VGG16, MobileNetv2, and SqueezeNet are utilized in the work to classify whether tomato plant leaves are diseased or not. AlexNet, SqueezeNet, GoogLeNet, VGG16, and MobileNetv2 are also utilized in the study. Deep learning models are developed to fit the total number of classes that must be classified by the network model to the models used in the study. It achieved 99.17% accuracy when associated with GoogLeNet, MobileNetv2, AlexNet, SqueezeNet, and the VGG16 model outperformed them all.

The entire technique has been subdivided and detailed into its parts. Comparisons with a few other transfer learning models have been made, and corresponding graphs have been provided in the model's performance analysis. As the model's depth rises, new image data is necessary for the greatest potential simplification of the CNN model. After preprocessing, the data is enlarged to increase the size of the dataset. Finally, DL was applied to a popular pretrained model, ResNeXt50 + LSTM. Many trials have been conducted to develop ResNeXt50 + LSTM and optimize the suggested model for this particular dataset.

11.6.3 Future Work

India's population is projected to exceed 1.6 billion by 2030. One is predicting a significant increase in demand for agricultural goods because of the enormous population growth. As a consequence of the growth of the service industry, there has been a considerable movement in the workforce away from the primary sector and toward the tertiary sector. These techniques have the potential to be the most important aspects of any future solutions to crises. DL algorithms for plant disease diagnostics were explored within the scope of this study. Furthermore, a range of visualization methodologies and mappings have been given to identify diseases of plants. There has been significant progress in the previous three to four years. The collective method or the hybrid method is utilized to merge the judgments of distinct class models. DL algorithms are used to discover differentiating traits that lead to enhanced categorization.

References

1 Bhattarai, S. (2018). New Plant Diseases Dataset, https://www.kaggle.com/datasets/vipoooool/new-plant-diseases-dataset

2 Vamshi, J., Tanwar, S., Thatipudi, J.G. et al. (March 2024). Novel plant leaf disease detection approach using hybrid deep learning strategy. *2024 5th International Conference on Intelligent Communication Technologies and Virtual Mobile Networks (ICICV)*, pp. 266–273. IEEE, 11–12 March 2024, Tirunelveli, India.

3 Tanwar, S. and Singh, J. (2023). ResNext50 based convolution neural network-long short term memory model for plant disease classification. *Multimedia Tools and Applications* 82 (19): 29527–29545.

4 Kalaivani, S., Shantharajah, S.P., and Padma, T. (2020). Agricultural leaf blight disease segmentation using indices based histogram intensity segmentation approach. *Multimedia Tools and Applications* 79 (13): 9145–9159.

5 Trivedi, J., Shamnani, Y., and Gajjar, R. (2020). Plant leaf disease detection using machine learning. *International Conference on Emerging Technology Trends in Electronics Communication and Networking*, pp. 267–276. Springer, Surat, India, 7–8 February, 2020.

6 Elaraby, A., Hamdy, W., and Alruwaili, M. (2022). Optimization of deep learning model for plant disease detection using particle swarm optimizer. *Computers, Materials & Continua* 71 (2): 4019–4031.

7 Atila, Ü., Uçar, M., Akyol, K., and Uçar, E. (2021). Plant leaf disease classification using EfficientNet deep learning model. *Ecological Informatics* 61: 101182.

8 Wagle, S.A. (2021). A deep learning-based approach in classification and validation of tomato leaf disease. *Traitement du Signal* 38 (3): 699.

9 Kumar, S. (2021). Plant disease detection using CNN. *Turkish Journal of Computer and Mathematics Education (TURCOMAT)* 12 (12): 2106–2112.

10 Nigam, S. and Jain, R. (2019). Plant disease identification using deep learning: a review. *Indian Journal of Agricultural Sciences*.

11 Chohan, M., Khan, A., Chohan, R. et al. (2020). Plant disease detection using deep learning. *International Journal of Recent Technology and Engineering* 9 (1): 909–914.

12 Sujatha, R., Chatterjee, J.M., Jhanjhi, N.Z., and Brohi, S.N. (2021). Performance of deep learning vs machine learning in plant leaf disease detection. *Microprocessors and Microsystems* 80: 103615.

13 Dey, A.K., Sharma, M., and Meshram, M.R. (2016). Image processing based leaf rot disease, detection of betel vine (Piper Betle L.). *Procedia Computer Science* 85: 748–754.

14 Hridoy, R.H., Habib, T., Rahman, S., and Uddin, M.S. (2022). deep neural networks-based recognition of betel plant diseases by leaf image classification. *Evolutionary Computing and Mobile Sustainable Networks*, pp. 227–241. Springer: Singapore, 21 Mar 2022.

15 Shah, J.P., Prajapati, H.B., and Dabhi, V.K. (2016). A survey on detection and classification of rice plant diseases. *Proc. 2016 IEEE Int. Conf. Current Trends Advanced Computing (ICCTAC)*, 10–11 March 2016, Bangalore, India, pp. 1–8.

16 Singh, A. and Singh, J. (2024). Synthesis of affective expressions and artificial intelligence to discover mental distress in online community. *International Journal of Mental Health and Addiction* 22: 1921–1946. https://doi.org/10.1007/s11469-022-00966-z.

17 Rastogi, S. and Singh, J. (2023). Performance enhancement of vision based fall detection using ensemble of machine learning model. *Cluster Computing* 26: 4119–4132. https://doi.org/10.1007/s10586-022-03818-6.

18 Kumar, A. and Jha, R. (2022). IoT in smart farming: a boon to farmers. In: *Internet of Everything: Smart Sensing Technologies* (ed. T. Kavitha, V.A. Devi, S.N. Pari, and S. Ramanathan), 185–206. Nova Publisher.

19 Singh, D. and Kumar, A. (2024). A deep recurrent neural network for plant disease classification. *SN Computer Science* 5 (8): 1053.

20 Kumar, A., Singh, D., and Vohra, R. (2023). Improving learning abilities using AI-based education systems. In: *AI-Assisted Special Education for Students With Exceptional Needs* (eds. A. Kumar, A. Nayyar, R. Sachan, and R. Jain), pp. 137–155. IGI Global.

21 Kumar, A., Walia, G.S., and Sharma, K. (2020). Real-time multi-cue object tracking: benchmark. *Proceedings of International Conference on IoT Inclusive Life (ICIIL 2019)*, NITTTR Chandigarh, India (pp. 317–323). Springer Singapore.

12

FarmTechAI: Artificial Intelligence-Based Modern Farmer Management System

Murat Can Cardak, Muhammed Golec, and Sukhpal Singh Gill

School of Electronic Engineering and Computer Science, Queen Mary University of London, London, UK

12.1 Introduction

In parallel with the increase in population around the world, there has been an increase in the need for food and crop products [1]. In addition to solutions such as genetically modified food, technological developments continue to create revolutions in the field of agriculture. Advances such as artificial intelligence (AI)/machine learning (ML) and the Internet of Things (IoT) are being used to aid better decision-making processes and operational efficiencies in agriculture [2]. Studies show that ML techniques have transformative effects in providing predictive insights into crop yields and fully automated farming applications [3]. These technologies provide farmers with opportunities to make the most of resources with minimum waste and therefore meet the increasing crop demand [4]. This type of technological integration into farming practices not only supports sustainable progress but also moves the entire industry toward more data-driven and insightful tools like the farm management dashboard (FMD). Farmers face challenges in tracking dynamically changing information in real time, such as weather, finances, and crop demands inherent in agriculture [5]. Therefore, there is a need to draw useful conclusions from the huge amounts of data produced alongside industrialization in agriculture. Although some work has been done to meet this need, it is still difficult to use these systems due to reasons such as interface complexity [6]. A comprehensive platform combining various aspects of farm management is sorely needed to overcome these challenges.

Emerging Smart Agricultural Practices Using Artificial Intelligence, First Edition.
Edited by Ashish Kumar, Jai Prakash Verma, and Rachna Jain.

Published 2025 by John Wiley & Sons, Inc.

12.1.1 Motivation and Our Contributions

With the combination of technological developments and traditional agriculture, promising developments have emerged that increase the operational efficiency of farms. However, there is still a need for new systems that keep interface complexity at a minimum and allow farmers to make quick decisions based on data. In this article, we present a modern farmer management system called FarmTechAI with a user-friendly interface that can perform inventory tracking, financial analysis, crop monitoring, and ML-based real-time data analysis to meet this need. The main contribution of this article can be listed as follows:

- To integrate data from sources such as weather Application Programming Interfaces (APIs), sensors, and financial databases.
- To develop a user-friendly interface for farmers to access and interpret real-time data.
- To implement features for crop monitoring, inventory tracking, and financial analysis.
- Allow users to customize the dashboard based on their specific needs, preferences, and the type of crops they cultivate.
- Design the system to be scalable, accommodating the growing needs of farms as they expand in size or diversify their agricultural activities.
- Develop user-friendly guides and provide ongoing support to ensure that farmers can effectively utilize the dashboard and its features.
- Integrate monitoring systems that track crop growth, detect diseases, and predict harvest times, contributing to improved crop yield and quality.

The content of the article continues as follows. Section 12.2 discusses similar studies found in the literature. Section 12.3 describes the architecture and system design of the proposed system. Section 12.4 evaluates the performance of the use of AI/ML technique and the control panel. Section 12.5 concludes the chapter with a discussion of conclusions and future work.

12.2 Related Works

In this section, existing systems and a comparison of existing systems and the proposed system are examined.

12.2.1 Existing Systems

The envisioned FMD is a paradigm leap in agricultural technology, in contrast to current farm management systems. Traditional technologies frequently display drawbacks such as disjointed functionality, intricate user interfaces (UIs), and a deficiency in real-time capabilities [7]. Usually, these systems concentrate on discrete areas of farm management, which results in a fragmented user experience. On the other hand, our FMD (FarmTechAI) offers a unified and up-to-date picture of farm operations by smoothly integrating with a variety of data sources [8]. Our idea places a high priority on an intuitive interface design, which ensures accessibility for farmers with varied degrees of technological skill, an area where traditional technologies fall short [9]. They concluded that the main challenge encountered when applying ML in arable farming is the lack of high-quality datasets, which directly affects the performance of the models [7, 8]. Table 12.1 compares the existing farming systems based on important key parameters.

12.2.2 Critical Analysis

After a comprehensive examination of existing farm management systems, it can be revealed that inherent limitations highlight the need for the proposed FarmTechAI. Traditional systems frequently take a fragmented approach, addressing specific aspects of farm management but failing to provide a cohesive and holistic solution. The lack of real-time data analysis capabilities in these systems is a major bottleneck that prevents farmers from making timely and informed decisions. This study will provide insights into how digital tools can influence farm performance with integrated data sources and technologies within the purview of its FMD [10]. Furthermore, complex interfaces contribute to usability issues, making it difficult for farmers who may lack advanced technological expertise to fully exploit the potential of these systems.

On the other hand, the FarmTechAI aims to solve these drawbacks by providing a comprehensive, feature-rich, and user-friendly solution. The dashboard aggregates information from multiple sources, such as financial databases and weather APIs, to give farmers a real time, comprehensive picture of their operations. To ensure accessibility for users with different levels of technological literacy and to close the gap between technology and agriculture, the critical analysis highlights the significance of intuitive design principles in the UI.

In addition, this work recognizes that the existing systems are fragmented and aims to reshape the environment by incorporating advanced functionalities. To meet the changing demands of contemporary farmers, the integration of ML algorithms for predictive analytics and mobile accessibility for use in the field is a progressive strategy. The system's ability to not only lessen current difficulties but also establish a new benchmark for farm management technology and promote a paradigm shift in favor of sustainable practices and precision agriculture is highlighted by the critical analysis.

Table 12.1 Comparison of the existing farming systems.

Application	Rating (out of 5)	Features Identified	Limitations
Farmforce[a]	4.1	• Farm-level traceability • Yield forecasting • Input management and usage • Deforestation monitoring • Harvest collection and adding farmers	• The software is missing an ability for field staff to review what data has been entered • It can be expensive to use this application • Farmforce only works with an internet connection
Granular[b]	4.7	• Ability to track every aspect of the farming operation • Very good product support • It also integrates well with various manufacturers' equipment • Fairly easy to use	• The onboarding process is brutal. • It is missing a lot of data, so you don't get to see many numbers on the mobile app • Setting up an account and to get it work is taking a lot of effort and time at first
Kizeo Forms[c]	4.5	• It is easy to use and customizable by the user • Very efficient for farmers to track their crop progress • Farmers can see the weather conditions so they can act in time according to the weather	• No personal login account • Its requiring users to update which makes extra cost
Traction[d]	4.9	• The users can create their own accounts with their own progress and dashboard • Farmers can water the harvest very easily with a simple button. • The setup of the application is very easy	• It is too expensive so it's not going to be economic • It is only cloud-based, so if there isn't a stable internet connection

[a] https://farmforce.com/.
[b] https://m.farms.com/agriculture-apps/technology/granular.
[c] https://www.kizeo-forms.com/en/.
[d] https://www.tractionag.com/.

12.3 FarmTechAI: Proposed System

In this section, the software, functional, and nonfunctional requirements required to deploy the framework are examined in Section 12.3.1. Then, the system design and architecture are explained in detail in Section 12.3.2. The section ends by giving implementation and deployment details in Section 12.3.3 so that future researchers can perform the experiment again.

12.3.1 System Design Requirements

The software, functional, and nonfunctional requirements required to deploy the framework are examined below:

12.3.1.1 Software Requirements

Table 12.2 provides a summary of the essential software and their respective versions needed for the application's deployment.

12.3.2 Functional Requirements

Table 12.3 provides a summary of the essential functional requirements needed for the application's deployment.

Table 12.2 Software requirements for the system.

Requirement ID	Software requirement	Description
SR1	Programming languages	Python 3.8 or later for Django backend, JavaScript ES6 for React frontend
SR2	Frontend framework	React.js 17.0 for building the user interface
SR3	Backend framework	Django 3.2 including Django REST framework for API development
SR4	Deployment	Docker 4.38 for deployment of the project
SR5	Database	PostgreSQL for production and for data storage
SR6	Development tools	Use Git for version control and Postman for API testing
SR7	Security	HTTPS for secure communication and Django's built-in security features
SR8	Code quality tools	ESLint for JavaScript, Flake8 for python, to ensure code quality and consistency

Table 12.3 Functional requirements for the system.

Requirement ID	Functional requirement	Description
FR1	User authentication	Secure Login and registration for farmers and administrators
FR2	Dashboard customization	Interactive charts and graphs display data on crop health, weather conditions, and financial statistics
FR3	Real-time data visualization	Interactive charts and graphs displaying real-time data on various metrics
FR4	Data integration	Automatic integration of data from IoT sensors, weather APIs, and financial databases
FR5	Notifications and alerts	Real-time alerts for critical conditions affecting farm operations
FR6	Crop management	Tools for tracking and scheduling crop cycles
FR8	Inventory management	Tracking of supplies, equipment, and inventory with restocking reminders
FR9	Financial management	Financial transaction recording and analysis
FR10	Reporting	Customizable reports on productivity, finances, and resource usage
FR11	Collaboration tools	Communication and data sharing tools for team members and advisors
FR12	Task management	Assignment and tracking of tasks with progress updates

12.3.3 Nonfunctional Requirements

Table 12.4 provides a summary of the nonfunctional requirements needed for the application's deployment.

12.3.4 System Design and Architecture

The system design and architecture are explained below:

12.3.4.1 Design Overview

FarmTechAI, an FMD, is carefully designed to be a cornerstone of modern agricultural management. Our system offers a rich, user-friendly interface that combines essential data from financial databases and meteorological APIs. When designing, we developed with the modern farmer management systems in mind; its goal is to

Table 12.4 Nonfunctional requirements for the system.

Requirement ID	Nonfunctional requirement	Description
NFR1	Performance	Quick load times and response, ensuring efficiency during use
NFR2	Usability	Intuitive navigation and easy operation for all user levels
NFR3	Scalability	Capability to handle increasing data and user loads smoothly
NFR4	Security	Strong data integrity, privacy measures, and user authentication
NFR5	Compatibility	Smooth operation across various browsers and responsive design
NFR6	Maintainability	Clean, well-documented codebase for easy updates and fixes
NFR7	Reliability	Minimal downtime with quick recovery from failures
NFR8	Accessibility	Adherence to accessibility standards for users with disabilities

not only display data but also convert it into useful insights that allow for quick, well-informed decision-making. The Model-View-Template (MVT) architectural pattern we get across serves as a basis for the dashboard, which improves its robustness, clarity, and extendibility so that it may grow with the needs of the agriculture industry [11]. The combination of various data sources in our system and attention to user-centric design principles will open the way for a future in which technology and agriculture will work together to promote sustainable growth and improved operational efficiency. Figure 12.1 illustrates the data flow diagram explains what section is doing what and how each section of the system is associated with the user.

12.3.4.2 Software Development Life Cycle (SDLC)

The Agile SDLC model [12] focuses on small, frequent releases, collaboration, and customer feedback through flexible approaches. Agile processes always allow for changing new requirements in time using cross-functional teams working with an iterative work cycle, which is known as a sprint or iteration, while traditional models, such as Waterfall, do not [13]. Agile works even better when the management goal and the user needs at the end are either not precisely defined or, in the real sense, must change. It enables the relationship between developers, stakeholders, and users to be more proactive, giving rise to a development process with

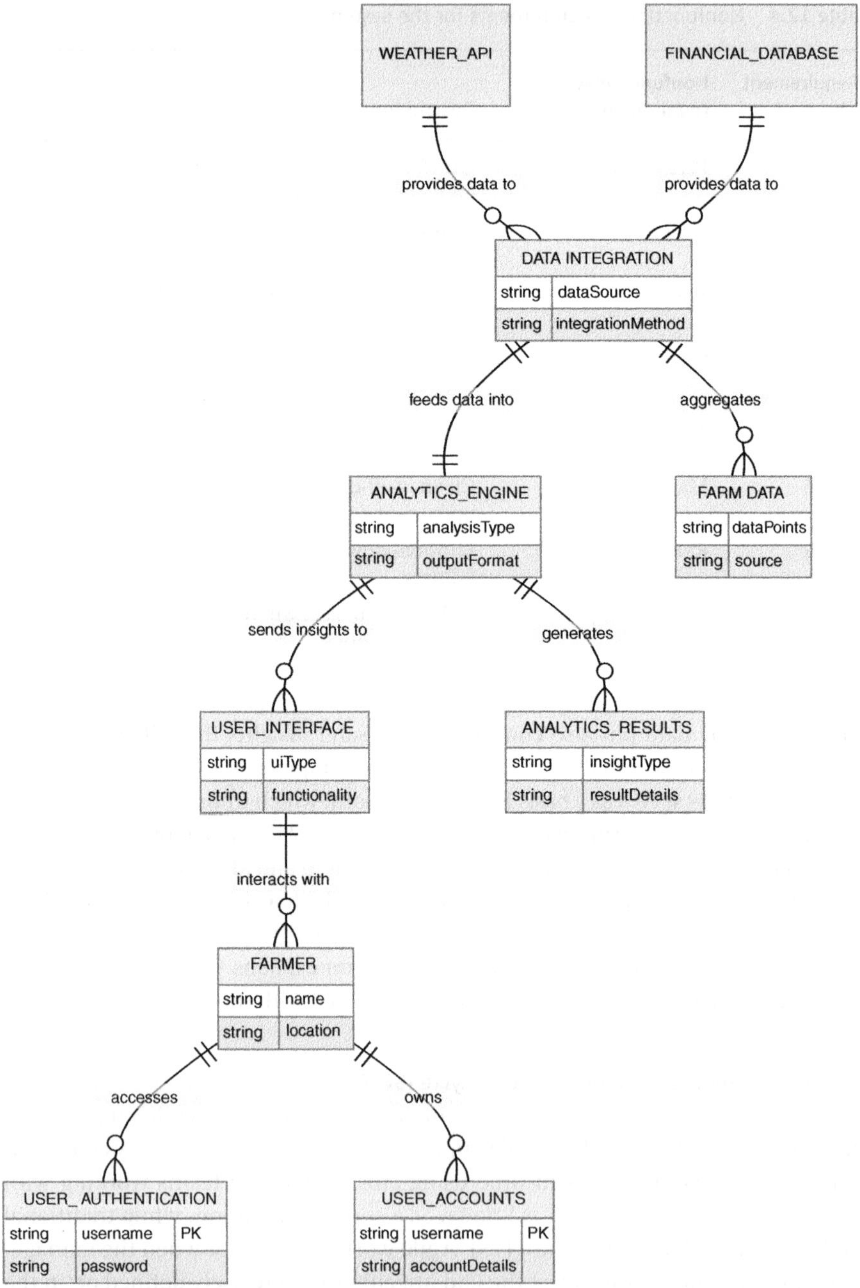

Figure 12.1 Data flow diagram for FarmTechAI.

much more user focus in which products are honed through many iterations based on continuous input.

The Agile SDLC model, together with Scrum practices, is the very key that evolved our Dashboard from concept to production. Sprint-based development was carried out with periodic assessments of timelines and deliverables, soliciting stakeholder feedback right after each release. In so doing, the process was both iterative and adaptive, ensuring that the dashboard reflected human needs on the one hand, and practical farm management demands of end users on the other.

12.3.5 Application Walkthrough

In this section, the application walkthrough is explained in detail.

12.3.5.1 Sign Up Page

This is what the users will see when they try signing up for the dashboard. A welcoming sign-up page prompts users to create an account (Figure 12.2). To ensure a safe onboarding process, this page asks the users for necessary information, including a username, email address, and a password, where it also asks to the user to confirm their password.

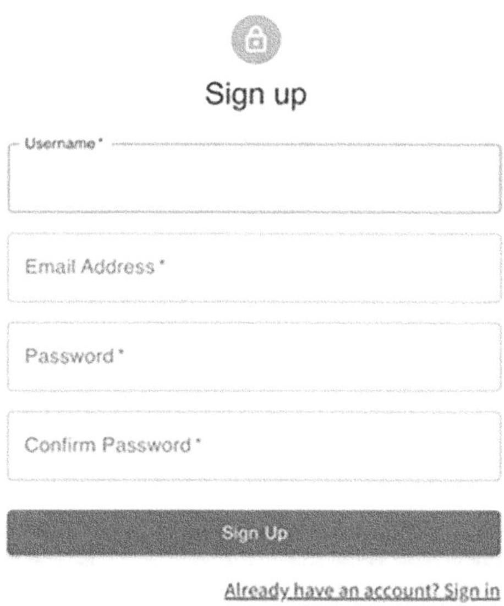

Figure 12.2 Sign-Up page in Farm TechAI.

12.3.5.2 Login Page

After successfully creating an account with the systems user can login into their dashboard using the sign in page (Figure 12.3). This is the page where users are greeted with a clean, straightforward login interface, encouraging a secure sign-in process with fields for email and password. Our login design also ensures ease of access with a "Remember me" option.

12.3.5.3 Dashboard

This is the dashboard page of the system where the users have the functionality to add new income entries via a simple interface (Figure 12.4). Adding an income feature ensures that it captures the essential details of the process. For example, it asks for the date, amount, and the description. After successfully adding the income the financial overview chart then gives a real-time visual representation of the farm's income trends over time. With the help of this integrated approach, users can monitor their financial health and make decisions based on the data displayed.

12.3.5.4 Weather

The weather feature is very helpful in planning agriculture activities because it provides localized updates (Figure 12.5). This feature is successfully done using weather API, and it tells the users the real-time data. With this page, having hourly and daily forecasts, farmers can plan ahead and successfully take advantage of the most recent weather patterns.

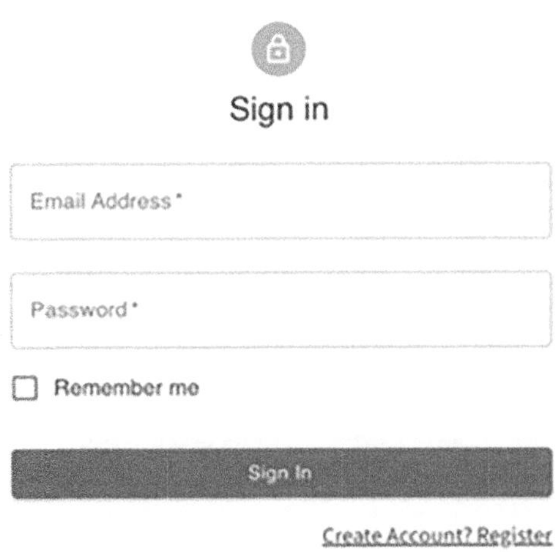

Figure 12.3 Login page in FarmTechAI.

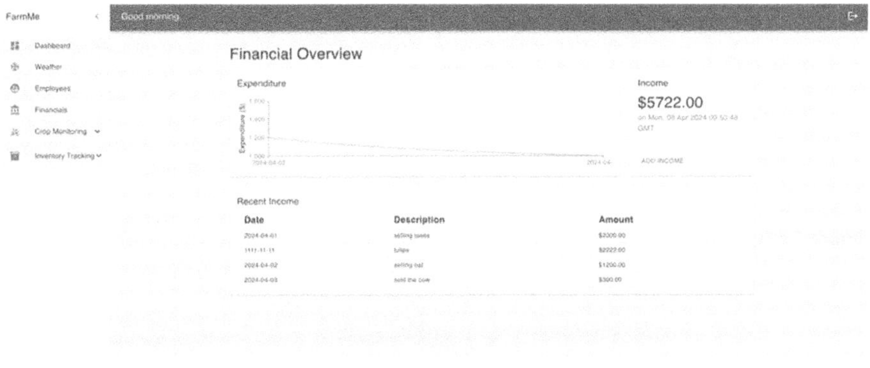

Figure 12.4 Main dashboard view in farm management system.

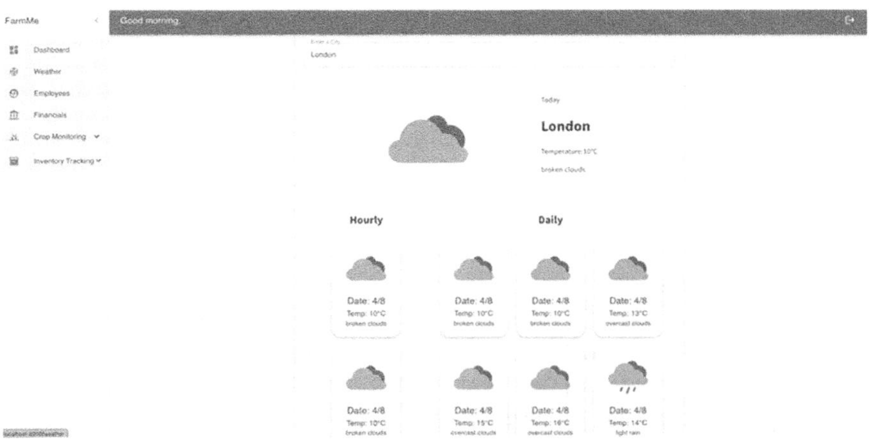

Figure 12.5 Weather page in the farm management system.

12.3.5.5 Employees

This is the employee page where the user can see all the current employees working in the company (Figure 12.6). This page also allows the user to add newly joined employees. Having this page will improve the maintaining of an organized record of all the employees and their contributions.

12.3.5.6 Add Employee

The user is asked by this page to provide basic information about the employee, including name, phone number, position of the employee, salary of the employee, and the performance (Figure 12.7). This added employee will then be shown up in the main employee list, which allows effective staff administration and supervision.

Figure 12.6 Employee page in the farm management system.

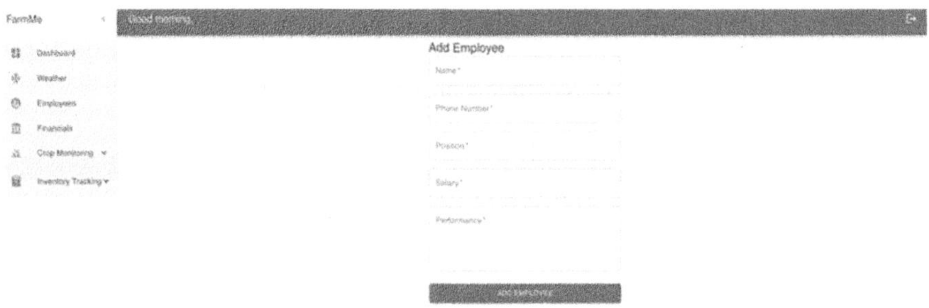

Figure 12.7 Process of adding employee page.

12.3.5.7 Financials

This is the page where the users can add their expenses to keep track of their financial status (Figure 12.8). This interface will require the user to input the expense amount, the description, and the date that the expense was made. This would be very useful for the user to keep track of how much money they are spending and to see what they are spending money on.

12.3.5.8 Crop Monitoring

We can add new crops and view the existing crop details.

Figure 12.8 Financial management page in the farm management system.

12.3.5.8.1 Add Crop

This interface of the system will be used to enter crop details into the system (Figure 12.9). It asks the user for specific information about the crop being added to the system. For example, the name of the crop, the variety of the crop and most importantly the key dates for planting and harvesting the crop. This function of the system will be very useful for the farmers as it will help them to keep track of and monitor their crops.

12.3.5.8.2 View Crop

Viewing the crops page of the system is very useful for the users (Figure 12.10). This is because this page will allow users to oversee the details of all the crops they have recorded. This page will display each crop, including its variety, planting date, and expected harvesting date. This page will give the users a quick view of their crop's progress. This page is also having update and delete options, which offer flexible crop management as well as ensure that the crop data is up to date. This function is very useful for the user as it will make sure that the user stays organized and makes timely decisions.

Figure 12.9 Adding crop page.

Figure 12.10 Viewing crop page.

12.3.5.9 Inventory Tracking

We can add new machines and livestock and view machine inventory details.

12.3.5.9.1 Add Machine

This interface of the system is for users to add machines that are being used in the company (Figure 12.11). The user is asked to input the plate number of the machine, the name of the machine and the purchase price along with the date of purchase. Adding the purchase price is important for asset management budgeting and maintaining scheduling.

12.3.5.9.2 Add Livestock

This page is used to add livestock so that the users can keep track of their animals (Figure 12.12). With the help of this page, users can add detailed information for each animal, which includes tag number, type, breed, age, weight, and, of course, the purchase price along with the purchase date.

12.3.5.9.3 View Inventory

This page allows users to see their machine inventory, which is a good feature because having all machine-related information in one place will mean that users can effectively manage the operational assets in the company (Figure 12.13). Also,

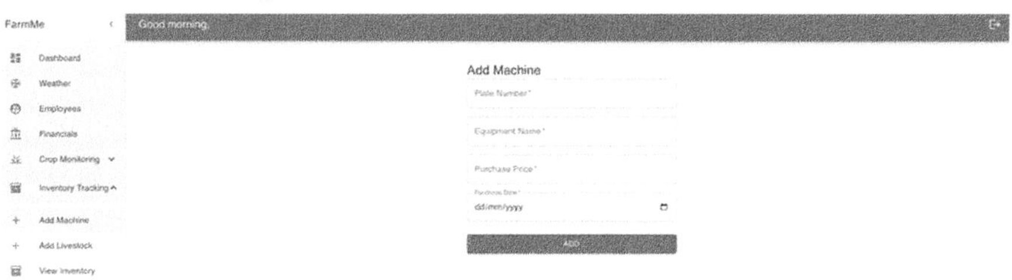

Figure 12.11 Adding machine page.

Figure 12.12 Adding livestock page.

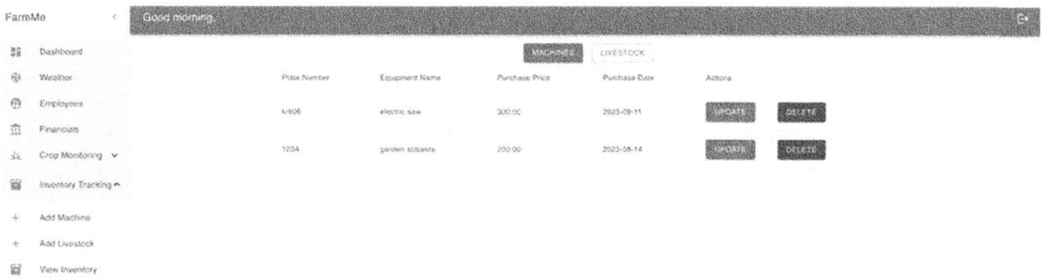

Figure 12.13 Viewing machine page.

Figure 12.14 Picture of the view livestock page.

this page will enable the users to update information about the machine and to delete the machine which provides full control of the resources.

This is the page showing the inventory of the livestock where users are again able to update the livestock information along with the feature of deleting the livestock if no longer needed (Figure 12.14). This feature is essential for planning and management because it enables users to maintain an up-to-date database of their livestock.

12.3.6 System Architecture

The design of the FarmTechAI is an MVT architectural pattern [11]. Specifically, the Model defines the structure and schema of the farm data, which refers to information related to the crops, livestock, and financial transactions. The View interacts with the Model to process requests from the user to fetch or update data and then decides the Template to render. After the view is running, it will display this to the users, enabling them to interact through the web interface with the dashboard. Such MVT architecture ensures good dashboard operation via a clean separation of concerns: Model undertakes data management; Views have business logic, and Templates are in charge of the presentation, yielding an intuitive and responsive user experience for the user.

12.3.6.1 Implementation and Deployment Details

We have used the following languages and frameworks in this work.

12.3.6.1.1 Languages

Python The backend development was critical with Python. Its simplicity in syntax and, at the same time, being powerful enough to handle even the most complex operation of data made it one of the most suitable to handle all server-side-related tasks. Data models, database queries, and business logic needed for farm management operations have all been done using Python within the dashboard.

JavaScript On the front end, JavaScript was used to make it responsive and interactive. It changed the DOM in real time, enabling users to get timely live feedback on their actions, hence not waiting for full-page reloads. Especially in the design of a single-page application that was to dynamically renew its interface with new information as the interactions take place, the Reacts JavaScript library was put in use.

12.3.6.1.2 Frameworks

Django The Django framework powers the FarmTechAI backend. It provides a powerful object-relational mapping (ORM), ensuring the most efficient interaction with the database, as it only defines models to include all key entities, such as details on crops, records for livestock, and even financial transactions within it. These views process the business logic from the frontend user input as per necessity and query the required data from the database. Powered by Django, the admin interface can have easy content control of the dashboard. User authentication and session management of security aspects are handled by the in-built modules of Django with expert perfection for maintaining any user data and dashboard integrity.

React On the frontend, the UI is rendered by declarative components in React, responding fast to user interaction. This avails, in React state, the prospect of the dashboard being updated automatically each time the user adds new information or makes some changes to their details, hence offering live visual feedback without a complete reload of the page. This ranges from the real-time updating of the charts in the overview of financials to the smooth management of crop and inventory data. Besides, the modular nature of React components would make its elements reusable, and the organization of the codebase would facilitate further development and maintenance of such.

Django REST Framework FarmTechAI is powered by Django REST Framework (DRF), which is supposed to run on a very high note and functions entirely. DRF helps develop RESTful APIs in the Django framework so that accessible communication of front and backend in the FMD is obtained. These APIs act as middleware between the frontend and the backend of the service, sending/receiving the data in JSON format. Thus, with them, the standardized handling of HTTP requests was introduced.

In other words, FMD uses DRF for the serialization of farm data, such as jsonify, like crop details, and records of livestock into JSON. This then exposes the serialized data to the JavaScript on the front end, which will pull and display it, allowing real-time updates without full-page reloads. This results in an agile and interactive experience for the farmer; he can see the immediate reflection of his actions on the dashboard, hence enhancing the decision processes and efficiency of management.

12.3.6.1.3 Application Programming Interfaces (APIs)

APIs stand for application programming interface and its important when developing a project like FarmTechAI because APIs are allowing the system to interact with data sources and external services.

Weather Data API Below is the basic code snippet for integrating real-time weather data into the FMD (Figure 12.15). make ApiCall is an async function operating fetching weather data from OpenWeatherMap API. This process starts

```
// Function to make API calls to fetch weather data
const makeApiCall = async () => {
  try {
    setLoading(true);
    setError(null);

    // Fetch daily weather data
    const dailyResponse = await fetch(
      `https://api.openweathermap.org/data/2.5/forecast?q=${inputCity}&cnt=16&appid=6557810176c36fac5f0db536711a6c52`
    );
    const dailyData = await dailyResponse.json();

    // Fetch hourly weather data
    const hourlyResponse = await fetch(
      `https://api.openweathermap.org/data/2.5/forecast?q=${inputCity}&cnt=8&appid=6557810176c36fac5f0db536711a6c52`
    );
    const hourlyData = await hourlyResponse.json();

    // If both API calls are successful, update state with new data
    if (dailyData.cod === "200" && hourlyData.cod === "200") {
      setLoading(false);
      updateState(dailyData, hourlyData);
    } else {
      console.error("API call unsuccessful:", dailyData, hourlyData);
      setError("City not found. Please try again.");
    }
  } catch (error) {
    console.error("Error making API call:", error);
    setError("Error making API call. Please try again.");
  } finally {
    setLoading(false);
  }
};
```

Figure 12.15 A snippet of the weather API code.

by setting the loading state to "true," which indicates the start of an API call and resetting any existing errors. After that, it makes two separate API requests, both of which are controlled by impunity: one to retrieve the hourly prediction and the other to retrieve the day weather data. These are built as JavaScript Fetch API calls using template literals to build the request URL in an inline and dynamic manner. If it is a "200" status code of success coming in the HTTP response, then it would parse the JSON response and call update state with fresh new weather data, which would then reflect on the dashboard. This enables the users to be able to receive updated weather forecasts, which are very critical for their daily farm management decisions. Otherwise, in the case of an error, the catch block logs the error and sets appropriate error messages for the user. This integration is useful because it shows how powerful the dashboard is for giving users helpful data that will guide them in maximizing their farm activities based on the prevailing weather.

Financial Data API The code snippet of the financial API shows the basic operations of the process of financial data submission within the FMD (Figure 12.16). The function handleSubmit listens for the event when a user submits the expenditure form. First, he stops the default form submission action to handle the form submission asynchronously. The form data and user token should be encapsulated into a single object and sent as a JSON string in the body of a POST request to the server endpoint "API/expenditure." This operation gives the dashboard the capability of collecting and logging the financial transactions attached to a particular user account. On successful completion of the request, the system will immediately provide feedback to the user, showing that expenditure data has been added, hence giving a sense of completeness in the user's mind and bettering user experience for being assured about the data integrity, fostering trust from the user's side. Alerts the user application for every error, maintaining transparency and making troubleshooting easier.

12.3.6.1.4 Core Functionalities

The following are the main functionalities of modern FarmTechAI.

Crop Monitoring Function An important function of the FMD is the crop monitoring function that helps improve decision-making in agriculture (Figure 12.17). The feature helps throw farmers into advanced details of various parameters in crop health, growth stages, and soil conditions. The farmers will be able to have data in real time on actual environmental factors and responses of the crops, hence being able to adjust their crop practice accordingly for optimum production and high crop yield while reducing the exposure to the risk of failure.

```
// Handle form submission
const handleSubmit = async (e) => {
  e.preventDefault();

  // Check for empty fields
  const emptyFields = Object.keys(formData).filter(
    (key) => !formData[key]
  );

  // If there are empty fields, set error state and return
  if (emptyFields.length > 0) {
    setError(`Please fill in all fields: ${emptyFields.join(", ")}`);
    return;
  }

  // Reset error state
  setError(null);

  // Prepare data for submission
  const raw = {
    ...formData,
    user_token: key,
  };

  try {
    // Make POST request to API
    const response = await fetch("http://127.0.0.1:8000/api/expenditure/", {
      method: "POST",
      headers: {
        "Content-Type": "application/json",
      },
      body: JSON.stringify(raw),
    });
```

Figure 12.16 A snippet of the financial API code.

The code snippet for the crop monitoring function is shown in Figure 12.17. It encapsulates the functionality of the form submission process for adding crop data. The handleSubmit function is triggered when a user submits the crop details form. It first prevents the default form submission event from managing submission asynchronously. It then checks for any empty fields and sets an error message if necessary, ensuring data integrity and user feedback.

If the data form is valid, the function will construct a payload with the form data and a user token for authentication. This data will then be sent to the backend API using a POST request, where it is processed and stored. The user will then be redirected to a view page upon successful submission, or an error message will be displayed in case of failure.

```
const handleSubmit = async (e) => {
  e.preventDefault();

  // Check for empty fields
  const emptyFields = Object.keys(formData).filter((key) => !formData[key]);

  if (emptyFields.length > 0) {
    setError(`Please fill in all fields: ${emptyFields.join(", ")}`);
    return;
  }

  setError(null);

  const raw = {
    ...formData,
    user_token: key,
  };

  const myHeaders = new Headers();
  myHeaders.append("Content-Type", "application/json");

  const requestOptions = {
    method: "POST",
    headers: myHeaders,
    body: JSON.stringify(raw),
    redirect: "follow",
  };

  console.log(raw);
  fetch("http://127.0.0.1:8000/api/crops/", requestOptions)
    .then((response) => response.json())
    .then((result) => {
      result && navigate("/view");
    })
    .catch((error) => alert(error));
};
```

Figure 12.17 A snippet of the crop monitoring function.

Having strong data management and user feedback methods are proved by this implementation. It makes sure that all required data is gathered precisely and gives user feedback in real time, improving the application's usability and user confidence.

Livestock Tracking Function Tracking the livestock allows the farmer to have full and complete tools assisting him in monitoring and management of health, breeding, and their locations from the FMD (Figure 12.18). These features are very key to the optimization of livestock productivity and welfare, since they enable farmers to have real-time data about the status, history, and performance of every single animal within their farm.

The code snippet shows the process of submitting livestock data taking place in the FMD (Figure 12.18). Further down, there is the handleSubmit function, which is

```
const handleSubmit = async (e) => {
  e.preventDefault();

  // Check for empty fields
  const emptyFields = Object.keys(formValues).filter(
    (key) => !formValues[key]
  );

  if (emptyFields.length > 0) {
    setError(`Please fill in all fields: ${emptyFields.join(", ")}`);
    return;
  }

  setError(null);

  const raw = {
    ...formValues,
    user_token: key,
  };

  const myHeaders = new Headers();
  myHeaders.append("Content-Type", "application/json");

  const requestOptions = {
    method: "POST",
    headers: myHeaders,
    body: JSON.stringify(raw),
    redirect: "follow",
  };

  console.log(raw);
  fetch("http://127.0.0.1:8000/api/livestocks/", requestOptions)
    .then((response) => response.json())
    .then((result) => {
      console.log(result);
      result && navigate("/viewInventory");
    })
    .catch((error) => alert(error));
};
```

Figure 12.18 A snippet of the livestock tracking function.

essential in controlling user interactions in such a way that an addition of a new livestock entry is done. It will begin by preventing the default form submission behavior from handling the process asynchronously. It will check the fields for emptiness and give an alert to the user if there is such necessary information absent, assurance of data completeness. When successfully completing all fields, the function then compiled form data into a JSON payload, including a user token used for authentication. This data is then sent to a backend API endpoint using a POST request. This further smoothens the process of transitioning directly to the inventory view page when the data is successfully submitted and, therefore, improved the user experience with immediate visual feedback using the useNavigate hook from React. The implementation substantiates the effectiveness of the farm management system, strong error handling by React, and secure dynamic data interactions indispensable to real-time livestock management, only to name a few that are of main interest to minimize usability problems and increase the data quality.

Financial Management Function The financial management function in the FMD is very key to the effective recording and tracking of financial and resource allocation. Farmers will, therefore, be able to view and record all their financial transactions concerning their agricultural activities, such as income from their sales, supplies of expenses, and investment costs. All these comprehensive financial controls assist in planning the budget more appropriately, tracking profitability, and making informed financial decisions.

This code snippet is responsible for submitting the financial records, an important feature for the FMD (Figure 12.19). The function handleSubmit is basically the heart of how the system interacts with the users, collecting, validating, and submitting expenditure data, all through a flowy and easy-to-use interface.

This function is used to check the form submission behavior; it makes the process of them being submitted programmatical. It checks to ensure that all fields that are required, are filled and, in case some of the fields have not been filled, immediately gives feedback to users. From the above, this means that such users are prevented from submitting incomplete data, therefore ensuring the addition of newly added data is complete before the addition.

The form data is validated and then compiled with the user authentication token (user_token) to make sure this action is done by an authenticated user. Finally, the data is posted to the server using a POST request sent to the respective API endpoint. The server's response is handled in such a way that in either of the cases, it has a notification showing the user a successful submission of data and redirection to the main page. If the submission has failed, the user should be alerted with some kind of error.

Weather Forecasting Function The weather forecasting function in this FMD is an important feature of the dashboard, as it enables farmers to access the right amount of weather updates needed in time and accuracy for scheduling and planning daily activities for the farm. This is a function that gives forecasts on temperature, precipitation, wind conditions, and others that would basically help a farmer make informed decisions about how to protect his crop or livestock from hostile weather and the usage of available resources.

The first step that was taken when developing the forecasting function was the use of a comprehensive and reliable source of weather data. We integrated weather APIs that provide both meteorological real time and forecast data. The user interface (UI) component was then responsive and updated dynamically per the data received; it would be assured the farmer gets the latest weather

```
// Handle form submission
const handleSubmit = async (e) => {
  e.preventDefault();

  // Check for empty fields
  const emptyFields = Object.keys(formData).filter(
    (key) => !formData[key]
  );

  // If there are empty fields, set error state and return
  if (emptyFields.length > 0) {
    setError(`Please fill in all fields: ${emptyFields.join(", ")}`);
    return;
  }

  // Reset error state
  setError(null);

  // Prepare data for submission
  const raw = {
    ...formData,
    user_token: key,
  };

  try {
    // Make POST request to API
    const response = await fetch("http://127.0.0.1:8000/api/expenditure/", {
      method: "POST",
      headers: {
        "Content-Type": "application/json",
      },
      body: JSON.stringify(raw),
    });

    // Handle the response as needed
    if (response.ok) {
      // Data successfully submitted
      alert("Expenditure added successfully");
      navigate("/")
    } else {
      // Handle error response
      alert("Error adding expenditure");
    }
  } catch (error) {
    console.error("Error adding expenditure", error);
  }
};
```

Figure 12.19 Snippet of the financial management function.

information. The City Input component in the Weather Forecasting Function allows users to input a City Name and fetches the data when the user presses “Enter.”(Figure 12.20). It validates the city name by a regex, indicating loading while fetching data, and then updating the UI with either success data on the weather or an error message. This is an example of smooth user interaction with excellent React state management and event handling. Then, we make the API call async and update the UI to keep the UI informed in such a manner that it will be responsive enough to provide a glitch-free user experience for our weather forecast feature.

```
const CityInput = (props) => {
  const [loading, setLoading] = useState(false);

  // Function to handle the Enter key press
  const onKeyPressHandler = async (e) => {
    e.persist();
    const eventKey = e.which ? e.which : e.keyCode;
    const city = e.target.value;

    // check if input contains only letters after Enter was pressed
    if (eventKey === 13) {
      if (/^[a-zA-ZäöüÄÖÜß ]+$/.test(city)) {
        e.target.classList.add("loading");
        setLoading(true);

        // Make the API call and update the state
        if (await props.makeApiCall(city)) {
          e.target.placeholder = "Enter a City...";
        } else {
          // If city was not found, display an error message
          e.target.placeholder = "City was not found, try again...";
        }

        setLoading(false);
      } else {
        e.target.placeholder = "Please enter a valid city name...";
      }

      e.target.classList.remove("loading");
      e.target.value = "";
    }
  };
```

Figure 12.20 Code snippet of the city input for forecasting function.

12.3.6.1.5 Deployment

The deployment of the FMD represents the characteristics of modern ways in which applications should be deployed reliably and efficiently. Docker, on its part, applies the containerization technology to the encapsulation of the application in an environment where everything is controlled so that it is exactly as required [14]. The domain and dedicated server, on the other hand, reinforce accessibility so that the dashboard enjoys a permanent residence on the web. The deployment, in this sense, would mean more than just access; it is the realization of the pragmatically feasible use in the provision of tools to end users to enhance their efforts in agriculture. This deployment is a landmark for not only the maturity of this project but, in fact, an attestation to readiness toward real-world application, scalability, and lifecycle management that is on a continual basis.

Deployment of the FMD was streamlined with the use of Docker, a powerful platform supporting consistent and isolated environments through containerization. This began with the writing of a Dockerfile, which describes how the building of the application's Docker image is to be done and ensures that all necessary components are bundled up. Then, the image was managed with Docker Compose, which allows simplified coordination of multi-container applications through a simple file written in YAML.

Having verified with local testing that the containers have run smoothly, images are pushed to the Docker registry. The image was pulled down from Docker, and the containers were spun up on a server acquired along with a domain. Then, the DNS for the domain was pointed to the IP of the server so that the dashboard was accessible to the public. This approach underscored the benefits of Docker's "build once, deploy anywhere" [15] Approach: a deployment that is robust and easily reproducible, and which made the modern DevOps practices being followed in the project.

12.4 Performance Evaluation and Testing

In this section, the performance test results of the framework are evaluated. The results of the methods of evaluation applied at different stages of development and testing of the FMD.

12.4.1 User Acceptance Testing

User acceptance testing (UAT) for the FarmTechAI will ensure the capability of delivering to specification and providing usability at the end user level. It provides the final verification that the project meets requirements and functionality. Table 12.5 shows the description of the UAT process and its results. Table 12.5 ensures every major aspect of the FMD is in operation and being accepted as required by the stakeholders' requirements. UAT is the process of testing real-world situations that verify the system can perform tasks it was designed for in readiness for production, laying emphasis on its functionality and UI. Success in these affirms that the FMD can be used for the intended purposes in a real-world setup and hence be able to support farm management.

12.4.2 Usability Evaluation

This usability evaluation section talks about the applied methods and results obtained regarding user-friendliness and efficiency of the FarmTechAI. This involves user testing sessions or questionnaires to check for user interaction and

Table 12.5 Table for user acceptance testing.

UAT ID	Test description	Expected outcome	Test result
UAT1	Verify login functionality with correct user credentials	Successful login and redirection to dashboard	Pass
UAT2	Check response to incorrect login credentials	Display error message and remain on login page	Pass
UAT3	Test update functionality in crop management module	Updates to crop details should be saved correctly	Pass
UAT4	Verify navigation between different modules	Seamless navigation without errors or delays	Pass
UAT5	Validate financial recording features	Accurate recording and display of financial data	Pass
UAT6	Test logout functionality	Successful logout and redirection to login page	Pass

satisfaction. The three criteria of functional accuracy, accessibility, and performance are going to be examined in the dashboard for a seamless and effective user experience in conducting their farm management tasks.

A feedback survey of the FarmTechAI was done to get candid and honest feedback. This method was preferred for the purpose of getting unbiased responses simply since with anonymity, there is an increase in the quality of participation and quality data is collected. The survey included, in general, qualitative and quantitative questions, covering a large view of user satisfaction and interfaces' usability. It was widely accessible and easy to respond to since its electronic distribution. The survey findings were instrumented with a clear understanding of user preferences and difficulties pinpointed, which must be worked for some improvements that would be carried in the design and functionality of the Dashboard. This approach includes user-centric development with continuous improvement based on live user feedback.

12.4.3 Analysis of User Satisfaction Results

We have tested the proposed system with a different set of users.

12.4.3.1 Navigation

Based on the survey results, the ease of navigation within the FarmTechAI is rated positively by the majority of users (Figure 12.21). Half of the users found the navigation to be "Easy," indicating that they experienced little to no difficulty in moving through the application. An additional 35.7% of users rated the navigation as "Very Easy," reinforcing the notion that the majority of users find the dashboard

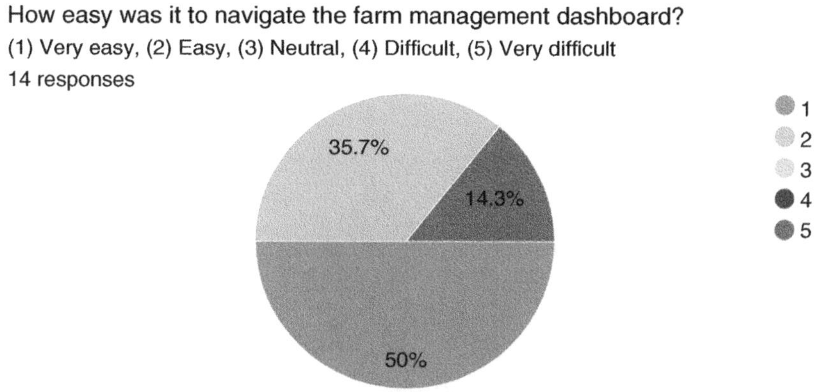

Figure 12.21 The results of the survey for navigation.

to be user-friendly and intuitive. This high percentage of positive responses suggests that the design principles guiding the navigation structure are well-aligned with user expectations and usability standards.

From the survey, users rated their satisfaction with the navigation at 14.3% as "Neutral." In essence, there were no big problems, but there was enough room for improvement. That is, this feedback is to unveil and fix subtle usability issues that hinder all users' satisfaction regarding to the navigation.

12.4.3.2 User Interface and Design

The survey results regarding the UI and design of the FarmTechAI reveal a generally positive sentiment among users, with a significant portion expressing satisfaction (Figure 12.22). Notably, 78.6% of respondents rated the UI and design as either "Good" or "Excellent." Specifically, 42.9% of users rated it as "Good," indicating a favorable impression of the dashboard's visual and interactive elements, while an additional 35.7% rated it as "Excellent," reflecting an even stronger endorsement of the design quality. This collective positive response suggests that the UI is generally well-received by users, with a majority finding it aesthetically pleasing and functional.

As seen from the report, 28.6% of users rated the UI and Design "Average" and "Poor"; it shows opportunities for improvement. This feedback, even if given by a minority, is necessary for fine-tuning specific areas that result in increased overall user satisfaction and usefulness of the dashboard.

12.4.3.3 Testing for Browser Compatibility

Table 12.6 shows the test results for browser compatibility testing. This test ensures that the FarmTechAI works equivalently on diverse web browsers and

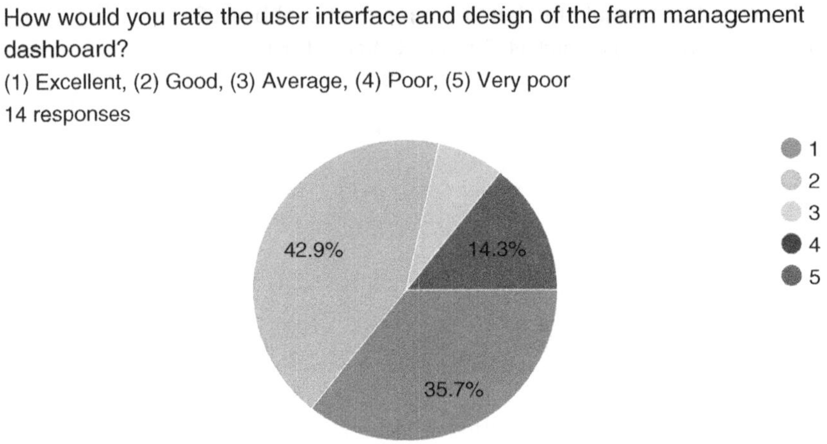

Figure 12.22 The results of the survey for user interface and design.

Table 12.6 Browser compatibility testing.

Browser test ID	Browser	Version	Description	Expected outcome	Test result
BT1	Safari	Latest	Test all main functionalities	All functions work as expected	Pass
BT2	Google Chrome	Latest	Test all main functionalities	All functions work as expected	Pass
BT3	Mozilla Firefox	Latest	Test all main functionalities	All functions work as expected	Pass
BT4	Microsoft Edge	Latest	Test all main functionalities	All functions work as expected	Pass
BT6	Mobile – Safari	Latest	Test on the latest iPhone	All mobile-specific functions work	Pass
BT7	Mobile – Chrome	Latest	Test on the latest Android	All mobile-specific functions work	Pass

devices. It will be the most critical testing; with this, one is assured that the application will be able to run comfortably by all users, no matter the kind of technology they decide to use. It identifies and helps in rectifying any discrepancies in functionality or display that would affect the user's satisfaction and, hence, the overall usability of the dashboard. Ensure that the application is accessible and reliable by testing its compatibility with the browsers. This will enable further user involvement with the app, and even from that, your software will have high levels of performance and standards that users can trust for their services.

Figure 12.23 displays the survey results regarding security, Figure 12.24 illustrates the usability of the current features, Figure 12.25 demonstrates the system's performance in terms of speed and response, Figure 12.26 presents the recommendation for the proposed system and Figure 12.27 showcases the proposed system's future usage. This feedback is very important, as it points to the very areas that need to be improved to further perfect user experience and ensure the dashboard becomes fully supportive to all categories of users in their respective farming operations.

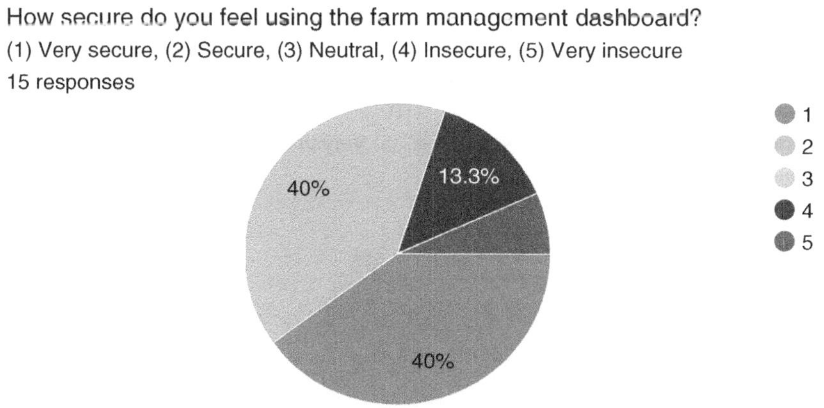

Figure 12.23 Survey results regarding security.

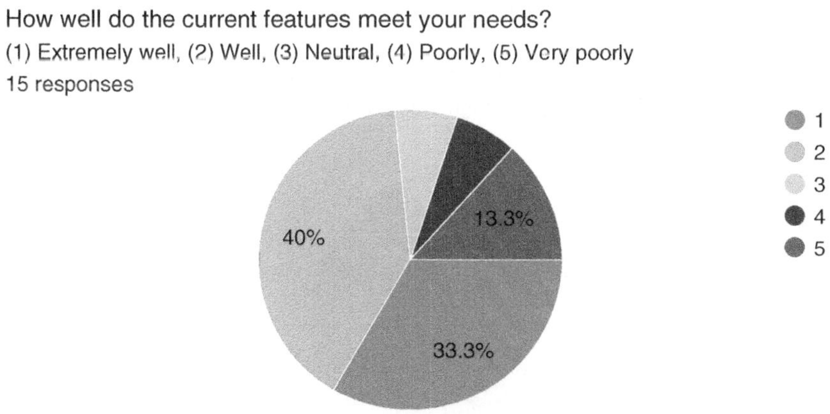

Figure 12.24 Survey results regarding the usability of the current features.

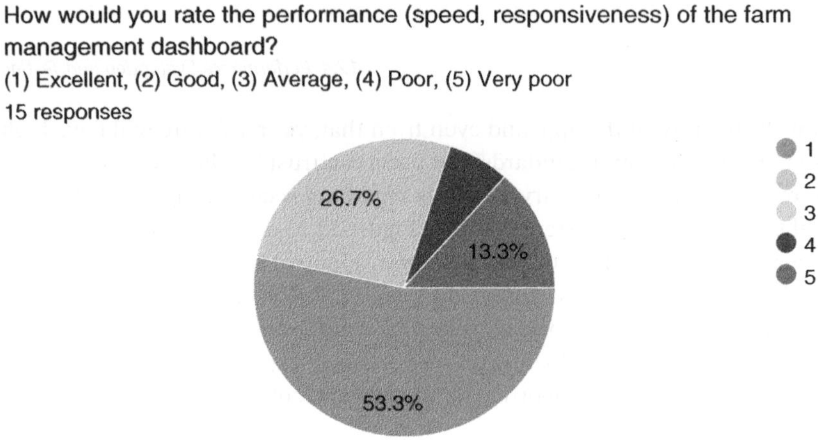

Figure 12.25 Survey results regarding the system's performance in terms of speed and response.

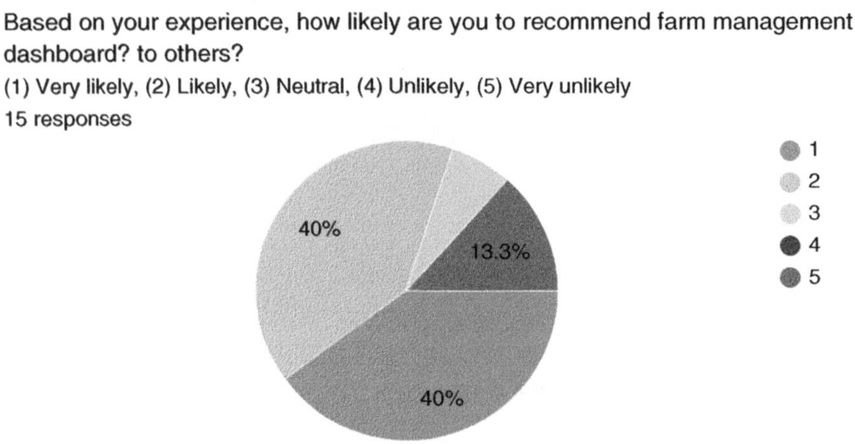

Figure 12.26 Survey results regarding the recommendation for the proposed system.

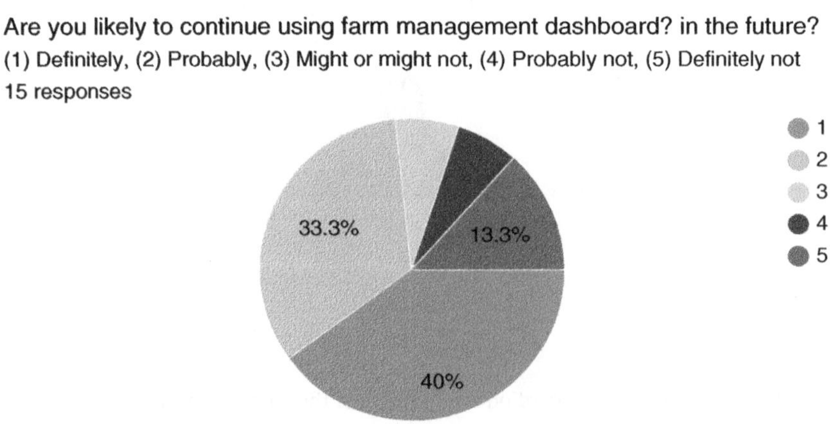

Figure 12.27 Survey results regarding the proposed system's future usage.

12.5 Legal, Social, Ethical, and Sustainability Issues

In this section, we discuss legal, social, ethical, and sustainability issues.

12.5.1 Legal Issues

As far as the issues concerning the law, the report FarmTechAI will be concerned with, among others, compliance with data protection laws like the General Data Protection Regulation (GDPR), which protects the data of the user by encryption and frequent audits. The dashboard respects intellectual property rights since it requires all software components to have the proper license. It also respects and ensures that the pertinent regulatory agricultural standards are respected. Pretty much every term of service was somewhat elaborately presented, and all the terms were set in a transparent user agreement, from ongoing legal reviews that would ensure the dashboard is continuously aligned with evolving legal requirements, to reinforce its commitment to the lawful operation and protection of its users.

12.5.2 Social Issues

In the discussion of social issues, this is the most critical part. Reflecting on how the FarmTechAI will affect the farming community, it might be useful to redefine labor dynamics by bringing automation to some tasks that require new skill sets, and it may displace tradition. So, the dashboard must play a role in building a network or, in a word, promoting community relations among the farmers to share knowledge and their best practices. Equally important is the social implication of disparities in data access so that technology not only serves a diversified kind of user but is socially inclusive for the agricultural sector.

12.5.3 Ethical Issues

As for ethical issues, the FarmTechAI places much emphasis on ethical data management practices, whereby the users' feedback is collected anonymously through a survey so that the responses are highly honest and not biased. This ethical stance also extends to transparent uses of the firm's ML algorithms, where applicable, to ensure decisions emanating from the system are fair and auditable. This project respects ethical norms by first taking consent from the user to utilize their data and observes privacy to ensure all are treated equally, regardless of their technical skills or domicile.

12.5.4 Sustainability Issues

Consistent with the study, the FarmTechAI is an aid in promoting sustainable agriculture practices. The dashboard minimizes wastage of resources by facilitating efficient management and optimizes water, fertilizers, and energy. The dashboard further supports sustainable decision-making. It provides real-time data that will enable the farmer to change his practices according to environmental crop health and conditions, hence productivity. More so, in promoting precision agriculture, the dashboard allows for the small environmental footprints that come from farming operations, all in line with the major objectives of enhancing the ecological sustainability of agriculture and guaranteeing the long-term productivity of farming resources.

12.6 Conclusions and Future Work

FarmTechAI, the FMD has fully achieved the first set of goals, integrating emerging technologies like real-time data analytics and ML to help make farmers' decisions and operations way more effective. Its user-friendly designs have made it be embraced by any farmer, even those that are not tech-savvy, in the process of democratizing the use of agricultural tools. The project will not only improve resource management and crop yields but also create a large collaborative farming community. Future development will focus on further increasing accessibility and effectiveness through expanded AI capabilities and mobile integration. Even with the issues of scalability and data security that need to be tackled, it is expected that the FarmTechAI already has quite a strong basis that can be further developed toward sustainable agriculture.

Promising future improvements can be listed as follows.

12.6.1 Having Advanced Analytics

Furthermore, the predictive and prescriptive insights tend to help the FarmTechAI in bringing improvement in the decision-making. Combining the analytics, predictive analytics can predict crop yields, forecast pests, and increase the optimization of resources [16]. Prescriptive analytics can help the respective farm states change their crops or irrigation routines, for example. This will involve data integration from sources, carrying an outline on how best to present, and preferably putting in place a UI with feedback mechanisms for real-time farm management and improved accuracy [17].

12.6.2 The Usage of IoT

Integration of the FarmTechAI with IoT would revolutionize farming. The real-time data from sensors and smart devices can synchronize crop and soil sensors with weather data, hence making irrigation and fertilization more impactful [18]. Second, it would reduce the downtime of equipment because of the system alerts and prediction when it requires maintenance. The second major advantage of using FarmTechAI is that it would have a UI showing live data and insights to help take action that will ensure enhanced farm efficiency, resource optimization, and environmental sustainability [19].

12.6.3 Integrating E-commerce Capabilities

Centralization of the management of the farm, sales, and marketing are some of the abilities that would come with incorporating e-commerce into the farm [20]. A farmer would do such things as add products, maintain his inventory, and even plan and monitor deliveries or order inputs using dynamic pricing tools tied to the market price, and analyze clients' details for tailored marketing all in one system [21, 23]. This would streamline sales, expand market reach, and reduce dependency on intermediaries, who also delay market responsiveness. Finally, quantum computing can provide a large amount of computation to manage real-time data efficiently [22].

Software Availability

We released FarmTechAI, which is available for free as open source. All code, datasets, and result reproducibility scripts are publicly available and can be accessed from GitHub: https://github.com/MuhammedGolec/FarmTechAI

Acknowledgments

We declare that this work has been submitted as a BSc project dissertation in partial fulfillment of the requirements for the award of the degree of Bachelor of Science submitted in the School of Electronic Engineering and Computer Science of the Queen Mary University of London, UK, is an authentic record of research work carried out by Murat Can Cardak (first author) under the supervision of Sukhpal Singh Gill (last author) and refers other researcher's work which is duly listed in the reference section. This BSc project dissertation has been checked using Turnitin at Queen Mary University of London, UK and submitted dissertation has been stored in repository for university record.

References

1 Dunwell, J.M. (2013). Global population growth, food security and food and farming for the future. *Successful Agricultural Innovation in Emerging Economies: New Genetic Technologies for Global Food Production* 100: 23.
2 Candoğan, K. and Özdemir, G. (2021). Novel approaches for a sustainable meat production. pp. 408–427.
3 Haval, A.M. and Rahman, F. (2024). Application of machine learning techniques and the internet of things for smart, sustainable agriculture. *BIO Web of Conferences.* Vol. 82. EDP Sciences, Moscow, Russian Federation, 5–6 December, 2023.
4 Gill, S.S., Chana, I., and Buyya, R. (2017). IoT based agriculture as a cloud and big data service: the beginning of digital India. *Journal of Organizational and End User Computing (JOEUC)* 29: 4.
5 Sengupta, A. (2021). Mobile edge computing based internet of agricultural things: a systematic review and future directions. Mobile Edge Computing.
6 Stočes, M. (2016). Internet of things (IoT) in agriculture-selected aspects. *Agris on-line Papers in Economics and Informatics* 8: 1.
7 Klompenburg, V., Thomas, A.K., and Catal, C. (2020). Crop yield prediction using machine learning: a systematic literature review. *Computers and Electronics in Agriculture* 177: 105709.
8 Fountas, S. (2015). Farm management information systems: current situation and future perspectives. *Computers and Electronics in Agriculture* 115: 40.
9 Brugger, F. (2018). Reducing transaction costs in contract farming arrangements: the case of farmforce. In: *Digital Technologies for Agricultural and Rural Development in the Global South* (ed. R. Duncombe), 54–67. Wallingford UK: CAB International.
10 Gumbi, N., Gumbi, L., and Twinomurinzi, H. (2023). Towards sustainable digital agriculture for smallholder farmers: a systematic literature review, MDPI. https://www.mdpi.com/2071-1050/15/16/12530 (accessed 24 April 2024).
11 Vidvan, T. (2021). All you need to know about django MVT architecture, TechVidvan. https://techvidvan.com/tutorials/djangos-mvt-architecture/ (accessed 05 January 2024).
12 Chaudhary, L. (2023). Agile in SDLC: characteristics and functionality of agile in SDLC, EDUCBA. https://www.educba.com/agile-in-sdlc/ (accessed 18 April 2024).
13 Aken, A. (2008). CHUNK: an agile approach to the software development life cycle. *Journal of Internet Commerce* 7 (3): 313–338. https://doi.org/10.1080/15332860802250385.
14 Docs, D. (2024). Docker overview, docker overview. https://docs.docker.com/get-started/overview/ (accessed 05 January 2024).

15 Macvittie, D. (2022). Build once, deploy anywhere, DevOps.com. https://devops.com/build-once-deploy-anywhere/ (accessed 05 January 2024).

16 Singh, S., Chana, I., and Buyya, R. (2020). Agri-Info: cloud based autonomic system for delivering agriculture as a service. In: *Internet of Things: Engineering Cyber Physical Human Systems*, vol. 9, 1–16. Elsevier Press.

17 Gia, T.N. (2019). Edge AI in smart farming IoT: CNNs at the edge and fog computing with LoRa. *2019 IEEE AFRICON*. IEEE, Accra, Ghana, 25–27 September 2019.

18 Gill, S.S. (2024). Modern computing: vision and challenges. *Telematics and Informatics Reports* 13: 1–38. Elsevier.

19 Gill, S.S. (2022). AI for next generation computing: emerging trends and future directions. *Internet of Things* 19: 100514.

20 Singh, R. and Gill, S.S. (2023). Edge AI: a survey. *Internet of Things and Cyber-Physical Systems* 3: 71–92.

21 Singh, D. and Kumar, A. (2024). A Deep Recurrent Neural Network for Plant Disease Classification. *SN Computer Science* 5 (8): 1053.

22 Golec, M. (2024). Quantum cloud computing: trends and challenges. *Journal of Economy and Technology* 2: 190–199.

23 Kumar, A. and Jha, R. (2022). IoT in smart farming: A boon to farmers. In: *Internet of Everything: Smart Sensing Technologies* (eds. T. Kavitha, V. Ajantha Devi, S. Neelavathy Pari, and S. Ramanathan), pp. 185–206. Nova Publisher.

13

Livestock Monitoring and Welfare

V. Kanakaris[1,2], E. Vrochidou[1], and G. A. Papakostas[1]

[1] *MLV Research Group, Department of Informatics, Democritus University of Thrace, Kavala, Greece*
[2] *Department of Economics, Democritus University of Thrace, Komotini, Greece*

13.1 Introduction

The advent of digital technologies introduced multiple sophisticated sensors, allowing for increased levels of automation in livestock farming, introducing precision livestock farming (PLF) [1]. PLF aims to improve the efficiency of production by simultaneously increasing animals' welfare by employing advanced technologies, targeted use of resources, and precision in the control of all stages of the production processes [2]. PLF is based on constant livestock monitoring for the observation of animals and their environments at all times. Monitoring livestock and guaranteeing their welfare are crucial components of contemporary animal husbandry procedures focused on maintaining the health and well-being of animals in agricultural environments [3].

Technological breakthroughs in sensors, artificial intelligence (AI), machine learning, computer vision (CV), and the Internet of Things (IoT) can provide easily accessible, affordable, accurate, and powerful tools, transforming the monitoring and management of livestock [4]. Sophisticated sensors and smart tags attached to animals include cameras, accelerometers, gyroscopes, temperature sensors, microphones, and more, aiming to replace traditional human visual and tactile observation tasks, in need of human expertise, that could be time-consuming and prone to errors, especially in large farms. Sensors are gathering a vast number of measurements in real time that need to be stored and processed. The latter is enabled by wireless communications and IoT, offering cloud storage

Emerging Smart Agricultural Practices Using Artificial Intelligence, First Edition.
Edited by Ashish Kumar, Jai Prakash Verma, and Rachna Jain.

Published 2025 by John Wiley & Sons, Inc.

for big data steaming from constant monitoring. Advancements in deep learning algorithms are able to process big data in real time and conclude in efficient extraction of information from the data and targeted decision-making [5]. Thus, farmers can be warned in real-time about the health and other issues of farmed animals and act immediately to prevent or resolve all rising problems. All these tools are combined in PLF to provide immediate monitoring of animal behavior, health metrics, and environmental factors, enabling early identification of problems and timely intervention. Figure 13.1 illustrates an overview of monitoring systems in PLF. For example, rumination cattle monitoring could help control mastitis and ketosis, while sensory systems are able to identify temperature and enable timely identification of animals with underlying diseases. Automated monitoring systems combined with efficient hardware and intelligent software, as well as with data analytics, can extract information from a broad range of sources and provide significant insights into animal care, production, and farm efficiency for both farmers and academics [6]. Therefore, PLF approaches can improve animal welfare, promote sustainable agricultural practices, and boost animal health results.

As a result, PLF is a subject of ever-increasing scientific interest. Conducted research on the Scopus scientific database using the general term "precision livestock farming," returned 1.291 documents, ranging from 1979 to date

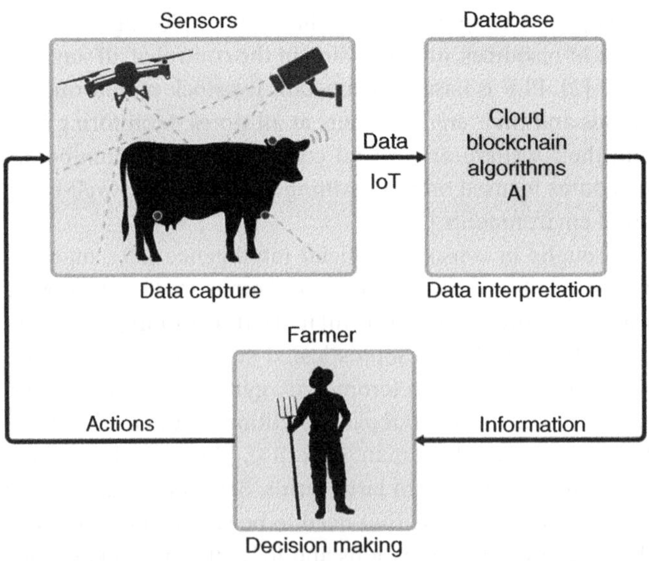

Figure 13.1 Livestock monitoring in a PLF system is empowered by sensors capturing data from animals, such as vision sensors and data from drones. Data are processed and interpreted, empowering farmers for more efficient decision-making.

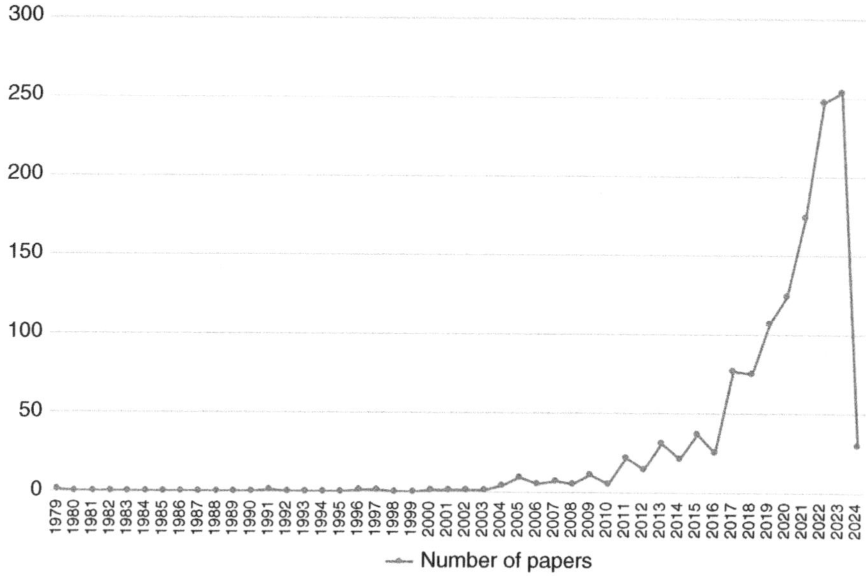

Figure 13.2 Number of papers per year on PLF, according to the Scopus database (research conducted on February 24, 2024).

(February 24, 2024). The distribution of articles over the years is illustrated in Figure 13.2, indicating the vivid research interest in the subject, especially within the last decade, with 30 articles already published by February 2024.

On top of the increased scientific interest, livestock monitoring market dynamics additionally indicate the economic projections of real-time livestock monitoring; the livestock monitoring market is expected to reach UDS 3.7 billion in 2030, at a compound annual growth rate (CAGP) of 11% from 2023 to 2030 [7], as illustrated in Figure 13.3. A major factor driving this increase in livestock market size is attributable to the adoption of cost-saving livestock monitoring solutions enabled by IoT and AI.

Livestock monitoring is applied to various types of animal husbandry, including but not limited to pig farming [8, 9], cattle farming [10, 11], poultry farming [12, 13], goat farming [14, 15], and sheep farming [16, 17]. The latter are some of the common types of animal husbandry where livestock monitoring techniques are utilized for various purposes such as health monitoring, behavior tracking, and overall management. The distribution of articles on livestock monitoring per animal type, based on article keywords derived from Scopus, is illustrated in Figure 13.4.

Note that the larger extent of the literature focuses on cattle-related applications. This is due to the fact that cattle are responsible for the largest share of the

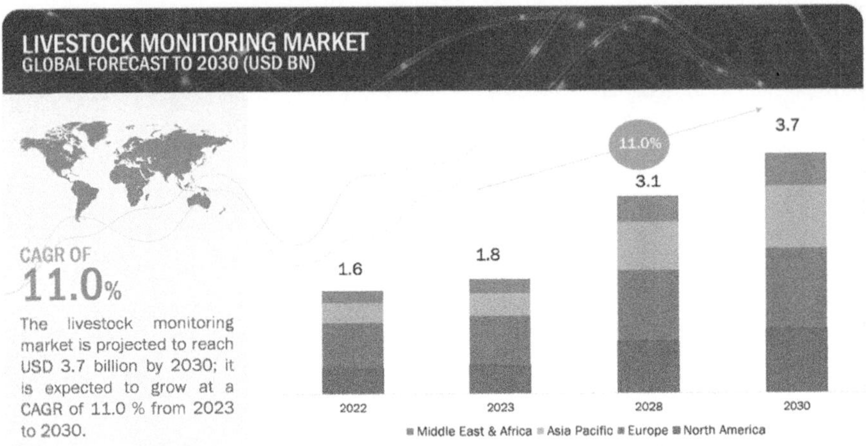

Figure 13.3 Livestock monitoring market size. *Source:* [7]/Markets and Markets Research Pvt. Ltd.

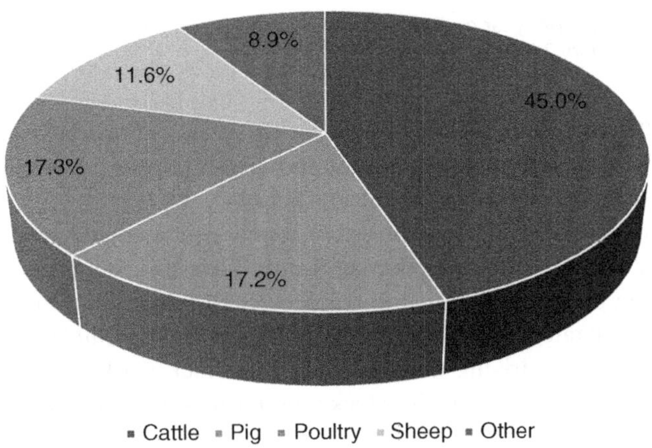

Figure 13.4 Percentage of papers on livestock monitoring per animal type.

livestock monitoring market, according to a report conducted on the market size by livestock type within the forecasting period 2022–2030 [7]. Many governments worldwide enforced strict regulations for the compulsory tagging of cattle, thus driving the adoption of livestock monitoring solutions for cattle, being on top of the market growth.

To this end, the objective of this chapter is to contribute to the existing body of knowledge by providing a comprehensive review of livestock monitoring technologies. The aim is to examine the literature, identify the current technological advancements in livestock monitoring, and discuss all related benefits in animal welfare and farming management. Additionally, livestock monitoring applications are discussed, covering the five most common types of animal husbandry. This work argues that to fully integrate PLF technologies into farming management systems, further developments and tests are required, as well as suitable farmers' training to better understand and, thus, be more willing to adopt modern approaches toward sustainable farming and increased animal welfare.

The rest of the chapter is organized as follows. Section 13.2 discusses the benefits of livestock monitoring. Section 13.3 presents advances in livestock monitoring technologies and methods, including current innovations in CV, the IoT, blockchain, and drones. The impact of livestock monitoring welfare is summarized in Section 13.4. The main findings of this work are discussed in Section 14.5, while Section 13.6 concludes the chapter.

13.2 Benefits of Livestock Monitoring

Livestock monitoring offers a multitude of benefits that encompass improved animal welfare, enhanced productivity, early disease detection, optimized resource management, and data-driven decision-making. In the following, the benefits of livestock monitoring are summarized, supported by references from the provided publications:

- **Improved Animal Welfare**: Animals are experiencing environmental stress and injuries that have an impact on their growth. Livestock monitoring systems enable real-time tracking of animal health parameters, behavior, and environmental conditions, leading to enhanced welfare outcomes. By utilizing sensor technologies and AI algorithms, farmers can proactively identify, and address issues related to animal health and well-being, ensuring a higher standard of care for the livestock [14, 18–20].
- **Enhanced Productivity**: Monitoring systems provide valuable insights into animal behavior, feeding patterns, and health status, allowing farmers to optimize feeding strategies, detect diseases early, and improve overall herd management practices. This leads to increased productivity and efficiency in livestock farming operations [15, 21–29].
- **Early Disease Detection**: Livestock monitoring technologies facilitate early detection of diseases through continuous monitoring of vital signs and behavior patterns. By leveraging IoT sensors and AI algorithms, farmers can identify signs of illness promptly, enabling timely intervention and reducing the spread of diseases within the herd [10, 18, 30–35].

- **Optimized Resource Management**: By monitoring environmental parameters such as temperature, humidity, and air quality, farmers can optimize resource utilization and create a more comfortable and healthier living environment for the livestock. This leads to improved feed conversion rates, reduced stress levels, and better overall performance of the animals [36]. The farmers are able to monitor the breeding and calving cycle of animals and, therefore, ensure the correct intake of nutrients. Moreover, the quality of food and water, can be accessed, hygiene levels and biosecurity of animals. Resources such as nutrients, food, water, medicine, and antibiotics are, thus, better handled [37]. Finally, surveillance tasks are automized, and farmers save labor costs related to visual monitoring and examination of animals, reducing veterinarian costs, and giving them more time to engage in additional farm activities. All the above result in optimized farm management, optimal reproduction processes, increased milk production, and weight gain of animals, allowing a feasible eco-farming at lower costs, saving resources, and reducing wastes without causing overall negative impacts on the environment. By optimizing resource utilization and reducing waste, farmers can minimize the environmental footprint of all their operations.
- **Data-Driven Decision-Making**: Livestock monitoring systems generate valuable data insights that empower farmers to make informed decisions regarding animal care, breeding programs, and herd management strategies. By analyzing data on animal behavior and health trends, farmers can implement targeted interventions to improve outcomes and drive sustainable practices [21].

Based on the above, it could be concluded that livestock monitoring technologies play a crucial role in promoting animal welfare, enhancing productivity, and ensuring sustainable agricultural practices. The use of advanced monitoring systems can enable farmers to optimize their operations, improve animals' health, and contribute to the overall efficiency and sustainability of the livestock farming industry.

13.3 Innovative Livestock Monitoring Technology Methods

In recent years, challenges such as climate change, increased consumer demands, and ever-increasing financial demands force livestock farmers to adopt cutting-edge technologies to maintain a sustainable future. Four main livestock technology methods that have witnessed significant advancements in recent years, introducing cutting-edge technologies that revolutionized livestock management practices, are the Internet of Things (IoT), vision-based systems, blockchain, and drones. Figure 13.5 illustrates the percentage of articles dealing with these four main

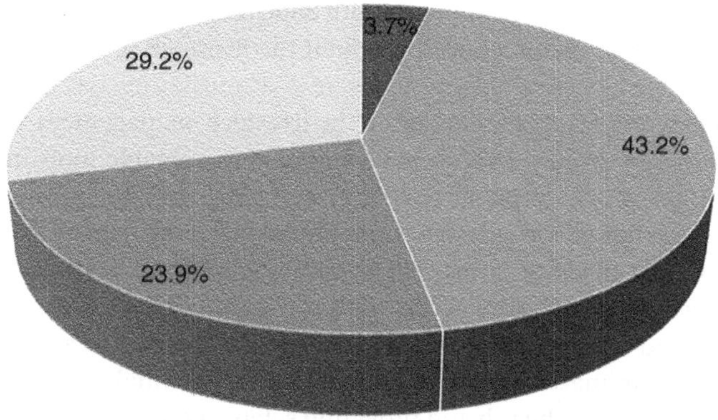

Figure 13.5 Percentage of papers per livestock monitoring technology method.

livestock technologies for monitoring, as concluded from their keywords on Scopus search, indicating as main contributors the IoT.

IoT technology holds a substantial role in livestock management by enabling real-time monitoring of animal health parameters, environmental conditions, and behavior patterns [10, 34, 36, 38, 39] by using sensors, devices and seamless data transmission. Data are gathered, transmitted, stored and analyzed, using appropriate technologies, algorithms and infrastructures. IoT devices such as sensors and wearables collect data on vital signs, feeding habits, and movement patterns, permitting farmers to make informed decisions and, thus, optimize herd management practices [10, 24, 34, 39]. Sensors are either attached to animals or located in key positions to continuously observe them. Real-time data is transmitted through wireless technologies and stored and analyzed in a central system, either in the cloud or in a database. Data analytics and efficient algorithms are used to interpret complex data and promote sustainable farming practices.

Vision-based systems, powered by AI and machine learning algorithms, offer advanced capabilities for livestock monitoring and behavior recognition. These systems use cameras and image processing techniques to track animal movements, detect anomalies, and ensure the well-being of livestock in real time [20, 40–43]. Cameras are installed in farm facilities aiming to gather precise training data. Efficient machine learning models are trained on collected data, to identify any kind of arising issues such as discomfort, stress, or pain.

Blockchain technology is increasingly being explored in livestock management for enhancing traceability, transparency, and security in supply chains. Blockchain is a decentralized recording system across a computer's network, able to ensure accountability and traceability of livestock-related data. By leveraging blockchain, farmers can create immutable records of animal health data, breeding history, and ownership, ensuring trust and transparency throughout the livestock production process [31, 44–46].

One of the most noteworthy applications of AI in livestock management is the exploitation of CV-based drones. Drones have emerged as valuable tools in livestock management, providing aerial surveillance, mapping, and monitoring capabilities for large-scale farms. Drones can autonomously survey substantial herds of animals across vast fields or farms. Their main contribution is in their capability to monitor vast areas and even access hard-to-reach locations. Drones equipped with cameras and sensors, combined with AI, enable farmers to assess pasture conditions, track animal movements, and identify potential risks or missing animals, protecting herds' health and providing safety [47–49].

The integration of all the abovementioned technologies represents a transformative shift toward data-driven and sustainable farming practices. By harnessing the power of these advanced technologies, farmers can enhance animal welfare, optimize productivity, and ensure the long-term sustainability of livestock operations. In what follows, these four technologies are further analyzed.

13.3.1 Computer Vision in Livestock Monitoring

CV techniques play a crucial role in livestock monitoring, enabling automated and efficient tracking, identification, and behavior analysis of animals. In what follows, an extensive summary of the most fundamental CV techniques used in livestock monitoring takes place.

Object Detection and Tracking: CV algorithms are employed to detect and track animals in real-time using visual data captured either from cameras or drones. Deep learning networks like You Only Look Once (YOLO) [50], Mask R-CNN [51], and fully convolutional networks (FCNNs) [52] are utilized in the literature for the accurate detection and tracking of livestock.

Individual Animal Identification: CV enables individual animal identification through techniques like back surface point clouds in 3D vision [52, 53], Kinect-sensed images, and cow face recognition using vision transformers [54]. These methods help in uniquely identifying animals for personalized monitoring and management.

Behavior Analysis: Continuous monitoring of animal behavior is essential for early disease detection and overall welfare. CV techniques can analyze movements, feeding patterns, and anomalies in behavior. This is achieved through methods like movement tracking, and monitoring canid scent marking using biologging and machine learning [52, 55].

Livestock Counting and Classification: CV is used for counting and classifying livestock from aerial images captured by drones. Techniques like Mask R-CNN [56] and convolutional neural networks (CNNs) are reported in the literature for accurate counting and classification of animals in the field [51, 52].

Health Monitoring: CV plays a role in monitoring the health of livestock by analyzing visual cues and anomalies. Techniques like the automated acquisition of top-view dairy cow depth image data and analysis of the drinking behavior of beef cattle are indicative examples of assessing the health status of animals based on their behavior and physical characteristics [52].

In general, CV enables PLF by providing real-time insights into animal behavior and health. Techniques like cow object recognition and double-camera fusion systems for animal-position awareness contribute to efficient and accurate monitoring of livestock in farming environments [52, 57, 58]. Using vision in livestock monitoring offers a range of benefits that enhance animal welfare, farm management, and overall productivity. Automated monitoring aims apart from monitoring livestock behavior, health, and activities, to reduce labor costs related to the need for manual observation and interventions [28]. CV can detect subtle changes in animal behavior or appearance that may indicate health issues, allowing for early intervention and disease prevention [15, 29]. Vision technology allows for the individual tracking and identification of animals, facilitating personalized care and management strategies [8, 57]. Thus, it can help optimize feed distribution, leading to better nutrition and resource utilization [37], and improve productivity and efficiency in livestock farming. Vision-based monitoring generates valuable data that can be analyzed to make informed data-driven decisions and targeted interventions regarding all aspects of animal care, breeding programs, and farm operations. Finally, vision systems can also be used to monitor environmental conditions in livestock facilities, such as temperature, humidity, and air quality, to ensure optimal living conditions for the animals.

The conducted review revealed that vision technology covers applications designated for various types of animal husbandry, each utilizing the technology for specific contributions: in poultry farming [12, 33, 59], cattle farming [10, 11, 16, 21–23, 25, 26, 34, 36, 39, 43, 44, 47, 52, 54, 57, 60–65], pig farming [9, 29, 53], and sheep farming [66–68]. Additionally, livestock monitoring applications for various other animals are also reported [14, 17, 22, 40, 64, 69–73] aiming to contribute to general farm operations. These applications demonstrate how vision technology enhances monitoring, management, and welfare practices across different sectors of animal husbandry, ultimately improving productivity and efficiency in livestock farming operations. Table 13.1 includes indicative applications from the literature for the four most common animal types, referring to the use of specific innovative technologies for livestock monitoring.

Table 13.1 Indicative publications including innovative technologies in livestock monitoring for the most common animal husbandry.

Farmed animal type	Computer vision	IoT	Blockchain	Drones
 Cattle	[10, 11, 16, 21–23, 25, 26, 34, 36, 39, 43, 44, 47, 52, 54, 57, 60–65]	[11, 24–26, 39, 60]	[44, 63, 74–76]	[47, 77–81]
 Poultry	[12, 33, 59]	[12, 33, 59]	[82–84]	—
 Pig	[9, 29, 53]	[9, 85–88]	[89]	[90]
 Sheep	[66–68]	[91–95]	[96]	[97]

13.3.2 IoT in Livestock Monitoring

IoT methods have transformed livestock monitoring by allowing immediate data gathering, examination, and control in animal farming. In what follows, a comprehensive overview of IoT methods employed in livestock monitoring is reported.

Sensor Integration: IoT systems integrate various sensors such as radio frequency identification (RFID) electronic labels, wireless sensor network (WSN) communication, and 3G remote communication technologies. These sensors collect data on parameters like body temperature, heart rate, respiratory rate, movement, and location of livestock [12, 30, 50, 98].

Data Transmission: IoT systems use wireless technologies like long range (LoRa) low-power wide-area network (LPWAN) for continuous monitoring and transmission of data from sensors to centralized platforms. This enables remote monitoring of livestock health and behavior [34, 50, 60, 64, 99].

Cloud Computing: Intelligent livestock farming systems leverage cloud computing for storing and processing large volumes of data collected from IoT sensors. Cloud platforms enable real-time analytics and decision-making for farmers [24, 30].

Machine Learning: IoT-based livestock monitoring systems incorporate machine learning algorithms for data analysis. These algorithms can detect patterns, predict health issues, and optimize livestock management practices [32, 100].

Historical Data Analysis: IoT systems capture and store historical data on livestock behavior and health parameters. Analyzing this data helps in identifying trends, optimizing feeding schedules, and improving overall animal welfare [12, 101].

Geofencing: IoT frameworks in livestock monitoring implement geofencing techniques for tracking and controlling the movement of animals within designated areas. IoT-enabled GPS location tracking systems provide real-time information on the whereabouts of livestock. This ensures better management of grazing spots, prevents theft and enhances security [25, 30, 32, 35, 52].

Using IoT technologies for livestock monitoring provides numerous advantages for farmers and animals. IoT technology provides immediate and precise monitoring of animals' physiological parameters, behavior, and environmental variables. IoT sensors can monitor environmental parameters in livestock facilities. This data assists in establishing ideal living circumstances for the animals, resulting in healthier animals. Instant data availability enables quick decision-making and intervention in case of any problems [33, 38, 39, 50, 59]. Timely identification of health problems and preventive treatment result in the improved general welfare of the livestock [32, 33, 39, 64, 100], preventing disease epidemics and reducing financial losses [10, 26, 30, 32–35], increasing productivity through feeding plans, breeding programs, and management procedures by utilizing data-driven insights [11, 100], and cost efficiency by optimizing operations, optimizing the use of resources and increased yields [24, 26, 59], eliminating human labor, and averting losses from health issues or theft [59, 64, 100]. IoT technologies provide remote monitoring of livestock, enabling farmers to check their animals' well-being from any location, improving convenience, and facilitating effective management of extensive animal operations [52, 59, 98, 101]. In general, IoT technologies support sustainable livestock production by improving resource efficiency, minimizing waste, and enhancing animal well-being. This leads to improved management practices and better outcomes in animal husbandry. This sustainability focus is in line with the increasing need for eco-friendly agriculture practices.

IoT technologies are utilized in various types of animal husbandry to enhance monitoring, management, and overall efficiency, included in Table 13.1: in poultry farming [12, 33, 59], cattle farming [11, 24–26, 39, 60], pig farming [9, 85–88], sheep farming [91–95], and more.

In conclusion, IoT technologies are applied across various animal husbandry sectors to improve monitoring, management, and productivity. Their contributions include early disease detection, precision farming practices, enhanced animal welfare, and sustainable agricultural operations in different livestock farming domains.

13.3.3 Blockchain in Livestock Monitoring

Blockchain technology is increasingly being leveraged in livestock monitoring systems to enhance data integrity, traceability, and security throughout the agricultural supply chain. Traceability can address the concerns of consumers, as well as make farming industries responsible for ethical and sustainable practices [102]. In the following, an extensive summary of how blockchain technology is utilized in livestock monitoring is presented, underlining its most crucial aspects.

Integrity of Data: Blockchain technologies could ensure the integrity of data through the creation of a decentralized and tamper-proof registry, aiming to record all transactions and data entries related to livestock monitoring in an immutable and transparent way [30, 103].

Security of Data: Blockchain technologies are capable of providing increased data security in livestock monitoring systems through the encryption of data. The latter can ensure data immutability and prevent unauthorized access and corruption of data. Blockchain, being a robust security framework, can, therefore protect sensitive information and maintain the integrity of livestock data [30, 51]. Blockchain technology can also employed to store and/or share livestock health data with adequate security, e.g. vaccination records, disease history, treatment information, etc. The latter can facilitate access to critical health data, detect underlying diseases quickly, and support decision-making for animal welfare based on historical information [30, 44, 63]. Livestock monitoring systems using blockchain technology could additionally incorporate identifiers, e.g. RFID tags or other sensors, associated with blockchain records, for enhanced accuracy and individual animal tamper-proof identification, thus, could aim to simplify tracking and other management tasks. [30].

Traceability: Blockchain can help to trace livestock products throughout all stages of the supply chain, by recording end-to-end information referring to breeding, feeding, as well as transporting and selling. Therefore, the origins of each livestock can be tracked, ensuring authenticity while at the same time preventing potential fraud [46].

Smart Contracts: Smart contracts refer to self-executing contracts implemented by following predefined rules written into code [104]. Smart contracts can be adopted on blockchain networks of livestock monitoring systems to automate processes such as payments, compliance checks, data sharing, and more [31, 65].

Data Sharing Security: Blockchain technologies permit secure data sharing between stakeholders (farmers, veterinarians, regulators, and consumers) in the livestock industry. Secure data sharing can facilitate real-time access to accurate information, thus improving the collaboration between stakeholders and facilitating decision-making [51, 63, 103].

Management of the Supply Chain: Blockchain can also improve supply chain management in livestock monitoring by providing a transparent and auditable record of transactions. The latter is essential to reduce frauds, optimize logistics, and ensure the following of regulations throughout the supply chain [103].

Compliance and Inspection: Blockchain technology can facilitate the obedience to regulations and auditing processes; the transparency of blockchain records can simplify compliance checks, certifications, and audits toward improved trust in the livestock industry [30].

Integrity of Food: Through all the above aspects related to the traceability of products, blockchain technology impacts both food integrity and safety standards; consumers are able to verify the origins and quality of products they buy, encouraging trust and transparency in the food supply chain [30, 44].

The adoption of blockchain in livestock monitoring is capable of offering multiple benefits, overall aiming to enhance transparency, traceability, security, and efficiency throughout all livestock-related processes. Summing up all the above, blockchain can ensure the integrity of livestock data by providing a safe and transparent registry, where all data and transactions could be recorded and stored with adequate security, and could not be altered retroactively in any way [31, 44, 51]. Moreover, blockchain, through the end-to-end traceability of livestock products, can allow stakeholders to track the origin, movements, and general handling of all individual animals throughout all stages of the supply chain. Consumers as well, could verify the origin and quality of the products they buy and reinforce the trust on the food supply chain [103]. Blockchain-related transparency can also help to prevent frauds such as fake products [46], while at the same time to maintain high safety standards by protecting data from unauthorized access and corruption [51]. Blockchain also facilitates the real-time sharing of data between stakeholders in the livestock industry, thus, improves the collaboration between them, and guides decision-making [46]. Blockchain enhances supply chain management in livestock monitoring by creating a transparent and auditable record of transactions. This helps optimize logistics, reduce fraud, ensure compliance with regulations, and improve overall supply chain visibility [31]. Smart contracts deployed on blockchain networks automate processes such as payments, compliance checks, and data sharing based on predefined conditions.

This streamlines transactions, reduces administrative overhead and ensures compliance with predefined rules [44]. Blockchain technology simplifies compliance checks, audits, and certifications in livestock monitoring by providing a transparent and verifiable record of transactions. This enhances trust, accountability, and regulatory compliance within the livestock industry [46].

Blockchain technology is utilized in various types of animal husbandry (Table 13.1): Poultry farming [82–84], cattle farming [44, 63, 74–76], pig farming [89], and sheep farming [96].

13.3.4 Drone in Livestock Monitoring

Drone technology has revolutionized livestock monitoring practices by providing farmers and ranchers with efficient, cost-effective, and real-time data collection capabilities. Drones, also known as unmanned aerial vehicles (UAVs), offer a wide range of applications in livestock monitoring, including tracking animal movements, assessing herd health, monitoring pastures, and enhancing overall farm management [105]. In the following, an extensive summary of drone technology used in livestock monitoring is reported.

Imaging and Remote Sensing: Drones equipped with high-resolution cameras and sensors, capture aerial images and videos of livestock, pastures, and infrastructure, aiming to provide valuable insights related to the behavior of animal, their health status, and pasture conditions. The latter enables farmers to conclude in targeted and efficient management decisions [47, 106–108].

Animal Counting and Tracking: Drones are widely used for counting and tracking livestock, such as cattle, sheep, and goats, when they move inside large feeding areas. Drones capture aerial data and empowered by advanced image processing algorithms, they can extract various information from images, such as individual animals' monitoring information, herd size, detection of anomalies from the movements of animals, locate missing animals, etc. [47, 49].

Monitoring of Pasture: Drones can be used to inform farmers regarding pasture conditions, vegetation, and water source conditions in large grazing areas. The captured information is analyzed to help farmers to optimize grazing patterns, assess available feeding areas, and identify areas in the farm that require maintenance or improvement [47].

Animal Health Evaluation: Drones are equipped with multiple sensors that can extract valuable information, such as thermal and multispectral cameras. Such sensors can detect changes in animals' body temperature and thus help farmers to identify indicators of either stress or illnesses and interfere in a timely to prevent potential health issues [30, 106].

Infrastructure Inspection: Drones, as already mentioned, are used for surveillance. More specifically, they can be employed to inspect livestock facilities, fences, and other infrastructures on the farm so as to identify needs for maintenance, locate potential security risks, and overall structural issues, safeguarding the well-being of both human workers and livestock [47, 97].

Monitoring of the Environment: Drones can also monitor environmental variables that may impact the livestock, such as weather conditions, vegetation, and water quality and availability. Farmers can use the captured information toward sustainable land management practices to guarantee the well-being of the herd [47, 108].

Accessibility and Scalability: Drones are both scalable and accessible, meaning that they can be easily employed in multiple livestock farming operations for farms of variable size and in different locations. Farmers can integrate drones into their existing monitoring practices without needing to make any interventions and adapt their use to their specific needs [30, 49].

Real-Time Monitoring: Drones can provide real-time monitoring through live in-field video streaming capabilities. Live monitoring can facilitate farmers to conclude in quick decisions and rapid responses to emergencies [47, 108].

Cost-Effectiveness: Drones constitute a cost-effective livestock monitoring solution, especially when compared to traditional methods of human-based inspections or integrated multisensory ground systems in the farm. Drones can reduce labor and operational costs while providing timely and accurate comprehensive data that can be used for efficient decision-making [106].

Based on the above main uses, drones can play a substantial role in PLF through the collection of individual animals' data, such as size, weight, and behavior. This data is used for personalized animal care, toward optimizing feeding schemes and improving breeding practices. The latter can increase the efficiency and productivity in livestock farm management. The employment of drones can offer multiple benefits to livestock monitoring. Sophisticated sensors are mounted on drones to collect high-resolution aerial images of livestock, pastures, and infrastructure. The collected data can provide valuable information for the behavior and health of animals, pasture conditions, and insights for the overall farm management [107, 108]. Drones can be used by farmers to monitor and track animals' movements inside large feeding areas. Current advances in GPS technologies and image processing algorithms, can be combined with drones for applications such as animals' count, individual animals' identification, detection of all kind of anomalies and of missing animals [47, 49]. Thermal and multispectral cameras can measure temperature and identify changes in animals' body temperature, indicating early signs of illness [106]. Environmental indicators can also be measured by appropriate sensors, such as weather conditions, vegetation and water quality, and their availability in the area. Thus, farmers can realize sustainable management of the sources in the area to ensure the well-being of their herd. Drone suggests a cost-effective solution for livestock monitoring and are used in several application as substitution additionally to traditional human-based surveillance [106]. Aerial surveillance and streamline data collection can optimize resources allocation, and improve farm management practices in general [49]. A key feature of drones is the capability of real-time monitoring of the field and farm, even from distant locations or difficult to reach. Real-time monitoring empowers quick decision-making and timely responses to emergencies [47]. Moreover, by using drones for livestock monitoring, farmers are able to minimize the need for physical contact with animals, reducing the risk of injuries for human workers. Safety hazards on the farm can also be identified from aerial monitoring, e.g. identification of areas that require maintenance or improvements, toward enhancing the overall farm safety for both workers and animals [106]. All these aerial information provides valuable data-driven insights for farmers, and can be used to optimize the use of resources, reduce environmental impact, and improve animal welfare, contributing to the long-term viability of livestock management operations [107, 108].

Drone technology has been widely utilized in various types of animal husbandry practices to enhance monitoring, management, and overall efficiency in cattle farming [47, 77–81], sheep farming [97], pig farming [90], and goat farming [109].

13.4 Impact of Livestock Monitoring Methods on Welfare

Livestock monitoring plays a crucial role in improving animal welfare by enabling farmers and animal caretakers to track, assess, and respond to various factors affecting the well-being of the animals. Some key factors monitored in livestock management that contribute to enhancing animal welfare are highlighted in the following.

Health Monitoring: Regular health monitoring through IoT sensors, wearable devices, and visual inspections allows early detection of illnesses, injuries, or abnormalities in livestock. Prompt identification and treatment of health issues improve animal welfare by ensuring timely care and reducing suffering [38, 44].

Nutritional Management: Monitoring feed intake, nutritional requirements, and body condition scores helps optimize the diet and feeding practices for livestock. Proper nutrition is essential for maintaining animal health, growth, and reproductive performance, thereby enhancing overall welfare [106].

Environmental Conditions: Monitoring environmental factors such as temperature, humidity, ventilation, and lighting in livestock facilities ensures optimal living conditions for animals. Providing a comfortable and stress-free environment promotes animal well-being and reduces the risk of heat stress, cold stress, or respiratory problems [40, 47].

Patterns of Behavior: Observation and analysis of behavioral patterns, social interaction, and level of activity of animals, indicates their mental and physical well-being. Any changes in the behavioral pattern of an animal, e.g. as aggression, lethargy, or abnormal movements, are indicating stress, pain, or discomfort, triggering timely interventions to improve the welfare of the animal [15, 43].

Reproductive Health/Management: Monitoring reproductive parameters, estrus cycles, mating behavior, and pregnancy status in livestock contributes to reproductive health management. Certification of correct breeding practices improves the animals' welfare through successful reproduction and offsprings' health [17, 100].

Biosecurity: Livestock monitoring data can help the realization of biosecurity protocols, disease surveillance, and vaccination programs for the prevention of infectious diseases inside the herd. The existence of biosecurity measures is essential to safeguard animal health and the welfare of the entire herd [10, 16, 17, 30, 31, 33, 34].

Social Well-being: Social dynamics, group interactions, and hierarchy within livestock populations can also be monitored. The latter aims to promote the social well-being of animals. Providing adequate space, enrichment activities, and socialization opportunities supports positive social behaviors and reduces stress in group-housed animals [43, 52, 60].

Pain Management: Monitoring pain indicators, postoperative care, and recovery processes in livestock undergoing procedures or treatments helps manage pain and discomfort effectively. Implementing pain management strategies improves animal welfare by addressing pain-related issues and promoting faster recovery [73].

Handling and Transport: Monitoring animal handling practices, transport conditions, and stress levels during transportation helps minimize stress, injuries, and discomfort, ensuring humane treatment and welfare throughout the process [18, 73].

Spatial Monitoring: Drones can provide spatial monitoring of livestock facilities, grazing areas, and infrastructure, enabling farmers to assess the layout and design of their operations for optimal livestock management and welfare [108].

Figure 13.6 illustrates the distribution of papers on livestock monitoring per welfare factor, as result from Scopus database, indicating health monitoring as a top priority task.

Table 13.2 summarizes the livestock monitoring methods that could provide the identified welfare factors in livestock management practices, toward enhancing the overall welfare, health, and quality of life of the animals under their care.

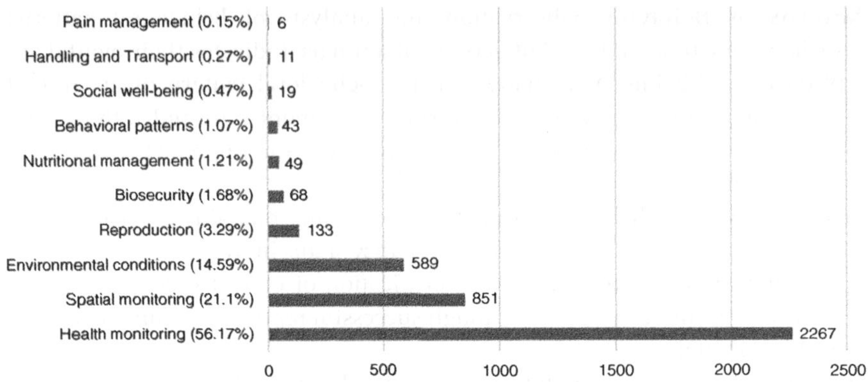

Figure 13.6 Number of papers on livestock monitoring per each defined welfare factor.

Table 13.2 Livestock monitoring methods that enhance the overall welfare.

	Livestock monitoring method			
Welfare factors	**Computer vision**	**IoT**	**Blockchain**	**Drones**
Health monitoring	✓	✓	✓	✓
Nutritional management		✓	✓	
Environmental conditions		✓	✓	✓
Behavioral observations	✓			✓
Reproductive health/management	✓	✓	✓	
Biosecurity measures		✓	✓	✓
Social well-being		✓		
Pain management	✓	✓	✓	✓
Handling and transport	✓	✓		
Spatial monitoring				✓

13.5 Discussion

While the benefits of livestock monitoring are evident, there are many challenges and limitations that need to be addressed. The complexity and high costs of technological innovations in livestock farming are major limitations to their adaptation. Farmers need to invest in adequate infrastructures, hardware devices, and corresponding software, to get trained to be able to use and interpret all derived information or to hire additional trained staff to deal effectively with monitoring systems. The latter is not always feasible, especially in small-scale farms. In the case of already existing infrastructures, integration of new technological systems with already existing ones may pose compatibility limitations, resulting in the need for additional investments toward adjustments and modifications. However, technological advancements foreseen in the coming years, are expected to lead to simplification and reduction of prices of livestock monitoring systems and wearable sensor technologies, making them accessible and easy to use for everyone.

Multisensory data information processing techniques able to handle big data are also in need, toward concluding in reliable extraction of information for precision farming decisions. Communication technologies supporting the recording and transferring of data through wireless modules, also are challenged to work efficiently for long-distance communications as well as on lighter energy consumption when integrated into wearable devices. Power supply of all devices also restricts the efficiency of wearable monitoring systems [5]; self-powered technologies driven by solar energy or by the energy of animals' movements need to be developed, especially for non-housed animals.

An additional challenge is related to the vast number of livestock monitoring data, including sensitive data such as health recordings and GPS locations, rising concerns about privacy, and security. While blockchain has the potential for sustainable livestock monitoring, the latter concerns still exist; data need to be safeguarded and processed thoroughly so as to comply with regulations on data privacy. As technological developments are ongoing, it is expected that blockchain technologies to be further enhanced, following ethical standards imposed by the farming industry. Transparency and awareness of all stages of livestock management are essential for the acceptance of technological interventions for farmers and consumers.

However, the global impact of livestock monitoring extends beyond animals' welfare, optimal farming management, and consumers' trust. It is associated with global sustainability goals. Livestock farming has a great environmental footprint, contributing to a large extent in greenhouse emissions, by almost 15% of all man-made emissions, it is responsible for deforestation actions, it is using about 10% of global agricultural and about 8% of the total available fresh water [7]. At the same time, the demands for milk and meat are increasing, raising serious concerns about the impact of livestock farming on the environment in the following years. Livestock monitoring has the potential to provide transformative tools toward sustainable livestock management.

Based on the above, future research directions on livestock monitoring could include the following aspects:

- Integration of advanced technologies, combining AI with IoT and sophisticated wearable sensors, toward smart large-scale farming.

- Advanced vision algorithms for animal health and environmental monitoring, toward early disease detection, automated nutrition decisions, waste reduction, and minimization of carbon footprint.
- Appropriate infrastructures for stable and efficient connectivity.
- Personalized animal care, through algorithms and applications focusing on individual animal identification and monitoring.

Overall, livestock monitoring can provide transparency, enhance traceability, improve animal welfare, and sustainable practices, suggesting a holistic solution to the challenges encountered by the livestock industry. In the future where livestock farming is sustainable and ethical, livestock monitoring emerges as a key factor in such a configuration [110, 111].

13.6 Conclusions

Livestock monitoring and welfare methods have been greatly affected by the contribution of technologies such as the IoT, CV, blockchain, and drones. IoT is essential for monitoring environmental factors, animal health data, and behavioral patterns in real time; farmers can therefore conclude informed decisions and optimize the use of farm resources. CV systems are valuable for the automatic detection, tracking, and analysis of animal behavior through image processing algorithms. Blockchain is employed to maintain data integrity, transparency, and traceability in livestock monitoring systems, encouraging the confidence between stakeholders, while reducing risks of any potential fraud. Drones facilitate aerial surveillance of farms, enabling distant monitoring of animals, pastures, and farm infrastructures, in real-time. In the future, livestock monitoring is expected to further advance through the foreseen advancements of these four key technologies. The incorporation of AI for predictive analytics, edge computing for real-time data processing, 5G connectivity for fast data transfer, and robotics for in-farm automations, are the most important trends that are expected to affect the future of livestock monitoring. The latter technologies are expected to transform the way farmers currently manage farms and care for animals, offering the potential for innovations, automation, and data-driven decision-making, toward more sustainable and efficient livestock management in the future.

Future research in the field could focus on identifying the impact of livestock monitoring on Environmental, Social, and Governance (ESG) regulation in the agricultural sector. In order to comply with ESG legislation, the following factors need to be addressed: reduction of environmental influence, promotion of social responsibility, assurance of transparency, enhancement of resource efficiency, and promotion of traceability in the supply chain. Future work on livestock monitoring could also focus on risk management issues, to prevent animal losses due to diseases or environmental factors, toward contributing to sustainability and resilience of the farming sector.

References

1 Buller, H., Blokhuis, H., Lokhorst, K. et al. (2020). Animal welfare management in a digital world. *Animals* 10 (10): 1779. Available from: https://www.mdpi.com/2076-2615/10/10/1779.

2 Banhazi, T.M., Lehr, H., Black, J.L. et al. (2012). Precision livestock farming: an international review of scientific and commercial aspects. *International Journal of Agricultural and Biological Engineering* 5 (3): 1–9.

3 Kleen, J.L. and Guatteo, R. (2023). Precision livestock farming: what does it contain and what are the perspectives? *Animals* 13 (5): 779. Available from: https://www.mdpi.com/2076-2615/13/5/779.

4 Brown-Brandl, T.M., Adrion, F., Maselyne, J. et al. (2019). A review of passive radio frequency identification systems for animal monitoring in livestock facilities. *Applied Engineering in Agriculture* 35 (4): 579–591. Available from: https://elibrary.asabe.org/abstract.asp?AID=50835&t=3&dabs=Y&redir=&redirType=.

5 Zhang, M., Wang, X., Feng, H. et al. (2021). Wearable internet of things enabled precision livestock farming in smart farms: a review of technical solutions for precise perception, biocompatibility, and sustainability monitoring. *Journal of Cleaner Production* 312: 127712. Available from: https://linkinghub.elsevier.com/retrieve/pii/S0959652621019302.

6 Berckmans, D. (2017). General introduction to precision livestock farming. *Animal Frontiers* 7 (1): 6–11. Available from: https://academic.oup.com/af/article/7/1/6/4638786.

7 Markets and Markets (2023). Markets and markets. livestock monitoring market size. Available from: https://www.marketsandmarkets.com/Market-Reports/livestock-monitoring-market-72634532.html (accessed 24 February 2024).

8 Tangirala, B., Bhandari, I., Laszlo, D. et al. (November 1 2021). Livestock monitoring with transformer. arXiv Prepr. arXiv:2111. Available from: http://arxiv.org/abs/2111.00801.

9 Psota, E., Mittek, M., Pérez, L. et al. (2019). Multi-pig part detection and association with a fully-convolutional network. *Sensors* 19 (4): 852. Available from: http://www.mdpi.com/1424-8220/19/4/852.

10 Moutaouakil, K.E.M., Jdi, H., Jabir, B., and Falih, N. (2023). Digital farming: a survey on IoT-based cattle monitoring systems and dashboards. *AGRIS on-line Papers in Economics and Informatics* 15 (2): 31–39. Available from: https://online.agris.cz/archive/2023/02/03.

11 Sulistiani, H., Yasin, I., Darwis, D. et al. (2023). IoT for monitoring system for cattle growth productivity. *2023 International Conference on Networking, Electrical Engineering, Computer Science, and Technology (IConNECT)*, pp. 180–183. IEEE, 25–26 Aug. 2023, Bandar Lampung, Indonesia. Available from: https://ieeexplore.ieee.org/document/10326758/.

12 Sriharee, G., Khongban, P., and Wongpipan, K. (2022). Toward IoT and data analytics for the chicken welfare using RFID technology. *2022 19th International Conference on Electrical Engineering/Electronics, Computer, Telecommunications and Information Technology (ECTI-CON)*, pp. 1–4. IEEE, 24–27 May 2022, Prachuap Khiri Khan, Thailand. Available from: https://ieeexplore.ieee.org/document/9795547/.

13 Grandin, T. (2020). Auditing and assessing the welfare of livestock and poultry during pre-slaughter handling and stunning. In: *The Slaughter of Farmed Animals: Practical Ways of Enhancing Animal Welfare*, 175–192. Wallingford UK: CABI http://www.cabidigitallibrary.org/doi/10.1079/9781789240573.0175.

14 Rao, Y., Jiang, M., Wang, W. et al. (2020). On-farm welfare monitoring system for goats based on internet of things and machine learning. *International Journal of Distributed Sensor Networks* 16 (7): 155014772094403. Available from: http://journals.sagepub.com/doi/10.1177/1550147720944030.

15 Bonneau, M., Godard, X., and Bambou, J.C. (2022). Assessing goats' fecal avoidance using image analysis-based monitoring. *Frontiers in Animal Science* 10: 3. Available from: https://www.frontiersin.org/articles/10.3389/fanim.2022.835516/full.

16 Herlin, A., Brunberg, E., Hultgren, J. et al. (2021). Animal welfare implications of digital tools for monitoring and management of cattle and sheep on pasture. *Animals* 11 (3): 1–20. Available from: https://www.scopus.com/inward/record.uri?eid=2-s2.0-85102434037&doi=10.3390%2Fani11030829&partnerID=40&md5=040b6c48551ffff0a88187b2301c637e.

17 Silva, S.R., Sacarrão-Birrento, L., Almeida, M. et al. (2022). Extensive sheep and goat production: the role of novel technologies towards sustainability and animal welfare. *Animals* 12 (7): 885. Available from: https://www.mdpi.com/2076-2615/12/7/885.

18 Neethirajan, S. (2023). Artificial intelligence and sensor technologies in dairy livestock export: charting a digital transformation. *Sensors* 23 (16): 7045. Available from: https://www.mdpi.com/1424-8220/23/16/7045.

19 Fayemi, P.O., Fayemi, O.E., Joel, L.O., and Ogungbuyi, M.G. (2021). COVID-19 syndrome: nexus with herbivory and exposure dynamics for monitoring livestock welfare and agro-environment. *Sustainability* 13 (22): 12381. Available from: https://www.mdpi.com/2071-1050/13/22/12381.

20 Guo, Y., Hong, W., Wu, J. et al. (2023). Vision-based cow tracking and feeding monitoring for autonomous livestock farming: the YOLOv5s-CA+DeepSORT-vision transformer. *IEEE Robotics and Automation Magazine* 30 (4): 68–76. Available from: https://ieeexplore.ieee.org/document/10303740/.

21 Lavanya, R., Abuthakir, S., and Ahamed, D.A. (2023). IoT-enhanced livestock monitoring for animal health and productivity. *2023 International Conference on Sustainable Communication Networks and Application (ICSCNA)*, pp. 504–509. IEEE, 15–17 Nov. 2023, Theni, India. Available from: https://ieeexplore.ieee.org/document/10370264/.

22 Anzai, H. and Hirata, M. (2021). Individual monitoring of behavior to enhance productivity and welfare of animals in small-scale intensive cattle grazing systems. *Frontiers in Sustainable Food Systems* 14: 5. Available from: https://www.frontiersin.org/articles/10.3389/fsufs.2021.694413/full.

23 Mendes Lima, L., Calebe Cavalcante, V., Guimaraes de Sousa, M. et al. (2021). Artificial intelligence in support of welfare monitoring of dairy cattle: a systematic literature review. *2021 International Conference on Computational Science and Computational Intelligence (CSCI)*, pp. 1708–1715. IEEE, 15–17 Dec. 2021, Las Vegas, Nevada, USA. Available from: https://ieeexplore.ieee.org/document/9799157/.

24 Kumarasamy, S. and Saraniya, S. (2018). Cloud IOT based novel livestock monitoring and identification system using UID. *Sensor Review* 38 (1): 21–33. Available from: https://www.emerald.com/insight/content/doi/10.1108/SR-08-2017-0152/full/html.

25 Aunindita, R.F., Shiam Misbah, M., Bin Joy, S. et al. (2022). Use of machine learning and IoT for monitoring and tracking of livestock. *2022 25th International Conference on Computer and Information Technology (ICCIT)*, pp. 815–820. IEEE, 17–19 Dec. 2022, Cox's Bazar, Bangladesh. Available from: https://ieeexplore.ieee.org/document/10055766/.

26 Ezhilarasan, K., Shilpa, A.N., Sridhar, N., and Lakshmi, V. (2023). Smart cattle farm infrastructure and monitoring cattle health status using IOT. *2023 International Conference on Sustainable Emerging Innovations in Engineering and Technology (ICSEIET)*, 14–15 Sept. 2023, ABES Engineering College, Ghaziabad, India, pp. 757–762. IEEE. Available from: https://ieeexplore.ieee.org/document/10303358/.

27 Tholhappiyan, T., Sankar, S., Selvakumar, V., Robert, P. (2023). Agriculture monitoring system with efficient resource management using IoT. *2023 Second International Conference on Augmented Intelligence and Sustainable Systems (ICAISS)*, pp. 1628–1633. IEEE, 23–25 Aug. 2023, Trichy, India. Available from: https://ieeexplore.ieee.org/document/10250720/.

28 Kadlec, R., Indest, S., Castro, K. et al. (2022). Automated acquisition of top-view dairy cow depth image data using an RGB-D sensor camera. *Translational Animal Science* 6 (4). Available from: https://academic.oup.com/tas/article/doi/10.1093/tas/txac163/6892921.

29 Yu, H., Lee, K., and Morota, G. (2021). Forecasting dynamic body weight of nonrestrained pigs from images using an RGB-D sensor camera. *Translational Animal Science* 5 (1). Available from: https://academic.oup.com/tas/article/doi/10.1093/tas/txab006/6102880.

30 Yang, L., Liu, X.Y., and Kim, J.S. (2020). Cloud-based livestock monitoring system using RFID and blockchain technology. *2020 7th IEEE International Conference on Cyber Security and Cloud Computing (CSCloud)/2020 6th IEEE International Conference on Edge Computing and Scalable Cloud (EdgeCom)*, 1–3 Aug. 2020, New York, NY, USA, pp. 240–245. IEEE. Available from: https://ieeexplore.ieee.org/document/9171001/.

31 Shen, L., Zhang, Z., Zhou, Y., and Xu, Y. (2023). Applying blockchain technology and the internet of things to improve the data reliability for livestock insurance. *Sensors* 23 (14): 6290. Available from: https://www.mdpi.com/1424-8220/23/14/6290.

32 Abidin, I.S.Z., Foo, Z.W., Ali, A.M., and Mansor, R. (2023). IoT–based livestock health monitoring and GPS location tracking system using radio frequency transmission. *2023 IEEE International Conference on Sensors and Nanotechnology (SENNANO)*, pp. 49–52. IEEE, 26–27 Sept. 2023, Putrajaya, Malaysia. Available from: https://ieeexplore.ieee.org/document/10352534/.

33 Joshi, K., Kumar, A., Anandaram, H. et al. (2023). Incorporating AI and IoT for smart poultry monitoring and health detection. *2023 Second International Conference on Augmented Intelligence and Sustainable Systems (ICAISS)*, pp. 76–81. IEEE, 23–25 Aug. 2023. Available from: https://ieeexplore.ieee.org/document/10250564/.

34 Balladares Ocana, A.M., Urgiles, R.I., Soria Perez, J.A. (2021). Design and implementation of a web platform prototype based an IoT gateway using raspberry Pi for livestock monitoring. *2021 IEEE Fifth Ecuador Technical Chapters Meeting (ETCM)*, pp. 1–6. IEEE, 12–15 Oct. 2021, Quito, Ecuador. Available from: https://ieeexplore.ieee.org/document/9590657/.

35 Jijesh, J.J., Lakshmi Manasa, B., Niranjana Vineeth, C. (2023). Design and implementation of smart livestock security system. *2023 International Conference on Distributed Computing and Electrical Circuits and Electronics (ICDCECE)*, pp. 1–6. IEEE, 29–30 April 2023, Ballar, India. Available from: https://ieeexplore.ieee.org/document/10150783/.

36 Vigneswari, T. (2021). Smart IoT cloud based livestock monitoring system: a survey. *Turkish Journal of Computer And Mathematics Education* 12 (10): 3308–3315.

37 Biglari, A. and Tang, W. (2022). A vision-based cattle recognition system using tensorflow for livestock water intake monitoring. *IEEE Sensors Letters* 6 (11): 1–4. Available from: https://ieeexplore.ieee.org/document/9924566/.

38 Muhamed, E.S., Ting, M., and SL, M.B. (2019). Livestock health monitoring using IoT technology for Ethiopia. *Journal of Advanced Research in Dynamical and Control System* 11 (11): 954–959. Available from: https://www.jardcs.org/abstract.php?id=4063.

39 Thakur, B.S. and Sheetlani, J. (2021). Analyzing a cattle health monitoring system using IoT and Its challenges in smart agriculture. *Proceedings of Intelligent System*

Design: INDIA 2019, November 1-2, 2019, Lendi Institute of Engineering & Technology, Vizianagaram, Andhra Pradesh, India. pp. 837–843. Available from: http://link.springer.com/10.1007/978-981-15-5400-1_79.

40 Nielsen, B.L. (2022). The role of precision livestock farming technologies in animal welfare monitoring: a review. *Veterinary Archives* 92 (3): 251–257. Available from: http://vetarhiv.vef.unizg.hr/papers/2022-92-3-4.pdf.

41 McDonagh, J., Tzimiropoulos, G., Slinger, K.R. et al. (2021). Detecting dairy cow behavior using vision technology. *Agriculture* 11 (7): 675. Available from: https://www.mdpi.com/2077-0472/11/7/675.

42 Hansen, M.F., Oparaeke, A., Gallagher, R. et al. (2022). Towards machine vision for insect welfare monitoring and behavioural insights. *Frontiers in Veterinary Science* 15: 9. Available from: https://www.frontiersin.org/articles/10.3389/fvets.2022.835529/full.

43 Islam, M.N., Yoder, J., Nasiri, A. et al. (2023). Analysis of the drinking behavior of beef cattle using computer vision. *Animals* 13 (18): 2984. Available from: https://www.mdpi.com/2076-2615/13/18/2984.

44 Gehlot, A., Malik, P.K., Singh, R. et al. (2022). Dairy 4.0: intelligent communication ecosystem for the cattle animal welfare with blockchain and IoT enabled technologies. *Applied Sciences* 12 (14): 7316. Available from: https://www.mdpi.com/2076-3417/12/14/7316.

45 Lopes, V., Pereira, N., and Alexandre, L.A. (2019). Robot workspace monitoring using a blockchain-based 3D vision approach. *2019 IEEE/CVF Conference on Computer Vision and Pattern Recognition Workshops (CVPRW)*, pp. 2812–2820. IEEE, 16–17 June 2019, Long Beach, California, USA. Available from: https://ieeexplore.ieee.org/document/9025368/.

46 Lopez, M.E. (2023). Immutable livestock tracking and compliance logs using blockchain. *2023 IST-Africa Conference (IST-Africa)*, pp. 1–11. IEEE, 31 May–02 June 2023, Tshwane, South Africa. Available from: https://ieeexplore.ieee.org/document/10187856/.

47 Liu C, Jian Z, Xie M, Cheng I. (2021). A real-time mobile application for cattle tracking using video captured from a drone. *2021 International Symposium on Networks, Computers and Communications (ISNCC)*, pp. 1–6. IEEE, October 31–November 2, 2021, Dubai, United Arab Emirates. Available from: https://ieeexplore.ieee.org/document/9615648/.

48 Katekar, V. and Cheruku, J.K. (2023). The application of drone technology for sustainable agriculture in India. *Current Agriculture Research Journal* 10 (3): 352–365. Available from: https://www.agriculturejournal.org/volume10number3/the-application-of-drone-technology-for-sustainable-agriculture-in-india/.

49 Li, X. and Xing, L. (2019). Reactive deployment of autonomous drones for livestock monitoring based on density-based clustering. *2019 IEEE International Conference on Robotics and Biomimetics (ROBIO)*, pp. 2421–2426. IEEE, 6–8 Dec. 2019, Dali,

Yunnan Province, China. Available from: https://ieeexplore.ieee.org/document/8961763/.

50 Nie, P., He, Y., Liu, F. et al. (2021). Livestock and aquaculture IoT systems. In: *Agricultural Internet of Things. Agriculture Automation and Control ((AGAUCO))* (ed. Y. He, P. Nie, Q. Zhang, and F. Liu). Cham: Springer https://doi.org/10.1007/978-3-030-65702-4_12.

51 Ortega, J.C.Ú., Rodríguez-Molina, J., Martínez-Núñez, M., and Garbajosa, J. (2022). A proposal for decentralized and secured data collection from unmanned aerial vehicles in livestock monitoring with blockchain and IPFS. *Applied Sciences* 13 (1): 471. Available from: https://www.mdpi.com/2076-3417/13/1/471.

52 Ilyas, Q.M. and Ahmad, M. (2020). Smart farming: an enhanced pursuit of sustainable remote livestock tracking and geofencing using IoT and GPRS. Anisi MH, editor. *Wireless Communications and Mobile Computing* 2020: 1–12. Available from: https://www.hindawi.com/journals/wcmc/2020/6660733/.

53 Zhou, H., Li, Q., and Xie, Q. (2023). Individual pig identification using back surface point clouds in 3D vision. *Sensors* 23 (11): 5156. Available from: https://www.mdpi.com/1424-8220/23/11/5156.

54 Zhao, H., Li, P., Chen, Z. et al. (2023). Cow face recognition with noise labels using vision transformer. *2023 IEEE 4th International Conference on Pattern Recognition and Machine Learning (PRML)*, 4–6 Aug. 2023, Urumqi, China, pp. 93–98. IEEE. Available from: https://ieeexplore.ieee.org/document/10348318/.

55 Bidder, O.R., di Virgilio, A., Hunter, J.S. et al. (2020). Monitoring canid scent marking in space and time using a biologging and machine learning approach. *Science Reports* 10 (1): 588. Available from: https://www.nature.com/articles/s41598-019-57198-w.

56 Xu, B., Wang, W., Falzon, G. et al. (2020). Livestock classification and counting in quadcopter aerial images using Mask R-CNN. *International Journal of Remote Sensing* 41 (21): 8121–8142. Available from: https://www.tandfonline.com/doi/full/10.1080/01431161.2020.1734245.

57 Bello, R.W., ASA, M., Talib, A.Z. et al. (2021). Computer vision-based techniques for cow object recognition. *IOP Conference Series: Earth and Environmental Science* 858 (1): 012008. Available from: https://iopscience.iop.org/article/10.1088/1755-1315/858/1/012008.

58 Huo, S., Sun, Y., Guo, Q. et al. (2022). Double-camera fusion system for animal-position awareness in farming pens. *Foods* 12 (1): 84. Available from: https://www.mdpi.com/2304-8158/12/1/84.

59 Ur Rehman, A., Arshad, J., Sadiq, M.T. et al. (2022). Implementation of an intelligent animal monitoring system using wireless sensor network and iot platform. *2022 International Conference on Cyber Resilience (ICCR)*, pp. 1–11. IEEE, 6–7 Oct. 2022, Al-Meydan Road, Nad Al Sheba, Dubai, UAE. Available from: https://ieeexplore.ieee.org/document/9996080/.

60 Nyamuryekung'e, S., Duff, G., Utsumi, S. et al. (2023). Real-time monitoring of grazing cattle using LORA-WAN sensors to improve precision in detecting animal welfare implications via daily distance walked metrics. *Animals* 13 (16): 2641. Available from: https://www.mdpi.com/2076-2615/13/16/2641.

61 Ahmad, M., Zhang, W., Smith, M. et al. (2023). Real-time livestock activity monitoring via fine-tuned faster R-CNN for multiclass cattle behaviour detection. *2023 IEEE 14th Annual Ubiquitous Computing, Electronics & Mobile Communication Conference (UEMCON),* 12–14 Oct. 2023, Columbia University, New York, USA, pp. 805–811. IEEE. Available from: https://ieeexplore.ieee.org/document/10316066/.

62 Sattar, F. (2022). A context-aware method-based cattle vocal classification for livestock monitoring in smart farm. *Chemistry Proceedings* 10 (1): 89. https://doi.org/10.3390/IOCAG2022-12233.

63 Leme, L., Medeiros, A., Srivastava, G. et al. (2020). Secure cattle stock infrastructure for the internet of things using blockchain. *2020 43rd International Conference on Telecommunications and Signal Processing (TSP),* pp. 337–341. IEEE, 7–9 July 2020, Milan, Italy. Available from: https://ieeexplore.ieee.org/document/9163494/.

64 Casas, R., Hermosa, A., Marco, Á. et al. (2021). Real-time extensive livestock monitoring using lpwan smart wearable and infrastructure. *Applied Sciences* 11 (3): 1240. Available from: https://www.mdpi.com/2076-3417/11/3/1240.

65 Hogeveen, H. and van der Voort, M. (2021). Advances in precision livestock farming techniques for monitoring dairy cattle welfare. In: *Understanding the behaviour and improving the welfare of dairy cattle*, 103–122. Burleigh Dodds Science Publishing https://shop.bdspublishing.com/store/bds/detail/product/3-190-9781801460927.

66 Alon, A., Shimshoni, I., Godo, A. et al. (2023). Machine vision-based automatic lamb identification and drinking activity in a commercial farm. *Animal* 17 (9): 100923. Available from: https://linkinghub.elsevier.com/retrieve/pii/S1751731123002203.

67 Hitelman, A., Edan, Y., Godo, A. et al. (2022). Biometric identification of sheep via a machine-vision system. *Computers and Electronics in Agriculture* 194: 106713. Available from: https://linkinghub.elsevier.com/retrieve/pii/S0168169922000308.

68 Himel, G.M.S., Islam, M.M., and Rahaman, M. (2024). Vision intelligence for smart sheep farming: applying ensemble learning to detect sheep breeds. *Artificial Intelligence in Agriculture* 11: 1–12. Available from: https://linkinghub.elsevier.com/retrieve/pii/S258972172300048X.

69 Zhu, Q., Ren, J., Barclay, D. et al. (2015). Automatic animal detection from kinect sensed images for livestock monitoring and assessment. *2015 IEEE International Conference on Computer and Information Technology; Ubiquitous Computing and*

Communications; Dependable, Autonomic and Secure Computing; Pervasive Intelligence and Computing, pp. 1154–1157. IEEE, 26–28 Oct. 2015, Liverpool, United Kingdom. Available from: http://ieeexplore.ieee.org/document/7363216/.

70 Li, N., Ren, Z., Li, D., and Zeng, L. (2020). Review: Automated techniques for monitoring the behaviour and welfare of broilers and laying hens: towards the goal of precision livestock farming. *Animal* 14 (3): 617–625. Available from: https://linkinghub.elsevier.com/retrieve/pii/S1751731119002155.

71 Berckmans, D. (2009). Automatic on-line monitoring of animal health and welfare by precision livestock farming. *European Forum; Livestock Housing for the Future.*

72 Mcloughlin, M.P., Stewart, R., and McElligott, A.G. (2019). Automated bioacoustics: methods in ecology and conservation and their potential for animal welfare monitoring. *Journal of the Royal Society Interface* 16 (155): 20190225. Available from: https://royalsocietypublishing.org/doi/10.1098/rsif.2019.0225.

73 Di Pasquale, A., Isocrono, E., Possenti, L. et al. (2009). Animal welfare monitoring and livestock traceability during transport. *Veterinaria Italiana* 45 (4): 577–586.

74 Meidayanti, K., Arkeman, Y., and Sugiarto (2019). Analysis and design of beef supply chain traceability system based on blockchain technology. *IOP Conference Series: Earth and Environmental Science* 335 (1): 012012. Available from: https://iopscience.iop.org/article/10.1088/1755-1315/335/1/012012.

75 Cao, S., Foth, M., Powell, W. et al. (2022). A blockchain-based multisignature approach for supply chain governance: a use case from the Australian beef industry. *Blockchain: Research and Applications* 3 (4): 100091. Available from: https://linkinghub.elsevier.com/retrieve/pii/S209672092200032X.

76 Adimabua Ojugo, A., Ogholuwarami Ejeh, P., Chukwufunaya Christopher, O. et al. (2023). Improved distribution and food safety for beef processing and management using a blockchain-tracer support framework. *International Journal of Informatics and Communication Technology* 12 (3): 205. Available from: https://ijict.iaescore.com/index.php/IJICT/article/view/20676.

77 Bello, R.W. and Oladipo, M.A. (2024). Mask YOLOv7-based drone vision system for automated cattle detection and counting. *Artificial Intelligence and Applications* https://doi.org/10.47852/bonviewAIA42021603. Available from: https://ojs.bonviewpress.com/index.php/AIA/article/view/1603.

78 Los, S., Mücher, C.A., Kramer, H. et al. (2023). Estimating body dimensions and weight of cattle on pasture with 3D models from UAV imagery. *Smart Agricultural Technology* 4: 100167. Available from: https://linkinghub.elsevier.com/retrieve/pii/S2772375522001319.

79 de Lima, W.F., de Moraes Weber, V.A., de Moraes, P.H. et al. (2023). Counting cattle in UAV images using convolutional neural network. *Remote Sensing Applications: Society and Environment* 29: 100900. Available from: https://linkinghub.elsevier.com/retrieve/pii/S2352938522002087.

80 Hwerbi, K., Amdouni, I., Laouiti, A. et al. (2023). Veterinary drone: blockchain-based system for cattle health monitoring. *2023 International Wireless Communications and Mobile Computing (IWCMC)*, pp. 1613–1618. IEEE, 19–23 June 2023, Marrakesh, Morocco. Available from: https://ieeexplore.ieee.org/document/10182480/.

81 Qiao, Y., Guo, Y., and He, D. (2023). Cattle body detection based on YOLOv5-ASFF for precision livestock farming. *Computers and Electronics in Agriculture* 204: 107579. Available from: https://linkinghub.elsevier.com/retrieve/pii/S0168169922008870.

82 Elham, M.N., Sabeghi, M.S., Al-Rasbi, F.S. et al. (2020). A preliminary study on poultry farm environmental monitoring using internet of things and blockchain technology. *2020 IEEE 10th Symposium on Computer Applications & Industrial Electronics (ISCAIE)*, pp. 273–276. IEEE, 18–19 April 2020, Penang, Malaysia. Available from: https://ieeexplore.ieee.org/document/9108820/.

83 Chouragade, G. (2023). Boosting poultry farm profits through blockchain technologies, AI, IoT, and machine learning. In: *Handbook of Research on AI and Knowledge Engineering for Real-Time Business Intelligence* (ed. K. Hiran, K. Hemachandran, A. Pise, and B. Rabi), 143–155. IGI Global Scientific Publishing. Available from: https://services.igi-global.com/resolvedoi/resolve.aspx?doi=10.4018/978-1-6684-6519-6.ch009.

84 Ibrahim, A., Kamoliddin, U., Yoo, J.H. et al. (2021). Blockchain-based poultry information management system design and implementation using hyperledger fabric. *Journal of the Chosun Natural Science* 14 (3): 107–115.

85 Neethirajan, S. (2023). AI in sustainable pig farming: IoT insights into stress and gait. *Agriculture* 13 (9): 1706. Available from: https://www.mdpi.com/2077-0472/13/9/1706.

86 Lee, S., Ahn, H., Seo, J. et al. (2019). Practical monitoring of undergrown pigs for IoT-based large-scale smart farm. *IEEE Access* 7: 173796–173810. Available from: https://ieeexplore.ieee.org/document/8911419/.

87 Su, J.J., Ding, S.T., and Chung, H.C. (2020). Establishing a smart farm-scale piggery wastewater treatment system with the internet of things (IoT) applications. *Water* 12 (6): 1654. Available from: https://www.mdpi.com/2073-4441/12/6/1654.

88 Hua, S., Han, K., Xu, Z. et al. (2021). Image processing technology based on internet of things in intelligent pig breeding. Tsai SB, editor. *Mathematical Problems in Engineering* 2021: 1–9. Available from: https://www.hindawi.com/journals/mpe/2021/5583355/.

89 Yuan, Y., Liu, X., Hong, K. et al. (2020). Trustworthy traceability of quality and safety for pig supply chain based on blockchain. In: *Cognitive Internet of Things: Frameworks, Tools and Applications. ISAIR 2018. Studies in Computational Intelligence*, vol. 810 (ed. H. Lu), 343–355. Cham: Springer. Available from: http://link.springer.com/10.1007/978-3-030-04946-1_34.

90 Friesenhahn, B.A., Massey, L.D., DeYoung, R.W. et al. (2023). Using drones to detect and quantify wild pig damage and yield loss in corn fields throughout plant growth stages. *Wildlife Society Bulletin* 47 (2): e1437. Available from: https://wildlife.onlinelibrary.wiley.com/doi/10.1002/wsb.1437.

91 Omarov, R.A., Isakhanov, M.Z., Kuder, K.M. et al. (2021). Experimental results of functional characteristics of IoT for free range sheep breeding. *International Journal of Agricultural Resources, Governance and Ecology* 3 (2–4): 301–315.

92 Odintsov Vaintrub, M., Levit, H., Chincarini, M. et al. (2021). Review: precision livestock farming, automats and new technologies: possible applications in extensive dairy sheep farming. *Animal* 15 (3): 100143. Available from: https://linkinghub.elsevier.com/retrieve/pii/S1751731120301452.

93 Alonso, R.S. RETRACTED CHAPTER(2021). Low-power distributed AI and IoT for measuring lamb's milk ingestion and predicting meat yield and malnutrition diseases. In: *Ambient Intelligence – Software and Applications. ISAmI 2020. Advances in Intelligent Systems and Computing*, vol. 1239 (ed. P. Novais, G. Vercelli, J.L. Larriba-Pey, et al.), 251–257. Cham: Springer. Available from: http://link.springer.com/10.1007/978-3-030-58356-9_26.

94 Nóbrega, L., Pedreiras, P., and Gonçalves, P. (2019). SheepIT, an IoT-based weed control system. In: *Information and Communication Technologies in Modern Agricultural Development. HAICTA 2017. Communications in Computer and Information Science*, vol. 953 (ed. M. Salampasis and T. Bournaris), 131–147. Cham: Springer Available from: http://link.springer.com/10.1007/978-3-030-12998-9_10.

95 Campiotti, V., Finozzi, N., Irazoqui, J. et al. (2023). Wearable device to monitor sheep behavior. *IEEE Embedded Systems Letters* 15 (2): 89–92. Available from: https://ieeexplore.ieee.org/document/9826898/.

96 Guo, Q., Sun, L., Liu, G. et al. (2021). MSBBS: a sheep breeding system based on blockchain. *Small Ruminant Research* 205: 106562. Available from: https://linkinghub.elsevier.com/retrieve/pii/S092144882100239X.

97 Westbrooke, V., Lucock, X., and Greenhalgh, I. (2023). Drone use in on-farm environmental compliance: an investigation of regulators' perspectives. *Sustainability* 15 (3): 2153. Available from: https://www.mdpi.com/2071-1050/15/3/2153.

98 Maria Anu, V. and Aroul, C.R. (2017). Livestock monitoring using RFID with R+ tree indexing. *Biomedical Research* 28 (6): 2407–2410.

99 Germani, L., Mecarelli, V., Baruffa, G. et al. (2019). An IoT architecture for continuous livestock monitoring using LoRa LPWAN. *Electronics* 8 (12): 1435. Available from: https://www.mdpi.com/2079-9292/8/12/1435.

100 K. Dineva, D. Parvanov, T. Atanasova, G. Mateeva, P. Petrov and G. Kostadinov, "Towards CPS/IoT System for Livestock Smart Farm Monitoring,"

2021 International Conference Automatics and Informatics (ICAI), Varna, Bulgaria, 2021, pp. 252–255, https://doi.org/10.1109/ICAI52893.2021.9639460. September 30–October 3, 2021, Varna, Bulgaria.

101 Yadavalli PK, Mutyala AK, Palla VK, Pappu A, Prathipati N. Smart IOT System for Monitoring and Controlling Livestock Parameters (January 19, 2020). *International Conference of Advance Research & Innovation (ICARI) 2020*, Available at SSRN: https://ssrn.com/abstract=3621946 or http://dx.doi.org/10.2139/ssrn.3621946. International Conference of Advance Research & Innovation (ICARI) 2020, January 25, 2020, Meerut, India

102 Alshehri, D.M. (2023). Blockchain-assisted internet of things framework in smart livestock farming. *Internet of Things* 22: 100739. Available from: https://linkinghub.elsevier.com/retrieve/pii/S2542660523000628.

103 Khan, P.W., Byun, Y.C., and Park, N. (2020). IoT-blockchain enabled optimized provenance system for food industry 4.0 using advanced deep learning. *Sensors* 20 (10): 2990. Available from: https://www.mdpi.com/1424-8220/20/10/2990.

104 Kumar, A., Sachan, R.K., Singh, D., and Jain, R. (2024). Analysis of block chain applications in asset backed securitization system. *Procedia Computer Science 233*: 147–153.

105 Dhifaoui, S., Houaidia, C., and Saidane, L.A. (2024). Computing paradigms for smart farming in the era of drones: a systematic review. *Annals of Telecommunications* 79 (1–2): 35–59. Available from: https://link.springer.com/10.1007/s12243-023-00997-0.

106 Sarantinoudis, N., Arampatzis, G., Valavanis, K.P., and Tsourveloudis, N. (2023). Unmanned aerial vehicles and livestock management: an application in western crete. *2023 International Conference on Unmanned Aircraft Systems (ICUAS)*, pp. 159–166. IEEE, 6–9 June 2023, Warsaw, Poland. Available from: https://ieeexplore.ieee.org/document/10155900/.

107 Kurniadi, F.A., Setianingsih, C., and Syaputra, R.E. (2023). Innovation in livestock surveillance: applying the YOLO algorithm to UAV imagery and videography. *2023 IEEE 9th International Conference on Smart Instrumentation, Measurement and Applications (ICSIMA)*, 17–18 Oct. 2023, Kuala Lumpur, Malaysia, pp. 246–251. IEEE. Available from: https://ieeexplore.ieee.org/document/10373473/.

108 Krul, S., Pantos, C., Frangulea, M., and Valente, J. (2021). Visual SLAM for indoor livestock and farming using a small drone with a monocular camera: a feasibility study. *Drones* 5 (2): 41. Available from: https://www.mdpi.com/2504-446X/5/2/41.

109 Sharma, D., Mishra, J., and Talukdar, N. (2023). Application of precision livestock farming for improved goat management and production. *Revista Electronica de Veterinaria* 24 (2): 461–470.

110 Kumar, A. and Jha, R. (2022). IoT in smart farming: a boon to farmers. In: *Internet of Everything: Smart Sensing Technologies* (ed. T. Kavitha, V.A. Devi, S.N. Pari, and S. Ramanathan), 185–206. Nova Publisher.

111 Singh, D. and Kumar, A. (2024). A deep recurrent neural network for plant disease classification. *SN Computer Science* 5 (8): 1053.

14

Smart Crop Management: Harnessing Green IoT Tomorrow

Shipra Raheja[1], *Vimal Gaur*[2], *and Rachna Jain*[3]

[1] *IT Department, Guru Tech Bahadur Institute of Technology, Delhi, New Delhi, India*
[2] *CSE Department, Maharaja Surajmal Institute of Technology, Delhi, New Delhi, India*
[3] *IT Department, Bhagwan Parshuram Institute of Technology, Delhi, New Delhi, India*

14.1 Introduction

Green IoT involves designing, developing, and deploying IoT devices, networks, and applications. Green IoT involves the usage of environmentally friendly technologies and practices throughout the lifecycle of IoT devices. Green IoT impacts the lifecycle of IoT devices, ranging from resource extraction and manufacturing to usage and disposal. Green IoT has vast potential in many fields, viz., smart homes, smart health, smart agriculture etc. The vast applications of green IoT play a great role in creating a resource-efficient world [1]. One of the most important applications of smart agriculture is smart crop management. As compared to traditional crop management, smart crop management helps in improving yields, reducing waste and finally making better decisions in managing farms. The purpose of smart crop management is to monitor, analyze, and optimize various aspects of farming operations. The process of smart crop management involves the usage of advanced technologies to make a more sustainable future. The components involved in managing crops smartly include sensor networks, IoT devices, remote monitoring, decision support systems, automation control, and weather monitoring. Each component of green IoT plays a key role in the management of crops. The deployment of sensor networks in agriculture greatly impacts the growth of crops. Sensors collect real-time data on various parameters, viz., soil moisture, humidity, temperature, and crop health. Sensors, when integrated with

Emerging Smart Agricultural Practices Using Artificial Intelligence, First Edition.
Edited by Ashish Kumar, Jai Prakash Verma, and Rachna Jain.

Published 2025 by John Wiley & Sons, Inc.

IoT devices provide valuable insight into conditions that enhance crop health. IoT devices allow farmers to access information remotely and make timely decisions. The data generated by IoT devices and sensors is analyzed using data analysis tools. These tools help to process and analyze large volumes of data. The resultant output is trends, patterns, and potential issues that affect the crops. Precision agriculture practices must be implemented to manage resources like water, fertilizers, and pesticides. With the advent of sensor networks, human intervention has been minimized. Automated health monitoring systems help farmers to predict weather conditions and hence planting and harvesting of crops are managed [2].

14.2 Greening Agriculture: Advancing with IoT Technology

Green IoT indicates a new era of innovation in agriculture practices, ensuring an efficient future for the agricultural sector. Green IoT solutions empower farmers in making informed about irrigation, fertilization, pest control, and resource allocation. These solutions help in minimizing resource usage and maximizing productivity. Green IoT benefits global food security and environmental conservational efforts. Green IoT for smart crop management requires the following steps.

1) **Planning and Design**: A complete assessment of site, technology, and system design is required. The first step is to identify specific farming challenges and goals (e.g. optimizing water usage, improving yield, and minimizing chemical inputs). The next step is to evaluate environmental factors (soil composition, climate, and water availability) and infrastructure requirements. Technology selection requires choosing appropriate sensors, data management platforms, and communication networks based on needs and budget. The most important step is to define data collection procedures, analysis methods, and control strategies for automation and optimization.
2) **Deployment and Installation**: Various issues need to be considered while deploying and installing sensor devices. To gather relevant data, sensors, greenhouses, and irrigation systems must be installed in fields. To ensure secure and seamless communication, a connectivity setup must be appropriate. Further, the accuracy of sensors functioning and data must be taken into consideration. Farmers must be trained with the knowledge and skills for operating and utilizing green IoT devices.
3) **Operation and Monitoring**: Real-time data acquisition is done using sensors for monitoring crop health and resource usage. Data gathered must be analyzed using advanced algorithms to identify trends and hence predict need of resources and detecting potential issues. Real-time alerts and recommendations for fertilization, irrigation, pest, and weed control must be generated. AI-driven decision support systems and automated systems must be implemented for crop management.

4) **Optimization and Adaptation**: To improve resource efficiency and crop yield data-driven decision-making must be used for adjusting and refining practices. Further, these data-driven approaches must be monitored continuously to identify areas of improvement. Since environmental conditions are continuously changing and so are the needs, hence models must be made to adjust these requirements. Finally, green IoT must be integrated with traditional farming practices to optimize agriculture processes.
5) **Maintenance and Sustainability**: Regular maintenance of sensors, communication networks, and data platforms must be done. Robust security measures must be implemented to protect sensitive data from being compromised. The devices must have long-term assurance in terms of biodiversity, environment sustainability. Finally, appropriate strategies must be developed for recycling sensor components at the end of their life span.

By following this life cycle approach, green IoT in smart crop management can ensure effective implementation, optimized performance, and long-term benefits for farmers, the environment, and society. Further, this can more clearly be explained as Figure 14.1 (process diagram) below:

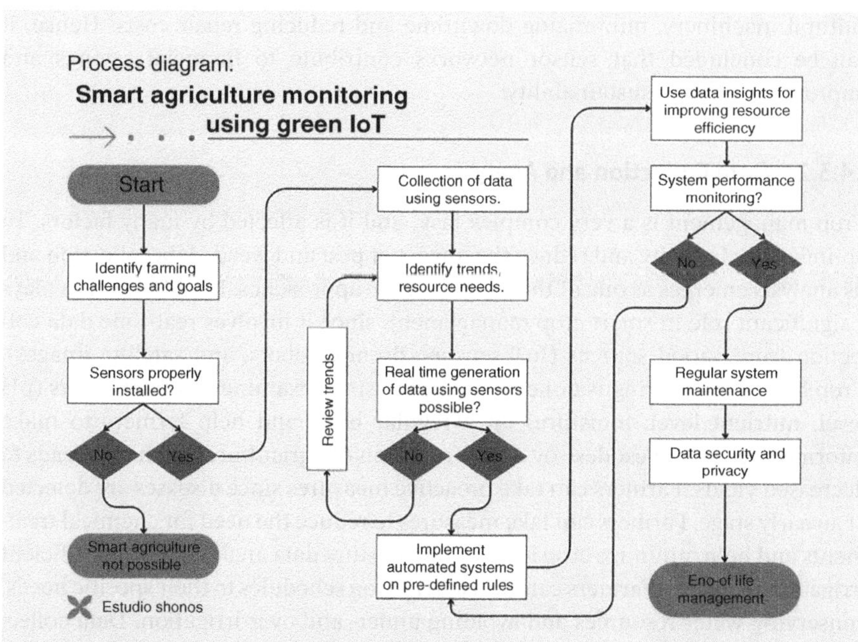

Figure 14.1 Smart crop management using green IoT.

14.3 Green IoT Key Components

The key components of green IoT for smart crop management have been listed below:

14.3.1 Sensor Networks

Sensors are deployed in agriculture for monitoring parameters like soil moisture, temperature, humidity, and crop health. Sensor networks have revolutionized traditional farming practices and hence contribute to increased efficiency, sustainability, and productivity. The key feature of sensor networks is to monitor real-time data, which helps in generating trends and patterns. The trends and patterns are analyzed carefully to reduce waste and improve resource efficiency. Sensors also assist in the determination of optimal irrigation schedules. Patterns indicate sign of diseases, pest infestations, irrigation schedules for crop in fields. The use of chemical treatment on crops is reduced if timely identification of diseases is done. Both under- and over-irrigation are avoided if trends are properly analyzed. Data received from sensors help the farmers to be resilient against climate-related challenges and make decisions accordingly. Sensor networks when integrated with actuators and automation systems remotely monitor and control agricultural processes. Sensor networks also play a major role in predictive maintenance of agricultural machinery, minimizing downtime and reducing repair costs. Hence, it can be concluded that sensor networks contribute to financial savings and improved economic sustainability.

14.3.2 Data Collection and Analysis

Crop management is a very complex task, and it is affected by many factors. To optimize productivity and reduce the impact of pest and weed, data collection and its analysis emerges as one of the most effective approaches. Data Collection plays a significant role in smart crop management, since it involves real-time data collection from varied sources (IoT sensors, drones, robots, and satellite images). Crop health monitoring is done by data analysts to examine soil parameters (pH level, nutrient level, moisture) on a regular basis and help farmers to make informed decision. Pest destroy a major portion of agriculture and hence leads to decreased yields. Farmers can take proactive measures since diseases are detected at an early stage. Farmers can take measures to reduce the need for chemical treatments and help minimize crop losses. Soil moisture data analysis informs efficient irrigation practices. Farmers can adjust irrigation schedules to their specific needs, conserving water resources and avoiding under- and over-irrigation. Data collection its analysis combines weather data, enabling farmers to anticipate and adapt to changing weather conditions. This strengthens climate-related challenges and supports informed decision-making. Analyzing large databases provides insights. Farmers can make decisions about planting, harvesting, resource allocation, and pest control, optimizing overall farm management. Analyzing data from remote sensors makes agricultural monitoring easier to automate and control. This increases operational efficiency, reduces manual labor, and improves overall housekeeping. Continuous data collection and analysis monitor plant health. This information can improve plant growth, thereby increasing yield and improving plant quality. Data analysis allows farmers to optimize energy use on farms. This includes efficient management of machinery and infrastructure, resource-saving, and contribution to sustainability. By simulating returns based on data analysis, farmers can reduce costs associated with excessive use of water, pesticides, and fertilizers. This contributes to financial savings and improved economic stability [3]. Predictive analysts allow to assess risks related to weather conditions, market conditions, and plant health. Farmers can improve overall sustainability and proactively plan and mitigate risk. As a result, data collection and analysis in crop management play a key role in changing agricultural practices. By using technology to collect, process, and interpret data, farmers can make informed decisions that optimize the use of resources, reduce waste, and lead to more sustainable and productive agriculture [4].

14.3.3 Precision Agriculture

The impact of precision agriculture on smart crop management is a highly efficient and targeted approach to farming practices. It enables precise use of resources such as water, fertilizers, and pesticides. This targeted approach reduces waste, conserves resources, and reduces environmental impact. This agronomy allows farmers to adjust practices according to specific conditions. This adaptation optimizes the growing conditions of crops, leading to increased yield and overall agronomic and, if necessary, chemical precision. Precision agriculture relies on sensors and real-time data monitoring equipment. This continuous flow of information helps farmers monitor crop health and take corrective action. Precision agriculture reduces the need for extensive chemical use by targeting specific areas affected by pests or diseases [5]. Reducing the use of chemicals contributes to environmental sustainability. It also uses advanced analytics to process and interpret large databases. This data-driven approach enables farmers to make informed decisions about planting, irrigation, and other important aspects of crop management. In addition, remote sensing technologies such as satellite imagery and drones will be used to collect detailed information about the fields. This helps in the effective control of large-scale agriculture. This agriculture uses efficient irrigation methods that take into account factors such as soil moisture, weather conditions, and plant water requirements. This targeted approach saves water and improves irrigation efficiency. In addition, targeted intervention in disease and pest management is possible. Farmers can reduce the need for extensive chemical treatments by checking areas affected by pests. Precision farming often requires the use of automatic equipment controlled by GPS and sensor technology [6]. This reduces the need for manual labor, optimizes efficiency, and reduces labor costs. Precision agriculture integrates IoT devices for seamless data connectivity. This connectivity enables real-time communication between sensors, machinery, and farm management systems. Hence, precision agriculture transforms conventional agriculture by incorporating technology to optimize resource use, increase productivity, and reduce environmental impact. The targeted and data-driven nature of precision agriculture contributes to more sustainable and efficient smart crop management practices.

The impact of smart management of water and crop water management is substantial, influencing various aspects of farming practices and contributing to improved efficiency, conservation of water resources, and sustainable agriculture. Smart crop management facilitates precision irrigation practices by using sensors to measure soil moisture levels. This allows farmers to optimize watering schedules and apply water precisely where and when it is needed. Through real-time monitoring and data-driven decision-making, smart agriculture helps minimize water waste. Farmers can avoid over-irrigation, leading to significant water conservation. Smart water management ensures an adequate and consistent supply of water to plants. This prevents water stress on the plants and helps them grow healthier and safer. Information from smart crop management allows farmers to allocate water resources more efficiently [7]. This targeted approach reduces water use, costs, and environmental impact. Smart water management systems integrate weather data to adapt irrigation operations based on changing weather conditions. This increases tolerance to climate change, ensuring plants receive adequate water levels. Smart agriculture helps control irrigation systems. Automated systems driven by real-time data increase operational efficiency and reduce the need for manual intervention. Monitoring soil moisture and actual weather conditions provides farmers with information to make decisions. This includes adjusting irrigation schedules, choosing drought-tolerant plants, or adopting water-saving techniques [8–10].

14.3.4 Crop Monitoring

Smart agricultural crop management significantly improves input efficiency, reduces environmental impact, and increases overall productivity. Real-time information on factors such as soil moisture, weather conditions and plant health help in taking decisions to reduce water and pesticides so as to increase yields. This data-driven approach helps in promoting farming practices, contributing to economic and environmental benefits. In smart agriculture, crop control can be easily managed by ensuring proper management of inputs. It improves yield forecasting, reduces waste, and promotes sustainable practices through data-driven insights. In addition, early detection of problems such as diseases or nutritional deficiencies allows for timely intervention, improving plant health and overall productivity.

14.3.5 Environmental Impact Reduction

Green IoT in agriculture helps in reducing the ecological footprint of agriculture by optimizing the use of resources, reducing chemicals, and using sustainable practices. Green IoT aim is to solve environmental problems and reduce environmental impact [11, 12]. These devices help in incorporating energy-efficient designs, low-power components, and optimized communication protocols also reduces the use of networks and IoT devices. IoT is used for smart grid systems that provide energy distribution, consumption control, and demand response mechanisms. This outcome reduced energy waste and lowered carbon emissions. IoT sensors are used for real-time environmental monitoring, collecting data on air and water quality, pollution levels, and weather conditions. This information helps in the early detection of environmental problems and facilitates timely action. In agriculture, green IoT enables precision agriculture by monitoring soil conditions, plant health, and weather conditions. As a result, resource use is optimized, chemical input is reduced, and crop yields are increased. IoT devices are also used in waste management systems to optimize waste collection routes, monitor landfill capacity, and improve recycling processes. This leads to a more efficient way of managing waste. Green IoT is integrated into smart building systems for energy-efficient lighting, heating, and cooling. Sensors adjust environmental conditions based on living conditions, saving energy and reducing environmental impact. IoT technology optimizes transportation routes, reduces fuel consumption, and improves logistics operations. This results in lower carbon emissions and increases the efficiency of supply chain management. Green IoT plays an important role in wildlife conservation by using sensors to monitor and track endangered species. This information helps create conservation strategies and protect ecosystems. By orienting IoT technology towards sustainability, green IoT contributes to more responsible and environmentally friendly practices in various industries, thereby reducing environmental parameters [13, 14].

14.3.6 Remote Monitoring and Control

Green IoT devices can remotely monitor and control energy consumption in buildings, factories, and other facilities. This allows optimization of lighting, heating, and cooling systems, which leads to reduced energy consumption. Remote monitoring and smart grid control enable efficient energy distribution, load balancing, and real-time adjustments based on demand. This contributes to energy savings and a more sustainable energy infrastructure [15]. Green IoT facilitates remote monitoring and control in precision agriculture. Farmers can remotely check crop conditions, moisture content in soil, and deploy automated systems to reduce irrigation, usage of water and energy consumption. In manufacturing, green IoT with remote monitoring and control enables efficient management of the production process by including optimizing device performance and reducing energy consumption. Green IoT systems can remotely monitor and manage waste levels in trash cans and landfills. This information helps us to optimize waste collection routes by reducing fuel usage and emissions associated with inefficient waste management practices [15]. Green IoT-enabled remote sensors monitor environmental conditions in real time. This includes air quality, pollution levels and climate data with timely responses to environmental changes and threats associated with them. Green IoT supports remote monitoring and control of renewable energy sources such as solar panels and wind turbines [16]. This improves performance and integration into the power grid, which contributes to a more sustainable energy landscape. Organizations and individuals can achieve greater energy efficiency by combining green IoT with remote monitoring and control capabilities [17, 18].

14.3.7 Smart Decision Support

The implementation of intelligent decision support in crop management improves agricultural efficiency, sustainability, and accuracy. For providing information to farmers intelligent decision support system with data analytics is used. This helps farmers to make decisions based on real time and historical data. Intelligent decision support enables farmers to use resources such as water, fertilizers, and pesticides more efficiently [19, 20]. This helps precision agriculture to reduce waste and increase crop production. A decision support system with timely intervention in crop management improves performance. Farmers can periodically respond to emerging challenges coming with diseases, pests and changing weather conditions and minimize possible damage due to this. Intelligent decision support uses predictive analytics to assess risks related to various factors, such as plant health due to weather and market conditions. Farmers can smartly reduce risk and plan for the uncertainty of the future well in advance [21, 22]. Decision support systems provide farmers with market intelligence including price trends for demand forecasting. This information helps in strategic decision-making for crop selection and market timing. Based on data-driven insights, intelligent decision support contributes to overall business efficiency by streamlining farming operations including optimal planting schedules harvest time and irrigation management. Decision support helps reduce waste by optimizing the use of resources such as fertilizers, water, and pesticides. This helps in saving resources and reduces the bad impact on the environment. Intelligent decision support helps improve crop yields with appropriate advice for planting, harvesting, and postharvest practices. This leads to increased productivity and profitability in the long run [23, 24]. Decision support systems help farmers in adapting these practices to ensure resilience against environmental change and climate changes. Decision support can play an important role in cooperation in agricultural supply chains. Integration with market stakeholders, suppliers, and distributors for improving efficiency [25]. Intelligent decision support systems provide feature user-friendly interfaces so that farmers can easily access and interpret and manage information. This increases the adoption of the technology in various agricultural contexts. Smart decision support in smart crop management contributes to more informed, efficient and sustainable agriculture. By using advanced analytics and data-driven insights, farmers can overcome challenges, optimize the use of resources, and make strategic decisions that lead to better agricultural results [26, 27]. The main goal of green IoT for smart crop management is to increase agricultural productivity while reducing negative environmental impacts. By using IoT technology, farmers can optimize the use of resources, reduce waste, and contribute to sustainable and ecological farming practices.

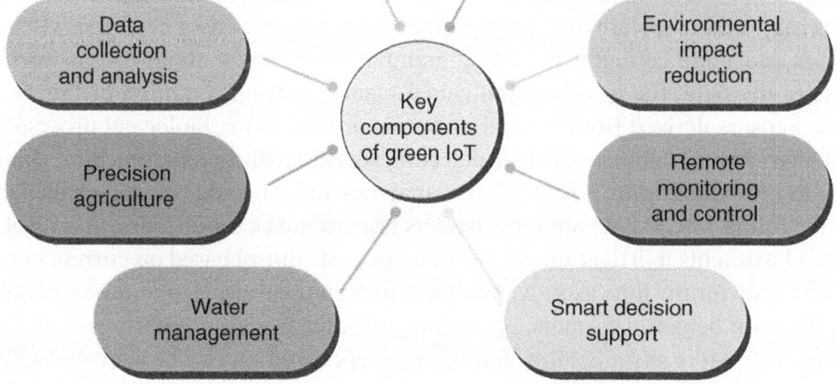

Figure 14.2 Key components of green IoT.

14.4 Future of AI in Agriculture

Advancement in agriculture technology helped farmers in doing farming efficiently and effectively and hence lowers the difficulties of farmers. Artificial intelligence plays a major role in sustaining agriculture and the food industry. Artificial intelligence (AI) in agriculture helps farmers by automating all the tasks with the use of agriculture robotics and computer vision techniques. AI techniques require little or no human intervention, and hence, humans can be made available at places where strategic planning is required [33, 34]. This is accomplished by advanced farming equipment and other AI-powered products [35, 36].

Furthermore, AI has transformed contemporary agriculture through real-time monitoring of crops, weather, customer requirements, market trends, availability of raw materials, etc. Once real-time monitoring of essential factors is done, farmers can take important decisions on crops. AI assists farmers right from crop selection to harvesting depending on soil quality, moisture content of soil, climatic conditions, etc. Sensors are deployed at different locations to monitor moisture content and quality of soil; important decisions are made about irrigation and pathogens. Simultaneously, weed detection and management are done using laser-based automated tools. With the advancement of AI, farmers are also transformed from laborers to administrators. AI is not a physical object but a collection of automated technologies that can learn from data and apply that knowledge to solve issues. AI plays a key role in industry revolution by training farmers about latest AI technologies [28, 29].

14.5 Conclusion and Future Aspects

The green IoT future for smart crop management is full of exciting possibilities, promising a more sustainable and productive agriculture. Sensor networks will be denser and more complex, providing granular information about the microclimate in the field, the health of individual plants, and the condition of the root zone. Sensors derived from natural materials or mimicking biological processes will offer new possibilities, such as detecting specific pathogens or tracking plant growth hormones. More advanced data analytics and AI could be a big challenge for the future [30, 31]. AI not only predicts but also takes action, suggesting real-time adjustments to irrigation, fertilization, or pest control based on current conditions and future forecasts. AI-powered systems can also make autonomous decisions on behalf of farmers, optimizing resource use even within predefined parameters. More sustainability and resource optimization can be achieved with sensors powered by renewable energy. The sensors are powered by solar, wind, or bio-harvesting technology, reducing their environmental footprint. Data transmission between sensors and the central system will be more efficient and reduce energy consumption in the network. Sensor materials and components will be designed to be reused and recycled, reducing waste and environmental impact [6, 32]. Future-proof agricultural data must be protected from cyber-attacks, and it is important to ensure responsible data use. Adaptation of green IoT technologies to small and medium farms is essential for widespread adoption. The potential for labor market disruption and ethical implications of AI-based decisions in agriculture should be carefully considered. Overall, the green IoT future for smart crop management is promising and offers the potential to create sustainable and resilient systems. However, it is necessary to manage the challenges and ensure responsible development to maximize benefits for farmers, consumers, and the environment.

References

1 Yan, C., Han, F., Yang, Y.-H. et al. (2014). Time-reversal wireless paradigm for green internet of things: an overview. *IEEE Internet of Things Journal* 1 (1): 81–98.

2 Simo, A. and Dzitac, S. (2023). Energy-efficient wireless sensor networks for greenhouse management. In: *Intelligent Method Systems and Applications in Computing, Communications and Control. ICCCC 2022. Advances in Intelligent Systems and Computing*, vol. 1435 (ed. S. Dzitac, D. Dzitac, F.G. Filip, et al.), 120–134. Springer.

3 Hassanalieragh, M., Page, A., Soyata, T. et al. (2015). Health monitoring and management using internet-of-things (IoT) sensing with cloud-based processing: Opportunities and challenges. *In Services Computing (SCC), 2015 IEEE International Conference on (2015)*, pp. 285–292, 27 June 2015–2 July 2015, New York, USA

4 Srisruthi, N. Swarna, G. M. Susmitha Ros, Edna Elizabeth. (2016). Sustainable agriculture using eco-friendly and energy efficient sensor technology. *Recent Trends in Electronics Information & Communication Technology (RTEICT) IEEE International Conference on.* IEEE. 20–21 May 2016, Bangalore, India

5 Srisruthi, S., Swarna, N., Ros, G.M.S., Elizabeth, E. (2016). Sustainable agriculture using eco-friendly and energy efficient sensor technology. *Proceedings of the 2016 IEEE International Conference on Recent Trends in Electronics, Information & Communication Technology (RTEICT)*, Bangalore, India (20–21 May 2016), pp. 1442–1446. IEEE: Bangalore, India.

6 Yang, X., Shu, L., Chen, J. et al. (2021). A survey on smart agriculture: development modes, technologies, and security and privacy challenges. *IEEE/CAA Journal of Automatica Sinica* 8: 273–302.

7 Wolfert, S., Ge, L., Verdouw, C., and Bogaardt, M.J. (2017). Big data in smart farming – a review. *Agricultural Systems* 153: 69–80.

8 Partha, R. (2017). Internet of things for smart agriculture: technologies practices and future direction. *Journal of Ambient Intelligence and Smart Environments* 9 (4): 395–420.

9 Lottes, Raghav Khanna, Johannes Pfeifer, Roland Siegwart, Cyrill Stachniss (2017). UAV-based crop and weed classification for smart farming. *IEEE International Conference on Robotics and Automation (ICRA)* (May 29–June 3, 2017). Singapore, Swiss Federal Institute of Technology, Zurich, Switzerland.

10 Kong, Q., Chen, H., Mo, Y.L., and Song, G. (2017). Real-time monitoring of water content in sandy soil using shear mode piezoceramic transducers and active sensing-a feasibility study. *Sensors* 17: 2395.

11 Mekala, M.S. and Viswanathan, P. (2017). A Survey: Smart agriculture IoT with cloud computing. *Proceedings of the 2017 International Conference on Microelectronic Devices, Circuits and Systems (ICMDCS)*, Vellore, India (10–12 August 2017), pp. 1–7. IEEE: Vellore, India.

12 Dholu, M. and Ghodinde, K.A. (May 2018). Internet of things (IoT) for precision agriculture application. *Proc. of the 2nd International Conference on Trends in Electronics and Informatics*, pp. 339–342, 11–12 May 2018, Tirunelveli, India.

13 Ferrández-Pastor, F., García-Chamizo, J., Nieto-Hidalgo, M., and Mora-Martínez, J. (2018). Precision agriculture design method using a distributed computing architecture on internet of things context. *Sensors* 18 (6).

14 Ahmed, N., De, D., and Hussain, I. (2018). Internet of things (IoT) for smart precision agriculture and farming in rural areas. *IEEE Internet of Things Journal* 5 (6): 4890–4899.

15 Elijah, O., Rahman, T.A., Torikumi, I. et al. (2018). An overview of internet of things (IoT) and data analytics in agriculture: benefits and challenges. *IEEE Internet of Things Journal* 5 (5): 3758–3773. https://doi.org/10.1109/JIoT.648890710.1109/JIOT.2018.2844296.

16 Tradigo, G., Vizza, P., Veltri, P. (2019). SISTABENE: an information system for the traceability of agricultural food production. *Proceedings of the 2019 IEEE International Conference on Bioinformatics and Biomedicine (BIBM)* (18–21 November 2019), pp. 2304–2309. San Diego, CA, USA.

17 Kim, J., Kim, S., Chanyoung, J., and Son, H.I. (2019). Unmanned aerial vehicles in agriculture: a review of perspective of platform, control, and applications. *IEEE Access* 7: 105100–105115. https://doi.org/10.1109/Access.628763910.1109/ACCESS.2019.2932119.

18 Shi, X., An, X., Zhao, Q. et al. (2019). State- of- the- art internet of things in protected agriculture. *Sensors* 19: 1833.

19 Ayaz, M., Ammad-Uddin, M., Sharif, Z. et al. (2019). Internet-of-things (IoT)-based smart agriculture: toward making the fields talk. *IEEE Access* 7: 129551–129583.

20 Haseeb, K., Ud Din, I., Almogren, A., and Islam, N. (2020). An energy efficient and secure IoT-based WSN framework: an application to smart agriculture. *Sensors* 20: 2081.

21 Kour, V.P. and Arora, S. (2020). Recent developments of the internet of things in agriculture: a survey. *IEEE Access* 8: 129924–129957. https://doi.org/10.1109/Access.628763910.1109/ACCESS.2020.3009298.

22 Lezoche, M., Hernandez, J., Diaz, M.D.M.A. et al. (2020). Kacprzyk Agri-food 4.0: a survey of the supply chains and technologies for the future agriculture. *Computers in Industry* 117: 103187–103201. https://doi.org/10.1016/j.compind.2020.103187.

23 López, I.D., Figueroa, A., and Corrales, J.C. (2020). Multi-dimensional data preparation: a process to support vulnerability analysis and climate change adaptation. *IEEE Access* 8: 87228–87242.

24 Shafique, K., Khawaja, B.A., Sabir, F. et al. (2020). Internet of things (IoT) for next-generation smart systems: a review of current challenges, future trends & prospects for emerging 5G-IoT scenarios. *IEEE Access* 8: 23022–23040.

25 Tripathy, P.K., Tripathy, A.K., Agarwal, A., and Mohanty, S.P. (2021). MyGreen: an IoT- enabled smart greenhouse for sustainable agriculture. *IEEE Consumer Electronics Magazine* 10: 57–62.

26 Bhat, S.A. and Huang, N.F. (2021). Big data and AI revolution in precision agriculture: survey and challenges. *IEEE Access* 9: 110209–110222.

27 Jiang, X., Zhang, H., Yi, E.A.B. et al. (2021). Hybrid low-power wide-area mesh network for IoT applications. *IEEE Internet of Things Journal* 8: 901–915.

28 Kumar, A. and Jha, R. (2022). IoT in smart farming: a boon to farmers. In: *Internet of Everything: Smart Sensing Technologies* (ed. T. Kavitha, V.A. Devi, S.N. Pari, and S. Ramanathan), 185–206. Nova Publisher.

29 Singh, D. and Kumar, A. (2024). A deep recurrent neural network for plant disease classification. *SN Computer Science* 5 (8): 1053.

30 Tao, W., Zhao, L., Wang, G., and Liang, R. (2021). Review of the internet of things communication technologies in smart agriculture and challenges. *Computers and Electronics in Agriculture* 189: 106352.

31 Oak Dusadeerungsikul, P. and Nof, S.Y. (2021). A cyber collaborative protocol for real-time communication and control in human-robot-sensor work. *International Journal of Computers Communications & Control* 16 (3): 4233. https://univagora.ro/jour/index.php/ijccc/article/view/4233.

32 Raj Kumar, G., Chandra Shekhar, Y., Shweta, V., and Ritesh, R. (2021). Smart agriculture – urgent need of the day in developing countries. *Sustainable Computing Informatics & Systems* 30: 100512.

33 Kumar, A., Walia, G.S., and Sharma, K. (2020). Robust object tracking based on adaptive multicue feature fusion. *Journal of Electronic Imaging* 29 (6): 063001.

34 Kumar, A. (2023). *Visual object tracking using deep learning*. CRC Press.

35 Vohra, R. and Kumar, A. (Eds.). (2024). *Advanced Geospatial Practices in Natural Environment Resource Management*. IGI Global.

36 Vohra, R., Kumar, A., and Rongali, G. (2024). Multi-scale extraction and spatial analysis of growth pattern changes in urban water bodies using sentinel-2 MSI imagery: a study in the central part of India. *Stochastic Environmental Research and Risk Assessment* 38 (6): 2397–2423.

15

Current Progress of Sustainable Smart Agriculture Using Internet of Things

Savita Kumari Sheoran[1], *Suraj Ranga*[1], *and Ghanapriya Singh*[2]

[1] *Department of Computer Science & Engineering, Indira Gandhi University Meerpur, Rewari, Haryana, India*
[2] *Department of Electronics & Communication Engineering, NIT, Kurukshetra, Haryana, India*

15.1 Introduction

Nowadays, IoT describes the billions of physical items with Internet-connected sensors that can collect and share data. This technology makes it possible to operate in agriculture, transportation, industrial automation, healthcare, and emergency reaction to natural and artificial disasters [1]. However, although most studies focus on technical debate, very few studies cover the IoT from a social standpoint. The UN established the sustainable development goals (SDGs) to solve several detrimental social and international concerns. The SDGs seek to eradicate poverty worldwide, safeguard the natural world, and promote peace and prosperity on a global scale. The Internet of Things (IoT) solves the most significant issues facing the modern world. The objectives, therefore, sit at the nexus of technology, sustainability, society, and nature. The United Nations 2030 Agenda laid out 169 targets and 17 SDGs to improve our current model of global development by the year 2030 [2].

SDG Goal 2, "**Zero Hunger**," aims to eradicate hunger by 2030, promoting food security, nutrition, and sustainable agriculture. It emphasizes sustainable production, resilient practices, and investments in rural infrastructure for small-scale farmers. By tackling hunger and food insecurity, SDG Goal 2 seeks to create a world where no one is left behind, fostering a healthier and more prosperous global population.

SDG 2, "**Zero Hunger,**" aims to eliminate global hunger and address poverty, health, education, inequality, and climate change. It focuses on sustainable

Emerging Smart Agricultural Practices Using Artificial Intelligence, First Edition.
Edited by Ashish Kumar, Jai Prakash Verma, and Rachna Jain.

Published 2025 by John Wiley & Sons, Inc.

agricultural practices and enhancing productivity for smallholder farmers and food producers in developing nations. Challenges include climate change impacts, rapid urbanization, and fostering sustainable practices. A holistic approach is needed to address SDG 2 and its complementary objectives, fostering equitable development and ensuring global food security. A synergistic approach is crucial for a more promising future. Undernourishment and hunger hinder productivity and income, affecting people's standard of living. Chronic malnutrition, affecting 148.1 million children under 5, increases the risk of infections and cognitive development. By 2050, there will be 2 billion more people on the planet but 25% less land for crop cultivation. Climate shocks and locust plagues pose a major challenge to ending hunger and food security. Small food producers face a disadvantage due to their lower productivity compared to large-scale producers [3, 4].

To reach SDG-3's "**Healthy Lives and Promoting Well-being**" the goal of making sure everyone is healthy and happy, a strong oath has been completed to end outbreaks of AIDS, TB, malaria, and other deadly diseases by 2030. Also, it wants to give everyone access to safe and effective medicines and vaccines and provide health care for everyone. Improving access to affordable drugs and paying for research and development of vaccines are both main parts of this procedure. Promoting health and well-being is one of the 17 global objectives that built up the 2030 Agenda for SD. An integrated strategy is required to achieve the various objectives [5].

In SDG-7, the objective of, **"Affordable and Clean Energy,"** and targeting Reasonable, consistent, and sustainable energy is critical for attaining SDGs, such as poverty abolition, water supply, healthiness, industrialization, education, and climate change mitigation. However, access to energy differs between nations, and the state of development is inadequate. Redoubled efforts are needed, especially for countries with high energy consumption deficits by 2030 [6]. Universal energy access will create employment and educational opportunities in the world's poorest communities, spur innovation, and impact local, regional, and national employment, with potential positive net impacts.

SDG-9's objective, "**Industry, Innovation, and Infrastructure**," aims to promote sustainable industrialization, build resilient infrastructure, and foster innovation. Global manufacturing recovered in 2021, with faster recovery seen in economies with diversified industrial sectors and robust infrastructure. By 2030, higher technology industries will have improved and recovered more quickly, underscoring the significance of technological innovation [7].

SDG 11 objective, **"Sustainable Cities and Communities,"** aims to make cities safe and human settlements that are inclusive, resilient, and sustainable. Most of the world's population lives in urban areas; by 2050, seven out of 10 people will call them home. Cities contribute over 80% of global GDP and 70% of greenhouse gas emissions. Sustainable urban development is crucial for generating inclusive

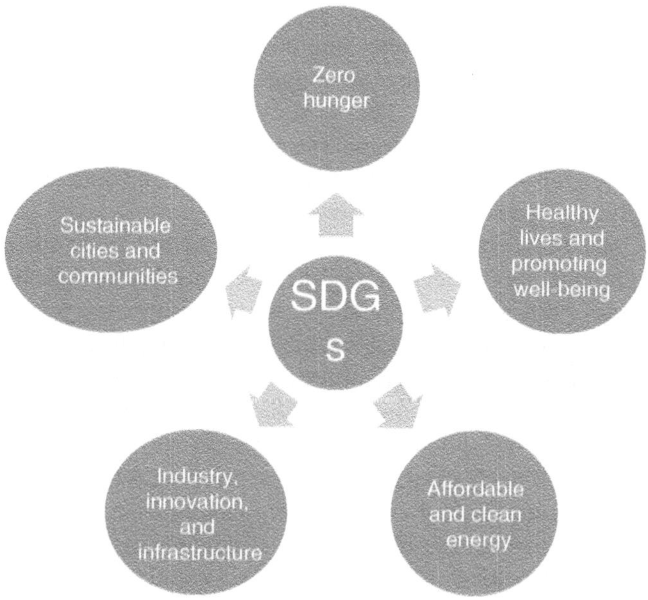

Figure 15.1 Relevant goals of SDGs.

prosperity, and strengthening preparedness and resilience is essential for responding to future crises [8]. The objective is to enhance inclusive and sustainable urbanization and human settlement planning and management in all countries by 2030.

15.2 Literature Review

This section reviews IoT applications such as sustainable agriculture, smart cities and grids, competent healthcare, innovative industries, and IoT–SDGs. The main focus of a literature review is to develop a new concept between IoT and SDGs and to summarize such IoT–SDG ideas. The analysis will concentrate on SDGs 2, 3, 7, 9, and 11 due to their relevance in IoT applications. This chapter will also focus on a systematic analysis of IoT–SDGs, as given below in Table 15.2.

15.2.1 Sustainable Agriculture

With an emphasis on Chinese universities and its applications in food supply chains, the research examines the usage of IoT technology in food safety. It was discovered that the first reports on IoT in food safety were published in 2011, with a preponderance of studies from Chinese colleges. The primary goals of the study were to outline opportunities and difficulties and to propose and create IoT systems for supply chains. Temperature, humidity, and position were measured by sensors in the majority of investigations; the most popular communication methods were RFID, WSN, and the Internet. The term "sustainable agriculture" describes a variety of approaches to solving numerous issues that affect the agricultural industry. Moreover, the term "sustainable" denotes a temporal dimension and the ability of an agricultural system to last forever [9].

According to projections, the global population will reach 9.7 billion by 2050, which means that food production efficiency and security will need to increase by 60% to meet the demands of this growing population. An agriculture that is sustainable is one that yields a large amount of food without destroying the environment or exhausting the planet's resources. Agriculture is the application of natural principles to the development of self-sustaining systems for livestock and crop production. In addition to being the agriculture of social values, sustainable agriculture is also the agriculture of lively rural communities, prosperous farm livelihoods for families, and nutritious food for all. However, as a collection of widely recognized methods or a role model of farm economy, sustainable agriculture is still in its infancy in the first ten years of the 21st century. Despite the fact that diminishing petroleum reserves, local food security, and the global economy are all related to sustainability in agriculture, its midwives [10].

Sustainable agriculture is a sustainable approach to food production that promotes eco-friendly practices, maintains soil quality, reduces soil degradation, saves water resources, improves land biodiversity, and ensures a healthy environment. Key accomplishments include crop rotation, controlling nutrient deficiency, pest control, recycling, and water harvesting. Sustainable agriculture is economically viable, providing basic human food and fiber needs and enhancing the quality of life for farmers and society. It operates within physical and biological resources, ensuring socio-economic viability and quality.

The importance of Artificial Intelligence Internet of Things (AIoT) in sustainable agriculture and food security is investigated in this paper. Because of environmental concerns and population expansion, there is a greater need than ever for innovative and efficient agricultural operations. Farmers and stakeholders may make informed decisions and maximize resource use with the help of real-time data and analytics from the AIoT. It covers AIoT in supply chain efficiency, intelligent irrigation, livestock management, and precision agriculture. The report also looks at edge computing and 5G, AIoT advancements, and AI and machine learning integration, as well as agricultural AIoT applications. It identifies and addresses the financial, technological, and data security challenges that arise with the deployment of AIoT. The findings highlight the relevance of AIoT in food security and sustainable agriculture. Agriculture might become more transparent, resource-efficient, and resilient with the use of AIoT and international collaboration [11].

SmartCrop is an IoT-based crop health monitoring system that combines cutting-edge technology with essential sensors to revolutionize contemporary agriculture. In addition to providing real-time data on temperature, humidity, and soil moisture, it excels at monitoring crop health and environmental conditions. It does this by using sophisticated algorithms for precise watering and quick danger identification. A user-friendly cloud platform provides farmers with access to this important data, allowing them to make well-informed choices for resource- and sustainability-efficient farming. By improving food security, lowering environmental impact, and fostering agricultural sustainability, SmartCrop stands as a ray of hope in a world where climate uncertainty and growing food demand are commonplace. Furthermore, SmartCrop supports a number of SDGs of the UN. SmartCrop demonstrates how technology may be used to create an agricultural future that is both commercially and ecologically sustainable [12, 13].

The IoT and smart sensors have the potential to significantly improve real-time data gathering and analysis for monitoring crop quality and yield, soil health, and water levels at a location. When combined, smart sensors and the IoT allow smart farming to replace traditional agricultural practices while increasing yields. IoT-enabled agricultural technology solutions aid in the evaluation of crop quality, soil fertility, soil health, soil erosion, and fertilizer needs. Additionally, IBM's "The Weather Company" The smart agricultural industry is growing thanks to the IoT, which makes it possible to monitor crop progress, seed quality, and optical irrigation. Water waste, energy usage, and cost savings may all be decreased with the use of automation, smart soil moisture sensors, and real-time data collecting. On the other hand, difficulties include privacy issues, security worries, cognitive difficulty, and subpar facilities [14, 15].

Precision agriculture uses optical sensors and IoT devices to optimize harvests and reduce agricultural hazards. This study presents an artificial swarm fish optimized naive bayes (ASFONB) method to monitor soil health and prevent diseases in cotton plants. The data is collected from environmental factors like temperature, humidity, light intensity, and chlorophyll content. The data is processed and analyzed by a central server, and the information is sent to farmers via an Android app. The proposed solution outperforms the status quo in disease identification [16].

Sustainable agriculture is a sustainable farming approach that balances food production with environmental health, economic profitability, and social equity. It uses renewable resources, conserves soil and water, and reduces chemical inputs. Techniques like crop rotation, agroforestry, integrated pest management, and organic farming enhance biodiversity and resilience. This model ensures food security for future generations while maintaining ecological balance and supporting farmers' livelihoods.

Sustainable agriculture is a sustainable approach to food production that promotes eco-friendly practices, preserving natural resources, halting biodiversity loss, and reducing greenhouse gas emissions. Key accomplishments include crop rotation, controlling nutrient deficiency, pest control, recycling, and water harvesting. Sustainable agriculture ensures a healthy environment by managing resources to meet human needs while conserving natural resources.

15.2.2 Smart City

This chapter focuses on the meaning of a smart city: collaborative advanced skills such as drones, AI, robotics, and IoT help us change our lifestyle, save people's lives, improve our quality of life, and sustain resources. This chapter describes applications in a smart city that show how IoT changes human beings' daily lives, such as data collection, privacy and security, environment protection, energy management, transportation management, waste management, e-waste reduction, disaster management, healthcare, delivery management, and other services management system [17].

This chapter confirms the existing state of science on Smart City SDGs development and establishes a theoretical framework for data integration. It compares Smart Cities with other cities and discusses the five Agendas of 2023, including people, planet, prosperity, peace, and collective participation. This article explores selecting indicators related to the current SDGs and their implications for Smart City development. It highlights the expected number of publications in 2022, nearly the same as in 2021, with 383 publications halfway through the year [18].

This chapter explores SDGs in smart cities, focusing on "soulmate" themes that promote human wellness and environmental sustainability. Both themes aim to improve collective and broad approaches to human well-being. A systematic literature search found no direct connections between smart cities and smart cities, leading to bibliographic mapping and a review of concepts and definitions. The literature review reveals that SCs primarily relate to technologies, specifically ICTs. The focus is on embracing all aspects, aiming for the welfare and wellness of human and social factors and the environment. Further research and evolution are needed to understand and develop SCs fully [19].

This chapter proposed a conventional attestation scheme for data source authentication based on threshold ring signature-trust certification. There are three steps in this scheme. The first step measures static and dynamic trust based on their behavior and then combines it. After that, the sensing nodes are classified for their trust value and level, and the last step is logical grouping, which is applied using a ring signature-based authentication system. This model identifies malicious nodes before the grouping and defends the trusted nodes, ensures the logical security of nodes, and works properly. The trusted measurement approach, motivated by the idea of a trust relationship in human society, ultimately considers the dynamism of the concurrent functioning of sensing nodes [20].

15.2.3 Smart Grid

The integration, applications, architectures, prototypes, taxonomies, IoT and non-IoT communications technologies, outstanding concerns, and future research directions have not been fully explored in surveys on IoT-aided Smart Grid systems. Significant facets of IoT-enabled SG systems are covered in this survey, including taxonomy discussions, current architectures, and prototypes. Applications, architectures, and prototypes are also covered. SG is a grid of the future that deals with problems including one-way information flow, energy waste, and rising energy demand. IoT technology makes Automation and tracking possible, providing connectivity and smart devices for grid monitoring, analysis, and control. This novel strategy provides dependability and security while addressing issues with the current electricity grid [21].

This article detailed the summary and role of the smart grid using IoT. It plays a crucial role in smart grids, connecting mechanisms, and fostering a soft relationship among people and society. It automates tasks and eliminates administrative burdens. However, cyber-security threats pose challenges in launching machine-to-human communication. IoT services in smart cities aim to improve human welfare and reduce operational costs. These services can be utilized in smart offices, grid stations, and workplaces for efficient energy usage management. This article discusses the role of IoT in smart grids, challenges, energy management systems, applications, security concerns, blockchain concepts, the Internet of Energy, and future directions. It also discusses the Internet of Energy and its potential impact on the future [22].

This research discusses the use of smart grid technology in Bangladesh, focusing on apparatus integration and domestic cross-platform communication. It also investigates the potential cost model for putting in a smart grid, considering the current state of the electricity industry. The processes needed to execute the smart grids in Bangladesh were better-understood thanks to research on India's smart grid. For an efficient power system, smart grid technology comprises cutting-edge features like distributed generation, smart metering, and a robust communication infrastructure. To address the country's expanding energy needs and bring its energy sector into line with the developed world, the proposal for a smart grid-enabled power sector in Bangladesh is presented in this chapter. It reports the electricity sector environment for smart grid implementation in Bangladesh, highlights changes required for successful implementation, and suggests a novel cost model [23].

The IoT has revolutionized power quality and reliability management in smart grid environments. A novel anonymous signature-based authenticated key exchange system called AAS–IoTSG has been developed to address security risks in these environments. With the help of this system, smart meters can safely receive services from service providers over unreliable pathways, resulting in a trustworthy and secure connection. AVISPA, formal security analysis, and informal security analysis have all been used to assess AAS–IoTSG thoroughly. AAS–IoTSG provides superior security and functionality features than existing clever grid system authentication mechanisms, requiring less communication and computation overheads. A thorough comparison research reveals this. This scheme has four phases: System setup, registration, authentication, and key agreement. The final phase is dynamic node addition, followed by security analysis [23].

15.2.4 Smart Health

This survey explores the application of emerging IoMT in health monitoring, proposing IoMT-based systems for individual, group, and disaster rescuers. The system focuses on individual connections, while group systems analyze key technologies like Markov models and game theory. Disaster rescuer systems address social attributes and routing repair. Future IoMT research will focus on integrating diverse networks, new technologies, and network security. The architecture, essential technologies, and research status of individual and group health monitoring systems for disaster rescue, along with the difficulties and potential paths for future research in health monitoring systems, are discussed in the study [24].

The impact of the SDGs on the uses of the IoT and intelligent systems is examined using a hybrid approach. It raises the question of whether IoT applications are essential for enhancing human health and the environment, and if so, if research and case studies demonstrating their effectiveness in improving public health can be used to evaluate the applications and projects. Indicators, objectives, and resources about building and deploying sensors, IoT, and intelligent systems within the context of the SDGs are examined in this article. It analyses the rules regulating the SDGs and merges technology, engineering, and the environment to enhance health and sustainable development [25].

A new architecture approach based on Dijkstra's theorem and the IEEE 802.15.6 CSMA policy. The algorithm aims to reduce retransmission processes, collisions, and query overhead and determine the shortest path. This improves the transmission quality of critical data, reduces failure probability, and reduces power consumption. Results show that distribution ensures system performance under general conditions compared to centralization and existing works. The chapter presents related works, models, and system architecture, including a new routing algorithm and a general architecture, with results and discussion [26].

15.2.5 Smart Industry or Industry 4.0

This chapter examines cutting-edge network automation technologies, use cases, and standardizations for IIoT in Industry 4.0. It also covers advancing the 6G-based Industry 5.0 blueprint-based next-generation network automation. IEEE P2668 is used to index network computerization for industrial applications to direct the transmission of h IIoT applications at vital moments.

Additionally, it covers potential obstacles when adopting the impending 6G-based Industry 5.0 and offers insights into next-generation system automation. This in-depth analysis of the advanced system, which is automation for IIoT powered by Industry 4.0, offers researchers pointers on supplying the new generation system, which is automation for Industry 5.0 powered by 6G. The infrastructure for network automation, standardizations, IIoT in Industry 4.0, new generation, 6G-based Industry 5.0, and probable difficulties and opportunities in the future of automation are covered in this article. It further emphasizes the necessity of 6G-based solutions [26].

It presents a protected fog-based IIoT structural design, incorporating security features and offloading tasks to fog nodes. This architecture reduces trust and burden on cloud and resource-constrained devices. Validation is accomplished through hypothetical overhead analysis, including simulation studies, practical experimentation, and testable implementation. 5G and 6G networking are becoming more prevalent in Industry 4.0 and Industry 5.0, with 5G and 6G being essential use cases. In Industry 5.0, 6G will replace traditional manual network administration, which accounts for 95% of management [27].

The article's main aim is that Industry 4.0 and Society 5.0 in Turkey influenced the academic staff at the Kafkas University to examine the country's orientation toward these SDGs. Because of the timely subject and technology components, the study is limited to academics at a state university in Kars, Turkey. Local governments, SMEs, managing natural disasters, managing human resources, and empowering women in the workforce can all benefit from Society 5.0. This chapter is divided into demographic questions, Industry 4.0 and Society 5.0 viability, two sub-dimensions, and 17 SDG-based questions. Participants' personal information is gathered in the first phase, and the viability of Industry 4.0 and Society 5.0 is examined in the second [28].

15.2.6 SDGs and IoT

Here, we will discuss how many goals we can achieve by combining IoT and SDG, and it shows what work is being done on which goals of SDG together with IoT. Table 15.1 discusses the nine collected articles and works on the meaning, title of the article, highlight on SDGs, and their focus; all these are given below in Table 15.1.

Table 15.1 List of articles studies with highlights on SDGs and focus.

Reference no	Highlight on SDGs	Focus
[29, 30]	1, 2, 3, 4, 5, 6, 7, 8, 9, 10, 11, 12, 13, 14, 15, 17	The chapter examines India's progress in implementing the sustainable development goals (SDGs) using OECD methodology. It highlights the wide variations in gaps and suggests international development cooperation for capacity building. The chapter emphasizes the need for timely SDG achievement
[31]	1, 2, 3, 4, 5, 6, 7, 8, 9, 10, 11, 12, 13, 14, 15, 17	India has launched the Digital India program, focusing on providing digital infrastructure, services on demand, and digital empowerment to all citizens using ICT4D, mobile, social media, Cloud, and data analytics
[32]	1, 2, 3, 4, 5, 6, 7, 8, 9, 10, 11, 12, 13, 14, 15, 17	India's extensive localization experience, driven by NITI Aayog, focuses on action at the sub-national level, offering valuable insights for other countries in implementing SDG localization
[33]	9	"SAMARTH – Udyog Bharat 4.0"
[34]	6, 7, 8, 9, 11, 12, 13	IT professionals, governments, and development organizations have a path toward a more sustainable society through technical innovation
[35]	3, 4, 7, 9, 10, 11, 12, 13	To promote the SDGs, this work analyses ground-breaking research that uses intelligent Internet of Things (IoT) systems in the hyper-connected, intelligent world. These systems will be created using IoT technology and advances in artificial intelligence
[36]	1, 2, 3, 4, 5, 6, 7, 8, 9, 10, 11, 12, 13, 14, 15, 17	This article shows the effect of COVID-19 on SDGs and who was more affected
[37]	11	To achieve SDG #11 targets, a more comprehensive approach to the governance of smart city initiatives is needed. This approach should incorporate social justice and sustainable development frameworks
[38]	11	The main goal is to determine smart governance's value in accomplishing the SDGs in urban areas
[39]	11	This chapter describes the need for smart cities through the work being done and demonstrates how IoT technology and smart city enablers may be deployed in urban contexts to support cities that improve their residents and get overall sustainable growth
[40]	9	This chapter presents an IoT architecture with increased security that is particularly beneficial to SDG #9
[41]	1, 2, 3, 4, 5, 6, 7, 8, 9, 10, 11, 12, 13, 14, 15, 16, 17	Role of IoT to achieve SDGs

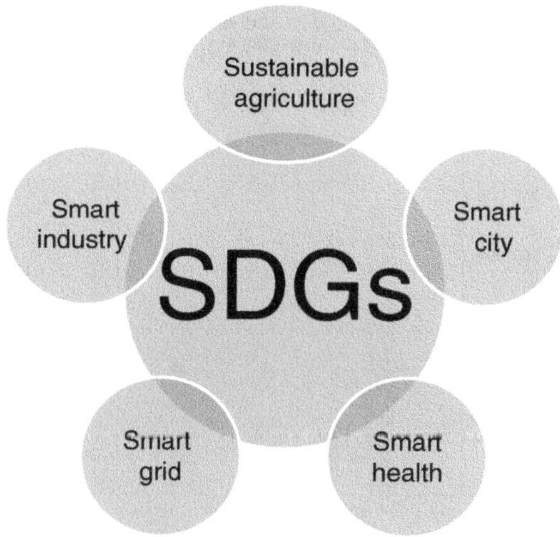

Figure 15.2 SDGs–IoT mapping.

15.3 Methodology

Every government implements policies for the benefit of its citizens through programs and initiatives. In August 2018, NITI-Aayog attempted to map the Central Sector Programmes and the Ministry of the Government of India with the SDGs. They divide the SDGs into 685 central sector initiatives and 36 officially funded programs. Additionally, the mapping carried out by the NITI Aayog has been appropriately modified to ensure that the particular initiative or program is mapped with an appropriate SDG rather than expanding the scheme's or project's purpose throughout various cross-cutting SDGs. The primary purpose of this budget analysis is to find how much budget allocations and expenditures have done under various government/nongovernment schemes/projects during the years, such as 2019, 2020, and 2021, covered our SDGs as well as some reports that were published from 2018 to 2022 to analyze works in the field of IoT–SDGs. In the end, the current status of India is according to the SDGs and where India is standing in the global current ranking. Also, we analyzed the urban city index, which helps to cover the SDGs.

15.3.1 Budget for Centrally Sponsored Schemes/Projects Under SDGs

A three-year action plan that spans the years 2017–2021 was published in 2017 and was focused on delivering "Ease of Living" to all of its residents and is built on the three themes of "Economic Development," "Caring Society," and "Aspirational India." Figure 15.3 shows the budget allocated by the Indian government for SDGs

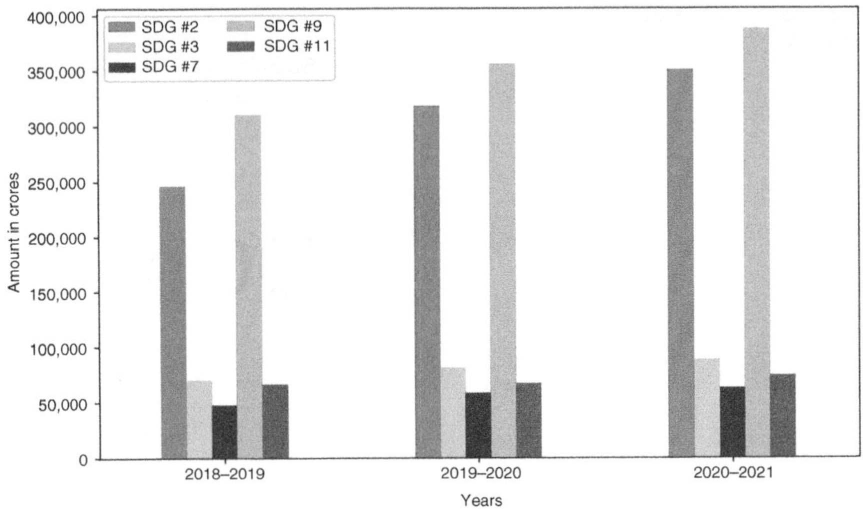

Figure 15.3 Budget for the years (2018–2021) in the form of various schemes/projects.

3, 7, 9, and 11 in the three-year financial years 2018–2019, 2019–2020, and 2020–2021, respectively. From 2018 to 2021, we have studied how much of the budget was allocated by the Indian government for SDG. Finally, we figured out that the highest budget was allocated for SDG 9. All values are in crores [42].

Figure 15.4 shows how much budget was allocated by the Indian Government for SDGs 3, 7, 9, and 11 from 2018 to 2019. It shows that from 2018 to 2019, 4.7% were allocated to SDG 3. In this way, if we talk about SDG 7, 3.2% were given, for SDG 9, 21% percent were issued, and for SDG 11, 4.4% were allocated, and the rest of SDG was 66.6% and 21% of the maximum budget was allocated for SDG 9.

Similarly, Figures 15.5 and 15.6 show how much budget was allocated for SDGs 3, 7, 9, and 11 in 2019–2020, from which it is clear that from the 2019 to 2020 budget allocation, 4.7% of the total was allocated to SDG 3 and 3.3% was allocated to SDG 7; if we talk about SDG 9, then 20.5% was given, SDG 11 was 3.9%, and the rest of the SDGs were 67.6%. It was found that 20.5% of the maximum amount was allocated for SDG 9.

15.3.2 Study of SDGs Reports

Table 15.2 shows the year-wise published reports on IoT and SDGs and the main focus of the reports. The table below contains data collected from the "Internet of Things Guidelines for Sustainability," "Technology and Innovation Report 2021", "IoT Week 2022 Report," and others, etc.

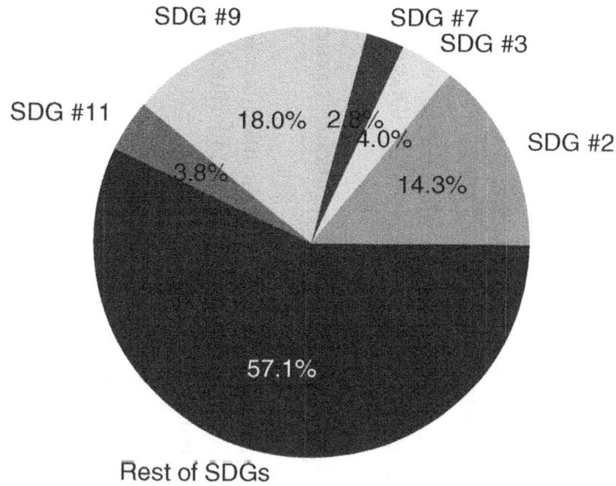

Figure 15.4 Budget for the year (2018–2019) in the form of various schemes/projects.

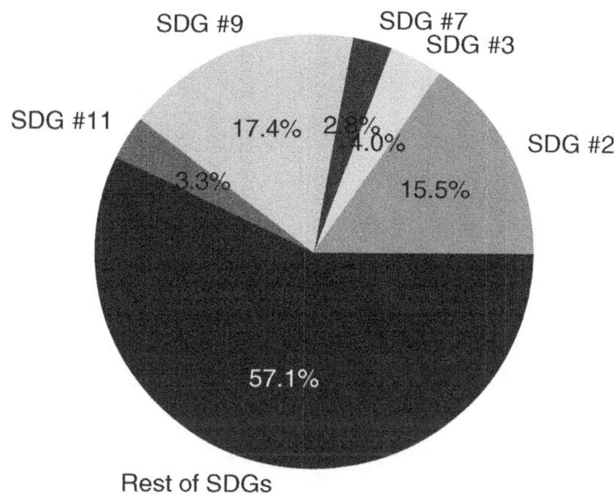

Figure 15.5 Budget for the year (2019–2020) in the form of various schemes/projects.

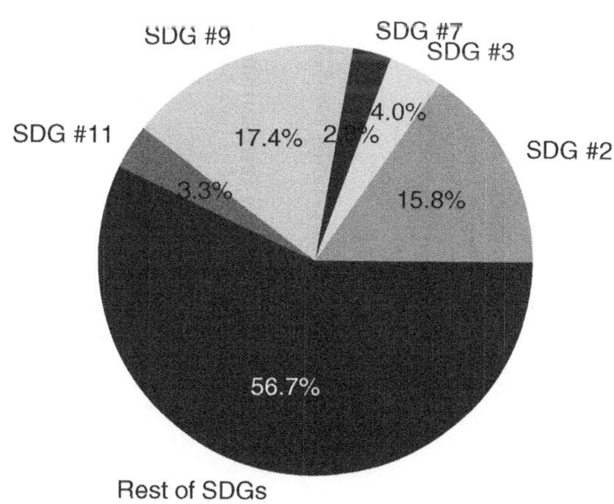

Figure 15.6 Budget for the year (2020–2021) in the form of various schemes/projects.

Table 15.2 List of SDGs reports studies on published year and focus.

Reference no.	Report name	Year	Focus
[43]	A connected world, a sustainable world		This report provides a comprehensive framework for achieving prosperity through various initiatives, including gender equality, digitalization, and sustainable practices
[44]	Internet of Things guidelines for sustainability	2018	This report highlights the need for entities, such as start-ups, small and medium enterprises, and large corporations, to analyze the impact of their IoT projects on the SDGs and consider external accreditation
[45]	5G, IoT, UN–SDG OMA LwM2M, IPSO	2018	This report discusses several methods for leveraging 5G and IoT technology to support some of these SDGs
[46]	ICDL 2019 Digital Transformation for SDGs	2019	The report discusses the potential of digitization to save energy in the manufacturing sector. It emphasizes the necessity for businesses to adopt digitalization and underlines the characteristics of Industry 4.0, which differentiates it
[47]	Technology and innovation report	2021	This describes the implications for achieving the SDGs
[48]	Frontier ICTs for sustainable development for digital leaders	2022	It provides an overview of IoT, including its key characteristics, current status, and potential government applications. The chapter also references the WEF report on the State of the Connected Era 2020, which likely provides further insights and findings
[49]	IoT week 2022 report	2022	Afforded people the chance to carefully investigate the state of IoT technology, applications, and upcoming problems, including how IoT solutions may help achieve the UN's sustainability goals
[50]	Sustainable development goals and fisheries sector progressing towards a sustainable future	2023	New methods and approaches are needed for sustainable fisheries and ecosystem implementation, with AI and advanced technologies like big data, IoT, and remote sensing playing crucial roles in achieving SDGs
[51]	IOT 4 SDGS	2020	IoT is a promising technology that can contribute to several SDGs by increasing efficiency and saving costs. To maximize the benefits of the IoT revolution while mitigating risks, policymakers need to adopt tailored policies and investments. The European Union should consider policy measures to boost IoT's potential for sustainability, in line with the European Green Deal and the upcoming 2021–2027 financial framework
[52]	SDG digital acceleration agenda	2023	Digital technologies, including mobile phones and AI, are transforming global economies, improving financial inclusion, enhancing government service delivery, and empowering citizens. They could potentially drive the SDGs

15.4 Current Status of SDGs (Global and Local) in Ranking

The rank is the number that shows the best performance among them for a particular parameter. In SDG, rank is assigned based on the SDG index score. The index score shows the status of a geographical area achieving SDGs in a particular duration. To calculate the SDG index scores, we first evaluate each goal's score using the formula for the mean of its indicators. The ultimate index score is then calculated by averaging these scores across all 17 SDGs. Figure 15.7 shows that four categories have been divided based on the score: first is our aspirant between 0 and 49, and then our performance from 50 to 69. Our frontrunner comes between 70 and 99; if it scores 100, it is the achiever.

As per Figure 15.8, the Indian Urban City SG, Shimla, scored 75.5, followed by Coimbatore with 73.3, Chandigarh with 72.4, and Trivainipuram with 72.4. Kochi has 72.3, Panji has 71.8, and Faridabad has 58.6. The Indian Urban SDG Index result analysis shows Shimla is in the first position in the Indian SDG Urban Index. [53].

Figure 15.9 represents the global SDG scores of different countries with their index and rank scores. Finland got an 86.8 score, followed by other countries such as Sweden (86.0), Denmark (85.7), Germany (83.4), Austria (82.0), and India (63.5). Finland ranks first out of 193 UNO members at the global level, and India's rank is 112. [54]

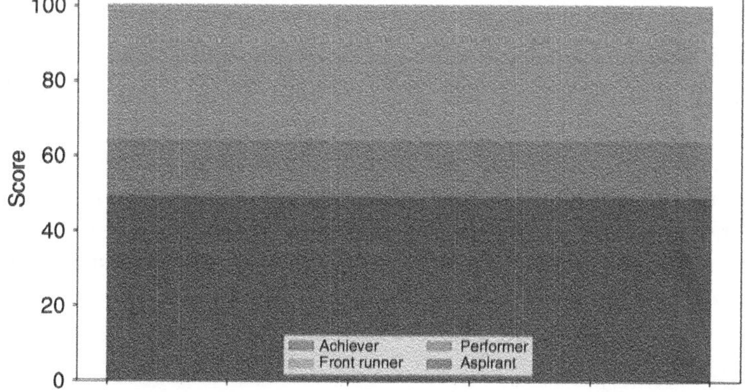

Figure 15.7 Score-based category.

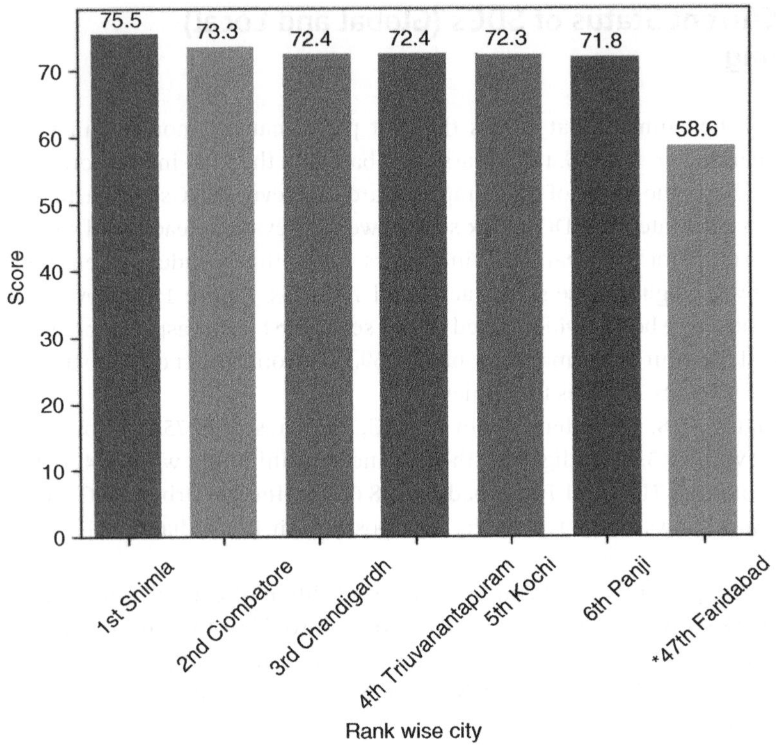

Figure 15.8 India SDGs city index and score.

15.5 Analysis

Table 15.3 contains the budget-wise details and description of budgetary provisions for the different SDGs discussed above in the above figures. From 2018 to 2021, the maximum budget allocation for various schemes and projects in India was directed towards SDG #9, focusing on industry, innovation, and infrastructure. In the fiscal year 2018–2019, SDG #9 schemes received nearly 21.00% of the total budget, followed by 20.50% in 2019–2020 and 20.70% in 2020–2021. In terms of urban SDG achievements within India, Shimla ranked first in the India SDGs City Index. On a global scale, Finland secured the top position in the Global SDGs Index among all 193 UN member states, while India was positioned at 112th.

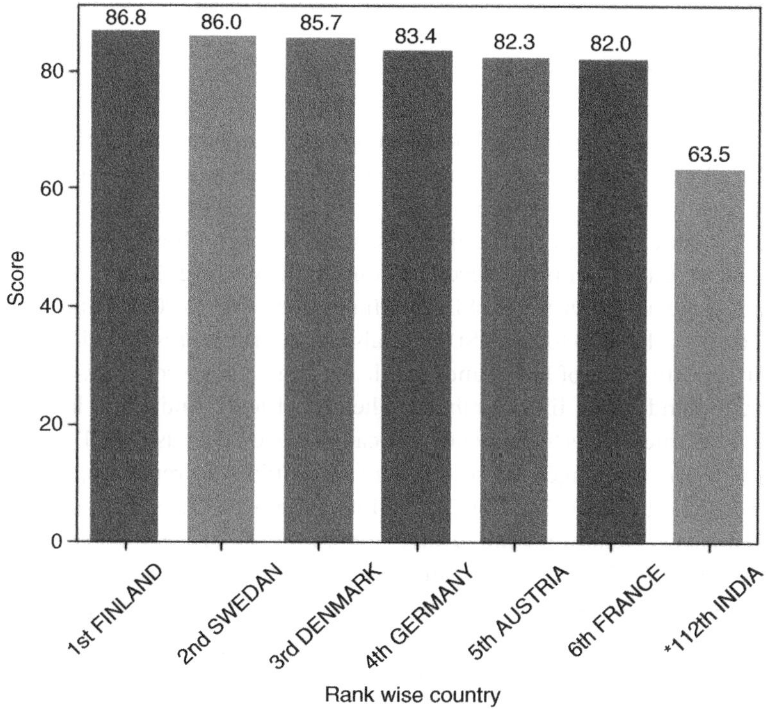

Figure 15.9 Global SDGs index and score.

Table 15.3 List of analysis.

Figure	Description	Analysis
15.3	Budget for the years (2018–2021) in the form of various schemes/projects	SDG #9(Schemes) allocated the maximum budget among them
15.4	Budget for the year (2018–2019) in the form of various schemes/projects	SDG #9(Schemes) allocated a maximum budget of nearly 21.00% for 2018–2019
15.5	Budget for the year (2019–2020) in the form of various Schemes/ projects	SDG #9(Schemes) allocated a maximum budget of nearly 20.50% for 2019–2020
15.6	Budget for the year (2020–2021) in the form of various schemes/projects	SDG #9(Schemes) allocated a maximum budget of nearly 20.70% for 2020–2021
15.8	India SDGs city index and score	Shimla got first rank in India's urban cities to achieve the SDGs
15.9	The performance of all 193 UN member states in SDGs (global SDGs index and score)	Finland got 1st rank in achieving SDGs globally, and India got 112nd

15.6 Conclusions

IoT and SDGs can work together to boost young people's interest in engineering and make it a lifetime hobby. After analysis, it is found that all SDGs are very important for daily life and to improve quality of life. This study aims to understand the IoT, tie it to realistic SDG objectives, and show how they relate. It also shows how good results can be achieved by combining SDGs (2, 3, 7, 9, and 11) and IoT. The budgetary provision for SDGs in India from 2018 to 2021 is analyzed with some reports related to IoT and SDGs. It also shows the scores of various cities in the Urban City Index of India, and Shimla has the highest score if discussed the country. Finland is the first number at the global level, and India has the 122nd rank. The article aims to show how we can achieve SDGs using IoT. In the future, it will open challenges for researchers to acquire the remaining SDGs using IoT. Predicting how IoT and SDGs can improve the world is also possible. This research is very helpful for the future, combining IoT and SDGs to impact our lives significantly. Sustainable farming systems utilize innovative science-based practices to enhance productivity and reduce environmental damage. These practices are suitable for all farms and produce various fuels, foods, and fibers, ensuring long-term productivity, profitability, and sustainability.

Funding

This work is funded by IIT Ropar-Technology and Innovative Foundation for Agriculture and Water Technology Development Hub (AWaDH) having grant File number: AWaDH-SpINe/IITRPR/2022/P1003 to Dr. Ghanapriya Singh.

References

1 Houhamdi, Z. and Athamena, B. (2020). Identity identification and management in the internet of things. *International Arab Journal of Information Technology* 17 (4A): 645–654. https://doi.org/10.34028/iajit/17/4A/9.

2 "Goal 3: Good Health and Well-being," Goal 3: Good Health and Well-being. https://www.jointsdgfund.org/sustainable-development-goals/goal-3-good-health-and-well-being (accessed 27 January 2025).

3 "SDG Zero Hunger." https://www.un.org/sustainabledevelopment/hunger/ (accessed 27 January 2025).

4 "Goal 2: Zero Hunger - the global Goals," The Global Goals, Jan. 23, 2024. https://www.globalgoals.org/goals/2-zero-hunger/ (accessed 27 January 2025).

5 Martin, "Infrastructure and Industrialization - United Nations Sustainable Development," United Nations Sustainable Development, Oct. 20, 2023. https://www.un.org/sustainabledevelopment/infrastructure-industrialization/ (accessed 27 January 2025).

6 Martin, "Cities - United Nations Sustainable Development Action 2015," United Nations Sustainable Development, Oct. 20, 2023. https://www.un.org/sustainabledevelopment/cities/(accessed 27 January 2025).

7 "Smart Cities Mission | IBEF," India Brand Equity Foundation. https://www.ibef.org/government-schemes/smart-cities-mission https://www.un.org/sustainabledevelopment/cities/ (accessed 27 January 2025).

8 "Goal 11: Sustainable cities and communities - The Global Goals," The Global Goals, Jan. 23, 2024. https://www.globalgoals.org/goals/11-sustainable-cities-and-communities/ /sustainabledevelopment/cities/ (accessed 27 January 2025)

9 Vishnoi, S. and Goel, R.K. (2024). Climate smart agriculture for sustainable productivity and healthy landscapes. *Environmental Science & Policy* 151: 103600. https://doi.org/10.1016/j.envsci.2023.103600. Available: https://www.sciencedirect.com/science/article/pii/S1462901123002496.

10 Bouzembrak, Y., Klüche, M., Gavai, A., and Marvin, H.J.P. (2019). Internet of things in food safety: literature review and a bibliometric analysis. *Trends in Food Science & Technology* 94: 54–64. https://doi.org/10.1016/j.tifs.2019.11.002. Available: https://www.sciencedirect.com/science/article/pii/S0924224419303048.

11 Shahab, H., Iqbal, M., Sohaib, A. et al. (2024). IoT-based agriculture management techniques for sustainable farming: a comprehensive review. *Computers and Electronics in Agriculture* 220: 108851. https://doi.org/10.1016/j.compag.2024.108851. Available: https://www.sciencedirect.com/science/article/pii/S0168169924002424.

12 Sai, P.S.S., Sunil, A., Mukesh, K. et al. (2024). SmartCrop: an IoT-powered crop health monitoring system. In: *Smart Trends in Computing and Communications*. SmartCom 2024, vol. 946 (ed. T. Senjyu, C. So-In, and A. Joshi), 175–187. Singapore: Springer. https://doi.org/10.1007/978-981-97-1323-3_14.

13 Tata Power Sector transformation: An SDG roadmap for electric utilities. https://www.tatapower.com/pdf/report-on-sdgroadmap.pdf sustainabledevelopment/cities/ (accessed 27 January 2025)

14 Balyan, S., Jangir, H., Tripathi, S.N. et al. (2024). Seeding a sustainable future: navigating the digital horizon of smart agriculture. *Sustainability* 16 (2): 475. https://doi.org/10.3390/su16020475.

15 Artificial Intelligence of Things for Achieving Sustainable Development GoalsAlaba, F.A., Jegede, A., Sani, U., and Dada, E.G. (2024). Artificial intelligence of things (AIoT) solutions for sustainable agriculture and food

security. In:, vol. 192 (ed. S. Misra, K. Siakas, and G. Lampropoulos), 109–126. Cham: Springer https://doi.org/10.1007/978-3-031-53433-1_7.

16 Durai, R.S.V., Vijayakumar, R., Lakshmisridevi, S. et al. (2024). IoT based optical sensor network for precision agriculture. *2024 International Conference on Optimization Computing and Wireless Communication (ICOCWC)**, pp. 1–7. Debre Tabor, Ethiopia, 29–30 Jan. 2024. https://doi.org/10.1109/ICOCWC60930.2024.10470879.

17 Alsamhi, S.H., Ma, O., Ansari, M.S., and Almalki, F.A. (2019). Survey on collaborative smart drones and internet of things for improving smartness of smart cities. *IEEE Access* 7: 128125–128152. https://doi.org/10.1109/ACCESS.2019.2934998.

18 Gong, B., Liu, J., and Guo, S. (2021). A trusted attestation scheme for data source of internet of things in smart city based on dynamic trust classification. *IEEE Internet of Things Journal* 8 (21): 16121–16141, 1 Nov.1. https://doi.org/10.1109/JIOT.2020.3006349.

19 Saleem, Y., Crespi, N., Rehmani, M.H., and Copeland, R. (2019). Internet of things-aided smart grid: technologies, architectures, applications, prototypes, and future research directions. *IEEE Access* 7: 62962–63003. https://doi.org/10.1109/ACCESS.2019.2913984.

20 Parra-Domínguez, J., Gil-Egido, A., and Rodríguez-González, S. (2022). SDGs as one of the drivers of smart city development: the indicator selection process. *Smart Cities* 5 (3): 1025–1038. https://doi.org/10.3390/smartcities5030051.

21 Visvizi, A. and Pérez-delHoyo, R. (2021). *Smart Cities and the Un SDGs*. Elsevier. https://doi.org/10.1016/C2020-0-01556-2.

22 Srinivas, J., Das, A.K., Li, X. et al. (2021). Designing anonymous signature-based authenticated key exchange scheme for internet of things-enabled smart grid systems. *IEEE Transactions on Industrial Informatics* 17 (7): 4425–4436. https://doi.org/10.1109/TII.2020.3011849.

23 Purohit, R.M., Verma, J.P., Jain, R., and Kumar, A. (2025). FedBlocks: federated learning and blockchainbased privacy-preserved pioneering framework for IoT healthcare using IPFS in web 3.0 era. *Cluster Computing* 28 (2): 1–15.

24 Maruf, M.H., ul Haq, M.A., Dey, S.K. et al. (2020). Adaptation for sustainable implementation of smart grid in developing countries like Bangladesh. *Energy Reports* 6: 2520–2530. https://doi.org/10.1016/j.egyr.2020.09.010.

25 Moutaib, M., Ahajjam, T., Fattah, M. et al. (2022). Application of internet of things in the health sector: toward minimizing energy consumption. *Big Data Mining and Analytics* 5 (4): 302–308, December. https://doi.org/10.26599/BDMA.2021.9020031.

26 Verdejo Espinosa, Á., Lopez Ruiz, J., Mata Mata, F., and Estevez, M.E. (2021). Application of IoT in healthcare: keys to implementation of the sustainable development goals. *Sensors (Basel)* 21 (7): 2330. https://doi.org/10.3390/s21072330.

27 Sengupta, J., Ruj, S., and Bit, S.D. (2021). A secure fog-based architecture for industrial internet of things and industry 4.0. *IEEE Transactions on Industrial Informatics* 17 (4): 2316–2324. https://doi.org/10.1109/TII.2020.2998105.

28 Chi, H.R., Wu, C.K., Huang, N.-F. et al. (2023). A survey of network automation for industrial internet-of-things toward industry 5.0. *IEEE Transactions on Industrial Informatics* 19 (2): 2065–2077. https://doi.org/10.1109/TII.2022.3215231.

29 Research and Information System for Developing Countries (RIS) Measuring the SDG progress in india with focus on gap analysis approach. https://www.ris.org.in/sites/default/files/Publication/DP-278-Krishan-Kumar-and-PK-Anand.pdf (accessed 27 January 2025).

30 United Nations Conference on Trade and Development (UNCTAD) Technology and Innovation Report 2021. https://unctad.org/system/files/official-document/tir2020_en.pdf. (accessed 27 January 2025)

31 United Nations SDG-vision-for-digital-India. https://www.local2030.org/library/674/SDG-Vision-for-Digital-India.pdf. (accessed 27 January 2025)

32 National Institution for Transforming India (NITI Aayog) Indian model of SDG localisation. https://www.niti.gov.in/sites/default/files/2022-07/The_Indian_Model_of_SDG_Localisation_13072022.pdf. (accessed 27 January 2025)

33 Beier, G., Niehoff, S., and Hoffmann, M. (2021). Industry 4.0: a step towards achieving the SDGs? A critical literature review. *Discover Sustainability* 2: 22. https://doi.org/10.1007/s43621-021-00030-1.

34 Berawi, M.A. (2019). The role of Industry 4.0 in achieving sustainable development goals. *International Journal of Technology* 10 (4): 644. https://doi.org/10.14716/ijtech.v10i4.3341.

35 Tasnim, S. and Ahad, M.D. (2022). Toward IoT enabled smart offices: achieving sustainable development goals. https://doi.org/10.48550/arXiv.2206.03556.

36 López-Vargas, A., Ledezma, A., Bott, J., and Sanchis, A. (2021). IoT for global development to achieve the united nations sustainable development goals: the new scenario after the COVID-19 pandemic. *IEEE Access* 9: 124711–124726. https://doi.org/10.1109/ACCESS.2021.3109338.

37 SenGupta, U. and SenGupta, U. (2022). SDG-11 and smart cities: contradictions and overlaps between social and environmental justice research agendas. *Frontiers in Sociology* 7: 995603. https://doi.org/10.3389/fsoc.2022.995603.

38 Giuliodori, A., Berrone, P., and Ricart, J.E. (2023). Where smart meets sustainability: the role of smart governance in achieving the sustainable development goals in cities. *BRQ Business Research Quarterly* 26 (1): 27–44. https://doi.org/10.1177/23409444221091281.

39 Singh, D. and Kumar, A. (2024). A deep recurrent neural network for plant disease classification. *SN Computer Science* 5 (8): 1053.

40 Abass, O.A. and Dawodu, A.A. (2021). S.D.G. Towards-9, an IoT Archit. Converging OT-It Ind. 4.0 and digital economy era. *University of Ibadan Journal of Science and Logics in ICT Research (UIJSLICTR)* 6 (1 and 2): 2021.

41 Philippe COUSIN, Franck Le Gall, Congduc PHAM et al. (2018). IoT standards for Africa and sustainable development goals (SDGs). *IST-Africa Week Conference (IST-Africa)*, Gaborone, Botswana, *09–11 May 2018*, pp. Page, 2018 1 of 8-Page 8 of 8. Available: https://cpham.perso.univ-pau.fr/Paper/ISTAfrica2018_Paper_ref_115.pdf

42 "Indian budget" (2022) https://www.intosaicommunity.net/document/articlelibrary/Budget_of_India_2020-2021_and_Sustainable_Development_Goals.pdf (accessed 27 January 2025).

43 José María Álvarez-Pallete, "A connected world, a sustainable world: An analysis of the direct and indirect impacts of Telefónica's activity on the United Nations Sustainable Development Goals," report, 2021. [Online]. Available: https://www.telefonica.com/en/wp-content/uploads/sites/5/2021/08/telefonica-sdg.pdf (accessed 27 January 2025)

44 World Economic Forum, "Internet of Things Guidelines for Sustainability," Jan. 2018. [Online]. Available: https://www3.weforum.org/docs/IoTGuidelinesforSustainability.pdf (accessed 27 January 2025)

45 P. Subramani, NOKIA, and IPSO, "5G, IoT, UN-SDG," May 2018. [Online]. Available: https://www.openmobilealliance.org/documents/whitepapers/lwm2m-5g-iot-un-sdg.pdf (accessed 27 January 2025)

46 ICDL (2019). Digital transformation for SDGs. (n.d.). https://www.teriin.org/events/icdl/sdg-files/pdf/Digital-Transformation-Brochure.pdf. (accessed 27 January 2025)

47 Technology and Innovation REPORT 2021. https://unctad.org/system/files/official-document/tir2020_en.pdf, (accessed 27 January 2025).

48 ICT (2022). https://www.unapcict.org/sites/default/files/2022-08/Frontier%20ICTs%20for%20Sustainable%20Development_Submodule%20C%20-%20IoT_0.pdf (accessed 27 January 2025).

49 I. Week (2022). https://iotweek.org/wp-content/uploads/2022/08/IoT-Week-2022-Report.pdf, (accessed 27 January 2025).

50 Ratheesh, K.R. Sustainable development goals and fisheries sector progressing towards a sustainable future. http://eprints.cmfri.org.in/17088/1/Aquaculture%20Spectrum_2023_Ratheesh%20Kumar%20R.pdf. (accessed 27 January 2025)

51 Renda, A. and Laurer, M. (2020). IoT 4 SDGs – what can the digital transformation and IoT achieve for Agenda 2030?. CEPS Papers, no. 26658, Centre for European Policy Studies, 2020. https://ideas.repec.org/p/eps/cepswp/26658.html (accessed 27 January 2025).

52 United Nations Development Programme (UNDP) SDG digital acceleration Agenda. https://www.undp.org/sites/g/files/zskgke326/files/2023-09/SDG%20Digital%20Acceleration%20Agenda_2.pdf. (accessed 27 January 2025)

53 Urban, S.D.G. (2024). Index. https://sdgindiaindex.niti.gov.in/urban/#/ranking. (accessed 27 January 2025)

54 "Sustainable Development Report 2024." https://dashboards.sdgindex.org/rankings (accessed 27 January 2025)

Index

Emerging Smart Agricultural Practices Using Artificial Intelligence, First Edition.
Edited by Ashish Kumar, Jai Prakash Verma, and Rachna Jain.

Published 2025 by John Wiley & Sons, Inc.